Biological Unit Processes for Hazardous Waste Treatment

BIOREMEDIATION

The *Bioremediation* series contains collections of articles derived from many of the presentations made at the First, Second, and Third International In Situ and On-Site Bioreclamation Symposia, which were held in 1991, 1993, and 1995 in San Diego, California.

First International In Situ and On-Site Bioreclamation Symposium

1(1)   *On-Site Bioreclamation: Processes for Xenobiotic and Hydrocarbon Treatment*

1(2)   *In Situ Bioreclamation: Applications and Investigations for Hydrocarbon and Contaminated Site Remediation*

Second International In Situ and On-Site Bioreclamation Symposium

2(1)   *Bioremediation of Chlorinated and Polycyclic Aromatic Hydrocarbon Compounds*

2(2)   *Hydrocarbon Bioremediation*

2(3)   *Applied Biotechnology for Site Remediation*

2(4)   *Emerging Technology for Bioremediation of Metals*

2(5)   *Air Sparging for Site Bioremediation*

Third International In Situ and On-Site Bioreclamation Symposium

3(1)   *Intrinsic Bioremediation*

3(2)   *In Situ Aeration: Air Sparging, Bioventing, and Related Remediation Processes*

3(3)   *Bioaugmentation for Site Remediation*

3(4)   *Bioremediation of Chlorinated Solvents*

3(5)   *Monitoring and Verification of Bioremediation*

3(6)   *Applied Bioremediation of Petroleum Hydrocarbons*

3(7)   *Bioremediation of Recalcitrant Organics*

3(8)   *Microbial Processes for Bioremediation*

3(9)   *Biological Unit Processes for Hazardous Waste Treatment*

3(10) *Bioremediation of Inorganics*

Bioremediation Series Cumulative Indices: 1991-1995

For information about ordering books in the Bioremediation series, contact Battelle Press. Telephone: 800-451-3543 or 614-424-6393. Fax: 614-424-3819. Internet: sheldric@battelle.org.

Biological Unit Processes for Hazardous Waste Treatment

Edited by

Robert E. Hinchee and Rodney S. Skeen
Battelle Memorial Institute

Gregory D. Sayles
U.S. EPA National Risk Management Research Laboratory

BATTELLE PRESS
Columbus • Richland

Library of Congress Cataloging-in-Publication Data

Hinchee, Robert E.
  Biological unit processes for hazardous waste treatment / edited by
  Robert E. Hinchee, Rodney S. Skeen, Gregory D. Sayles.
       p.   cm.
  Includes bibliographical references and index.
  ISBN 1-57477-010-1 (hc : acid free paper)
  1. Hazardous wastes—Biodegradation—Congresses. I. Hinchee,
  Robert E. II. Skeen, Rodney S. III. Sayles, Gregory D.
  TD1061.B53 1995
  628.4'2—dc20   95-32250
                                                                  CIP

Printed in the United States of America

Additional copies may be ordered through:
Battelle Press
505 King Avenue
Columbus, Ohio 43201, USA
1-614-424-6393 or 1-800-451-3543
Fax: 1-614-424-3819
Internet: sheldric@battelle.org

CONTENTS

Contents

v

FOREWORD

This book and its companion volumes (see overleaf) comprise a collection of papers derived from the Third International In Situ and On-Site Bioreclamation Symposium, held in San Diego, California, in April 1995. The 375 papers that appear in these volumes are those that were accepted after peer review. The editors believe that this collection is the most comprehensive and up-to-date work available in the field of bioremediation.

Significant advances have been made in bioremediation since the First and Second Symposia were held in 1991 and 1993. Bioremediation as a whole remains a rapidly advancing field, and new technologies continue to emerge. As the industry matures, the emphasis for some technologies shifts to application and refinement of proven methods, whereas the emphasis for emerging technologies moves from the laboratory to the field. For example, many technologies that can be applied to sites contaminated with petroleum hydrocarbons are now commercially available and have been applied to thousands of sites. In contrast, there are as yet no commercial technologies commonly used to remediate most recalcitrant compounds. The articles in these volumes report on field and laboratory research conducted both to develop promising new technologies and to improve existing technologies for remediation of a wide spectrum of compounds.

The editors would like to recognize the substantial contribution of the peer reviewers who read and provided written comments to the authors of the draft articles that were considered for this volume. Thoughtful, insightful review is crucial for the production of a high-quality technical publication. The peer reviewers for this volume were:

Carolyn M. Acheson, *U.S. Environmental Protection Agency*
Craig E. Armstrong, *Dames & Moore*
Boris N. Aronstein, *Institute of Gas Technology*
Steven D. Aust, *Utah State University*
Craig L. Bartlett, *DuPont Co.*
Netta Benazon, *CH2M Hill Engineering, Ltd.* (Canada)
Judith Bender, *Clark Atlanta University*
Philip C. Bennett, *The University of Texas at Austin*
Christopher J. Berry, *Westinghouse Savannah River Co.*
Richard Bleam, *Bioscience, Inc.*
Reinhard Bolliger, *EcoTerra* (Switzerland)
Robert Booth, *Environment Canada*
Annette Braun-Lüllemann, *Universität Göttingen*
Alec Breen, *U.S. Environmental Protection Agency*
Richard C. Brenner, *U.S. Environmental Protection Agency*
Kim Broholm, *Technical University of Denmark*
Thomas M. Brouns, *Battelle Pacific Northwest*
Mitchell D. Brourman, *Beazer East*

Gunter Brox, *TEKNO Associates*
Ronald J. Buchanan, *DuPont Co.*
Robert S. Burlage, *Oak Ridge National Laboratory*
Timothy E. Buscheck, *Chevron Research & Technology Co.*
Daniel Campos, *DuPont Co.*
Jason Caplan, *En Solve Biosystems, Inc.*
Adam Chen, *Burns & McDonnell Waste Consultants, Inc.*
Jong Soo Cho, *U.S. Environmental Protection Agency*
Wilson Clayton, *Groundwater Technology, Inc.*
Jerzy Dec, *Pennsylvania State University*
Kate Devine, *Biotreatment News*
Kim DeWeerd, *GE Corporate R&D Center*
Joel M. Dougherty, *U.S. Environmental Protection Agency*
Ronald Drake, *Dynamac Corp.*
Richard Egg, *Texas A&M University*
David Elmendorf, *University of Central Oklahoma*
Michael Enzien, *Argonne National Laboratory*
Patrick J. Evans, *AGI Technologies*
John F. Ferguson, *University of Washington*
John Flyvbjerg, *Water Quality Institute* (Denmark)
Chansheng Fu, *University of Cincinnati*
Clyde Fulton, *CH2M Hill Engineering, Ltd.* (Canada)
Jianwei Gao, *Battelle Pacific Northwest*
Phillip A. Gauglitz, *Battelle Pacific Northwest*
David J. Glass, *D. Glass Associates, Inc.*
Fred Goetz, *U.S. Navy*
Doug Goldsmith, *BioSystems Technology, Inc.*
Roberto Guzman, *University of Arizona*
Mark R. Harkness, *GE Corporate R&D Center*
M. H. Henssen, *Bioclear Environmental Biotechnology b.v.*
   (The Netherlands)
Pat Hicks, *Groundwater Technology, Inc.*
Ronald J. Hicks, *Groundwater Technology, Inc.*
Brian S. Hooker, *Battelle Pacific Northwest*
Wendy Huang, *Battelle Columbus*
Aloys Hüttermann, *Universität Göttingen*
Bo Neergaard Jacobsen, *Danish Water Quality Insitute*
Sytze Keuning, *Bioclear Environmental Biotechnology b.v.*
   (The Netherlands)
Richard Lee, *Skidaway Institute of Oceanography*
Robert Legrand, *Radian Corp.*
Tony Lieberman, *ESE Biosciences, Inc.*
Eric H Marsman, *TAUW Milieu b.v.* (The Netherlands)
John E. Matthews, *JMCO, Inc.*
Perry McCarty, *Stanford University*

Linda McConnell, *Logistics Management Institute*
Lawrence C. Murdoch, *University of Cincinnati*
David W. Ostendorf, *University of Massachusetts*
Sorab Panday, *HydroGeologic Inc.*
J. K. Park, *University of Southern California*
John R. Parsons, *University of Amsterdam*
Andrzej Paszczynski, *University of Idaho*
Rick Perkins, *DuPont Co.*
Brent M. Peyton, *Battelle Pacific Northwest*
Peter Phillips, *Clark Atlanta University*
Donald W. Phipps, *Orange County Water District*
Daniel F. Pope, *Dynamac Corp.*
J.H. Portwood, *Alpha Environmental, Inc.*
Hap Pritchard, *U.S. Environmental Protection Agency*
Santo Ragusa, *CSIRO* (Australia)
Laura E. Rice, *Institute of Gas Technology*
M. Marcel Ropars, *Institut Français du Pétrole*
John Sansucusino, *University of Tennessee*
Erwan Saouter, *Proctor and Gamble, ETC* (Belgium)
Edward D. Schroeder, *University of California, Davis*
Manish Shah, *Washington State University*
Rajesh Singh, *DuPont Environmental Remediation Services*
Lee Scott Stevenson, *Groundwater Technology*
Marleen Troy, *Groundwater Environmental Services, Inc.*
Michael J. Truex, *Battelle Pacific Northwest*
Jean-Paul Vandecasteele, *Institut Français du Pétrole*
F. Michael von Fahnestock, *Battelle Columbus*
Peter Wilderer, *Technische Universität München*
Patrick M. Woodhull, *OHM Remediation Services Corp.*
Ten-Fu Yen, *University of Southern California*

Finally, I want to recognize the key members of the production staff, who put forth significant effort in assembling this book and its companion volumes. Carol Young, the Symposium Administrator, was responsible for the administrative effort necessary to produce the ten volumes. She was assisted by Gina Melaragno, who tracked draft manuscripts through the review process and generated much of the correspondence with the authors, co-editors, and peer reviewers. Lynn Copley-Graves oversaw text editing and directed the layout of the book, compilation of the keyword indices, and production of the camera-ready copy. She was assisted by technical editors Bea Weaver and Ann Elliot. Loretta Bahn was responsible for text processing and worked many long hours incorporating editors' revisions, laying out the camera-ready pages and figures, and maintaining the keyword list. She was assisted by Sherry Galford and Cleta Richey; additional support was provided by Susan Vianna and her staff at Fishergate, Inc. Darlene Whyte and Mike Steve proofread the final copy. Judy Ward, Gina Melaragno,

Bonnie Snodgrass, and Carol Young carried out final production tasks. Karl Nehring, who served as Symposium Administrator in 1991 and 1993, provided valuable insight and advice.

The symposium was sponsored by Battelle Memorial Institute with support from many organizations. The following organizations cosponsored or otherwise supported the Third Symposium.

Ajou University–College of Engineering (Korea)
American Petroleum Institute
Asian Institute of Technology (Thailand)
Biotreatment News
Castalia
ENEA (Italy)
Environment Canada
Environmental Protection
Gas Research Institute
Groundwater Technology, Inc.
Institut Français du Pétrole
Mitsubishi Corporation
OHM Remediation Services Corporation
Parsons Engineering Science, Inc.
RIVM–National Institute of Public Health and the Environment
   (The Netherlands)
The Japan Research Institute, Limited
Umweltbundesamt (Germany)
U.S. Air Force Armstrong Laboratory–Environics Directorate
U.S. Air Force Center for Environmental Excellence
U.S. Department of Energy Office of Technology Development
   (OTD)
U.S. Environmental Protection Agency
U.S. Naval Facilities Engineering Services Center
Western Region Hazardous Substance Research Center–
   Stanford and Oregon State Universities

Neither Battelle nor the cosponsoring or supporting organizations reviewed this book, and their support for the Symposium should not be construed as an endorsement of the book's content. I conducted the final review and selection of all papers published in this volume, making use of the essential input provided by the peer reviewers and other editors. I take responsibility for any errors or omissions in the final publication.

Rob Hinchee
June 1995

Biodegradation of a Mixture of Aromatic Hydrocarbons and Heterocyclic NSO-Compounds

Erik Arvin, Jean-Pierre Arcangeli, and Anders T. Gundersen

ABSTRACT

The rate of biodegradation of specific compounds in a complex mixture of aromatic compounds in an aerobic biofilm system is presented. Two systems were investigated, one with a mixture of aromatic hydrocarbons (System A), the other with a mixture of aromatic hydrocarbons, phenols, and heterocyclic nitrogen-, sulfur-, and oxygen-containing compounds (NSO-compounds, System B). At total hydrocarbon concentrations below 0.1 mg/L, the compounds were degraded according to first-order kinetics. With increasing concentrations of total hydrocarbons (up to 1.8 mg/L), the pattern differed. The removal rates of naphthalene, biphenyl, phenanthrene, quinoline, phenol, and o-cresol increased with increasing total hydrocarbon concentration or reached a maximum (zero-order) level, whereas the removal rates of toluene, benzene, o-xylene, 1,4-dimethylnaphthalene, and indene reached a maximum and then decreased. Toluene and benzene are normally considered easily biodegradable under aerobic conditions based on experiments with single compounds. It was, therefore, surprising to observe that these compounds degraded slowly in complex mixtures at total hydrocarbon concentrations of only 1 to 1.8 mg/L. The inhibition of benzene and toluene degradation was particularly strong in System B, consisting of a complex mixture of aromatic hydrocarbons, phenols, and heterocyclic NSO-compounds.

INTRODUCTION

In situ or on-site bioreclamation of contaminated waste sites often involves treatment of complex mixtures of organic compounds, for example mixtures of hydrocarbons (HC), creosote compounds, or chlorinated hydrocarbons.

Most biodegradation studies focus on single compounds or simple mixtures. This implies a risk of overlooking phenomena that appear when dealing with multicomponent systems. Biodegradation of mixtures of compounds offers possibilities of both mutual inhibition or stimulation, for example by cometabolism.

The results of several studies have revealed that there may be complex inter-actions between the compounds in mixtures (Arvin et al. 1989, Alvarez and Vogel 1991, and Jensen and Arvin 1993). The purpose of this study was to investigate how individual aromatic compounds behave in complex mixtures that simulated the water-soluble fraction from creosote and oil pollution. A biofilm system was used because the biomass of many natural systems is fixed to solid materials.

MATERIALS AND METHODS

Biofilm Reactor and Culture Conditions

The experimental system was an annular biofilm reactor consisting of two concentric Plexiglas™ cylinders (surface area 0.16 m², volume 0.96 L), where the inner cylinder rotates at 200 RPM (Kristensen and Jansen 1980). The rotation ensures efficient mixing in the bulk liquid, and the effect of hydraulic sheer keeps the thickness of the biofilm uniform. The reactor operates without headspace. Experiments in combination with modeling of biodegradation has shown that sorption in the Plexiglas™ may be a problem only for the compounds phenanthrene, dibenzofuran, and dibenzothiophene at relatively high concentrations when the removal is zero order (not shown here).

The outer cylinder was equipped with removable slides opening inward for measurement of biofilm thickness and density. The temperature was 10 to 12°C, which is similar to the mean temperature of the groundwater in temperate climates. The reactor was inoculated with a mixed culture originally obtained from an oil-polluted sandy aquifer (Grindsted, Denmark).

The flow was adjusted in each experimental run, resulting in hydraulic retention times in the range of 0.4 to 16 h. The wide range of residence times was used in order to observe zero-order as well as first-order removal kinetics. At least 3.5 retention times were allowed from the adjustment to new experimental conditions before the next sampling was done. The oxygen concentration in the reactor was measured regularly by the Winkler method. The concentration was 3 to 8 mg/L, ensuring no oxygen limitation in all cases. This can be inferred from the biofilm theory of Harremoës (1978) on the basis of the concentrations of aromatic compounds and the oxygen concentration. The pH was not regulated, but the variation was small, around 7.

Substrate

Two different mixtures of aromatic compounds were studied. During one experimental run (System A, BIO1-4), a mixture of aromatic hydrocarbons were fed to the biofilm system: benzene, toluene, o-xylene, naphthalene, 1,4-dimethyl-naphthalene, biphenyl, and phenanthrene. Influent concentrations of each compound were 400 to 600 µg/L, except for phenanthrene which was 100 to 300 µg/L. In addition to these hydrocarbons, the other experimental run (System B, BIO2-4) contained the aromatic hydrocarbon indene, phenols, and heterocyclic

NSO-compounds: phenol (300 µg/L), *o*-cresol (300 µg/L), pyrrole (300 µg/L), quinoline (200 to 300 µg/L), dibenzofuran (200 to 250 µg/L), dibenzothiophene (150 to 250 µg/L), and fluorenone (300 µg/L). The concentrations of hydrocarbons were 250 to 300 µg/L, except for phenanthrene which was 150 to 250 µg/L.

The mixture of aromatics was dissolved in water and stored in a glass container. The substrate solution was changed regularly to avoid biological degradation. The concentrations were measured at the inlet to the biofilm reactor.

The mineral medium contained the following macroions per liter of distilled water: 147 mg $CaCl_2$, 0.4 mg $FeCl_3 \cdot 6H_2O$, 9.4 mg KCl, 50 mg $MgSO_4 \cdot 7H_2O$, 12.2 mg $NaNO_3$, 5.8 mg Na_2HPO_4, and 100 mg $NaHCO_3$.

The trace minerals were (per liter): 14 µg Mn(II), 14 µg Zn(II), 10 µg Mo(VI), 10 µg B, 8 µg Cu(II), and 8 µg Co(II). Analytical-grade chemicals were used.

Analytical Techniques

Analysis of organic compounds was performed with a Shimadzu GC-9A gas chromatograph (GC), equipped with a OCI-G9 on-column injector and a flame ionization detector (FID) connected with a Shimadzu C-R 3A computing integrator. The GC-column was a 20-m Chrompack WCOT fused-silica (liquid phase: CP-Sil-5 CB) 0.53 mm inner diameter (ID). Phenol and *o*-cresol were derivatized according to Flyvbjerg et al. (1993). All other analysis was done according to Arvin et al. (1989).

RESULTS AND DISCUSSION

The rates of removal of the individual aromatic hydrocarbons in the mixture versus total (aggregate) hydrocarbon concentration inside the reactor (= effluent concentration) in the experiment with hydrocarbons alone (System A) are shown in Figure 1. The rates are expressed in terms of mass removal per unit of surface area, mg/m²/d. In general, the removal was first-order at total hydrocarbon (HC) concentrations below 100 µg/L. The removal rate of naphthalene and biphenyl reached a constant level (zero-order) above concentrations of 800 to 1,000 µg/L. Phenanthrene removal increased in the whole concentration range, whereas the removal rate of benzene, toluene, *o*-xylene, and 1,4-dimethylnaphthalene decreased for total HC concentrations above 1,000 to 1,500 µg/L.

The corresponding picture of hydrocarbon removal in the mixture of aromatic hydrocarbons, phenols, and heterocyclic NSO-compounds (System B) is shown in Figures 2 and 3. The pattern from Figure 1 is repeated, except for two important aspects. First, the removal rates of individual compounds were much lower for naphthalene and biphenyl, by a factor of more than 2. Second, the decrease in benzene, toluene, *o*-xylene, and 1,4-dimethylnaphthalene degradation were much more pronounced. The maximum benzene removal was 10 times lower than in System A, and the removal of benzene was close to zero above a total HC concentration of 500 µg/L. The maximum removal of toluene was 4 times lower than in System A, and the toluene removal was also close to zero above

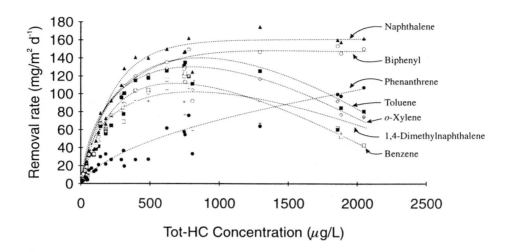

FIGURE 1. Removal rate of specific aromatic hydrocarbons in the mixture versus the total hydrocarbons concentration in the reactor (System A): naphthalene (▲); biphenyl (O); phenanthrene (●); toluene (■); o-xylene (◊); 1,4-dimethylnaphthalene; (+) benzene (□).

a total HC concentration of 1,500 μg/L. The removal of indene was fast up to a total HC concentration of 800 to 1,000 μg/L, and it then decreased.

The removal of heterocyclic compounds (except pyrrole) and phenols versus the total HC concentration is shown in Figure 4. The removal of quinoline, phenol, and o-cresol increased in the whole HC concentration range, whereas the removal of dibenzothiophene and dibenzofuran reached a constant level. For

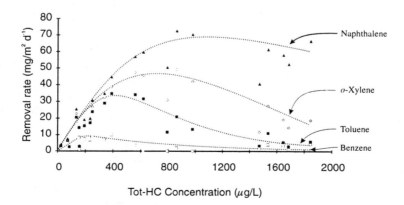

FIGURE 2. Removal rate of specific aromatic hydrocarbons in the mixture versus the total hydrocarbons concentration in the reactor (System B): naphthalene (▲); o-xylene (◊); toluene (■); benzene (□).

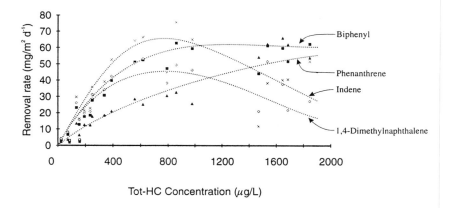

FIGURE 3. Removal rate of specific aromatic hydrocarbons in the mixture versus the total hydrocarbons concentration in the reactor (System B): biphenyl (■); phenanthrene (▲); indene (x); 1,4-dimethylnaphthalene (◊).

fluorenone, maximum removal was attained at a total HC concentration of 500 to 1,000 µg/L, after which the removal rate decreased. The removal of pyrrole is not shown because it was close to zero.

There was a pronounced difference in the degradability of the individual aromatic compounds. Naphthalene, biphenyl, phenanthrene, phenol, *o*-cresol, and quinoline were easily degraded, whereas the removal of compounds such

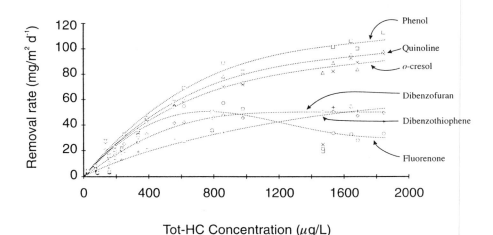

FIGURE 4. Removal rate of specific phenols, and heterocyclic NSO-compounds in the mixture versus the total hydrocarbons concentration in the reactor (System B): phenol (□); quinoline (△); *o*-cresol (x); dibenzofuran (◊); dibenzothiophene (+); fluorenone (O).

as benzene, toluene, *o*-xylene, 1,4-dimethylnaphthalene, indene, and fluorenone was inhibited to varying degrees at the highest total HC concentrations. This is particularly surprising for benzene and toluene, which are both easily degradable in many situations. However, in previous studies the recalcitrance of benzene has been attributed to the presence of pyrrole (Arvin et al. 1989).

It is not possible from the present data to infer whether the decrease in degradation rate of benzene, toluene, *o*-xylene, 1,4-dimethylnaphthalene, indene, and fluorenone at total hydrocarbon concentrations above 500 to 1,000 µg/L was due to competitive inhibition or any other type of inhibition. This has to be determined from experiments with a few compounds at a time.

CONCLUSIONS

1. All compounds except pyrrole were removed to some extent.
2. At total hydrocarbons concentrations of below 0.1 mg/L, the compounds were degraded according to first-order kinetics. With increasing concentrations of total hydrocarbons (up to 1.8 mg/L), the pattern differed. The removal rates of naphthalene, biphenyl, phenanthrene, quinoline, phenol, and *o*-cresol increased with increasing total hydrocarbon concentration or reached a maximum (zero-order) level, whereas the removal rates of toluene, benzene, *o*-xylene, 1,4-dimethylnaphthalene, and indene reached a maximum and then decreased.
3. The removal of benzene and toluene was strongly inhibited when the concentration of other aromatic compounds increased to between 1 and 1.8 mg/L, in particular in the mixture with phenols and heterocyclic NSO-compounds. Toluene and benzene are normally considered easily biodegradable under aerobic conditions judged on the basis of experiments with single compounds.

REFERENCES

Alvarez, P.J.J., and T. Vogel. 1991. "Substrate interactions of benzene, toluene and *para*-xylene during microbial degradation by pure cultures and mixed culture aquifer slurries." *Appl. Environ. Microbiol. 57*: 2981-2985.

Arvin, E., B. K. Jensen, and A. T. Gundersen. 1989. "Substrate interactions during aerobic biodegradation of benzene." *Appl. Environ. Microbiol. 55*(12): 3221-3225.

Flyvbjerg, J., C. Jørgensen, E. Arvin, B. J. Jensen, and S. K. Olsen. 1993. "Biodegradation of *ortho*-cresol by a mixed culture of nitrate-reducing bacteria growing on toluene." *Appl. Environ. Microbiol. 59*(7): 2286-2292.

Harremoës, P. 1978. "Biofilm kinetics." In R. Mitchell (Ed.), *Water Pollution Microbiology*, Vol. 2, pp. 71-109. John Wiley & Sons, Inc., New York, NY.

Jensen, K. B. and E. Arvin. 1993. "Aromatic hydrocarbon degradation specificity of an enriched denitrifying mixed culture." In R. E. Hinchee, B. C. Alleman, R. E. Hoeppel, and R. N. Miller (Eds.), *Hydrocarbon Bioreclamation*, pp. 411-417. Lewis Publishers, Chelsea, MI.

Kristensen, G. H., and J.L.C. Jansen. 1980. *Fixed Film Kinetics. Description of Laboratory Equipment*. Rep. 80-58. Department of Environmental Engineering and Science, Technical University of Denmark. DK-2800 Lyngby.

Residence Time Distribution Studies and Design of Aerobic Bioreactors

Manish M. Shah and David K. Stevens

ABSTRACT

The design of bioreactors for treating contaminated groundwater is normally based on degradation rate data and residence time distribution of the contaminant in the reactor. In our previous studies, the evaluation of the performance of fixed-film bioreactors (sparged, cocurrent, upflow) used for treatment of wood-preserving chemicals suggested that irrespective of water flowrates, the removal efficiency of naphthalene was the least affected among the mixture of PAHs. We subsequently reported that certain tracers, such as the polyaromatic Rhodamine B dye (which acts as a surfactant), had about half the residence time compared to KCl (no surfactant properties) because of its partitioning at the bubble interface in an upflow reactor. Later, we developed a model to understand the removal of contaminants because of their attachment and partitioning to the air bubbles. In the current communication, we are reporting the use of this model to understand the effect of bubble partitioning of pollutants in an aerobic fixed-film bioreactor during treatment of groundwater at a wood-preserving site. The study suggests that significant portions of polycyclic aromatic hydrocarbons (PAHs), especially naphthalene, having high Henry's law constant values would be removed from the reactor via volatilization. The results also suggest that the residence time of contaminants is independent of water flowrate but is dependent on the air flowrate and bubble diameter. The observed behavior of naphthalene removal efficiency with respect to water flowrate might be because the dominating pathway of naphthalene removal could be volatilization, which will not be affected significantly by changes in water flowrate.

INTRODUCTION

In recent years, bioremediation has received considerable attention as a cost-effective technology for treating contaminated groundwater, sludge, sediment, and soil. For the proper design of bioreactor, knowledge about the residence time distribution of a pollutant in the reactor is required (Hill 1977). In most cases, such studies are performed using chemically nonreactive tracers. The research reported herein demonstrates that the tracer studies, although useful, have serious limitations if the physicochemical properties of contaminants are much different than those of the tracer. The most important property is the ability of contaminants to partition at or inside the bubble surface.

Our previous studies evaluated the performance of field-scale (10,000 gal) fixed-film bioreactors (sparged, cocurrent, upflow) (Stevens et al. 1994). The evaluation of the performance of bioreactors suggested that, irrespective of water flowrate changes, the removal efficiency of naphthalene was the least affected among the mixture of PAHs. As a result, it was hypothesized that removal of naphthalene must be taking place via a pathway that is not affected significantly by changes in the water flowrate or residence time. We hypothesized that PAHs could attach and partition themselves at the air/water interface. We also hypothesized that, depending on the partitioning ability and volatility of the contaminants, they can be stripped out of the bioreactor. Consequently, we developed a model to understand the removal of contaminants because of their attachment and partitioning (Stevens et al. 1994). The potential for volatilization was estimated using a mass balance approach that incorporated the physical/chemical properties of the contaminant, reactor geometry, water flowrates, and air flowrates. Results of this model for naphthalene, phenanthrene, and benzene are reported here.

Our analysis suggests that the observed behavior of naphthalene removal efficiency with respect to water flowrate might be because the dominating pathway of naphthalene removal could be volatilization, which will not be affected significantly by changes in water flowrate. The residence time of the contaminant is independent of water flowrate but is highly dependent on the air flowrate and bubble diameter. Contaminants with high Henry's law constants such as naphthalene could be lost via volatilization in a sparged aerobic bioreactor.

RESULTS

Residence time distributions of Rhodamine B and KCl in the reactor were investigated in two aerated upflow reactors in series (Shah et al. 1993). It was concluded that the residence time of Rhodamine B was lower than that of KCl because of its partitioning at the bubble interface (Shah et al. 1993). The consequences of contaminant accumulation at the bubble interface could be serious, with two major consequences. First, the concentration of the contaminants throughout the reactor may be much less than near the top, where foam accumulates, as has been observed in the case of Rhodamine B (Shah et al. 1993). Second, contaminants having high Henry's law constants can be lost via volatilization from the reactor without degradation.

The potential for volatilization was estimated using a simple mass balance model that incorporates physical/chemical properties of the contaminants, reactor geometry, and air and water flowrates (Stevens et al. 1994). The model did not account for biomass. For development of the model, the reactor was assumed to be a continuously stirred tank reactor (CSTR) with respect to the aqueous phase and plug flow with respect to the air phase. The dimensions and conditions chosen for the reactor were based on an actual large-scale fixed-film bioreactor (sparged, cocurrent, upflow) in operation at a wood-preserving site and are shown in Table 1.

In the current communication, the model was used predict the removal efficiency of contaminant due to volatilization. Naphthalene is by far the most abundant of the PAHs at the wood-preserving site and the most volatile of the contaminants, and so it was used as an example. In addition, pentachlorophenol (PCP), also present at the site, was used as an example of a nonvolatile contaminant that has been shown to partition onto bubble surfaces (Valsaraj et al. 1991). The removal efficiency caused by volatilization is defined as

$$\text{Efficiency} = (1 - \frac{C_{we}}{C_{wo}}) \times 100\% \tag{1}$$

where C_{we} and C_{wo} are the effluent and influent concentrations. The relationship between the removal efficiency via volatilization and bubble diameter at various flowrates is shown in Figure 1. For values of bubble diameter less than 1 mm, bubble size no longer influences the air stripping efficiency. At the moderate flowrate of 25 scfm (0.71 m^3/min), approximately 30% of the naphthalene could be regarded as strippable. For larger diameters, the importance of stripping as a removal mechanism is diminished. For naphthalene, the importance of bubble partitioning is negligible compared to stripping (Stevens et al. 1994).

The volatilization removal efficiencies of phenanthrene, naphthalene, and benzene are shown in Figure 2 for various bubble diameters at a flowrate of 25 scfm

TABLE 1. Dimensions and conditions of the reactor considered for the model study.

No.	Item	Values
1	Volume (m^3)	37.65
2	Height (m)	4.27
3	Diameter (m)	3.35
4	Air Flowrate (m^3/min)	0.7073
5	Water Flowrate (m^3/min)	0.0454
6	Flow Conditions	Upflow, cocurrent
7	Packing Material[a]	BioDek™

(a) Munter BioDek 10960, Munters Inc., Ft. Myers, FL. The values of surface area and void ratio were 126 m^2/m^3 and 0.95, respectively.

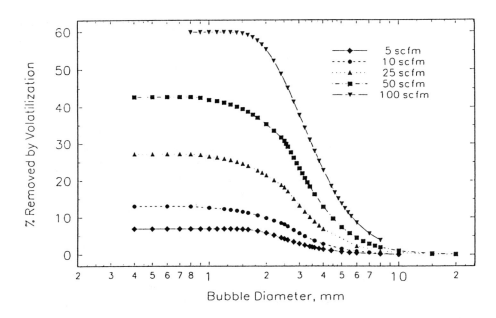

FIGURE 1. Removal efficiency due to volatilization as a function of flowrate and bubble diameter for naphthalene.

$(0.71 \ m^3/min)$. As can be expected, very little phenanthrene is removed by volatilization compared to naphthalene. However, the importance of bubble partitioning is greater because of its lower Henry's law constant values and higher bubble partitioning coefficient (Stevens et al. 1994).

In an actual large-scale bioreactor used for treating wood-preserving chemicals, it was found that the removal efficiency of naphthalene (measured by disappearance of the parent compound from the reactor) was least affected at a higher flowrate (Shah 1995). Table 2 compares the removal effect of flowrates on the removal efficiency of naphthalene, phenanthrene, and PCP in that reactor system. An important pathway of naphthalene removal could be volatilization, which will not be affected significantly by changes in water flowrate at the air:water ratio of 15:1 to 20:1 used.

CONCLUSIONS AND FUTURE STUDIES

In our previous studies (Stevens et al. 1994), the evaluation of bioreactor used for treatment of wood-preserving chemicals suggested that, irrespective of water flowrates, the removal efficiency of naphthalene was the least affected among the mixture of PAHs. The results reported here suggest that, depending on the volatility of the contaminants, they can be stripped out of the bioreactor.

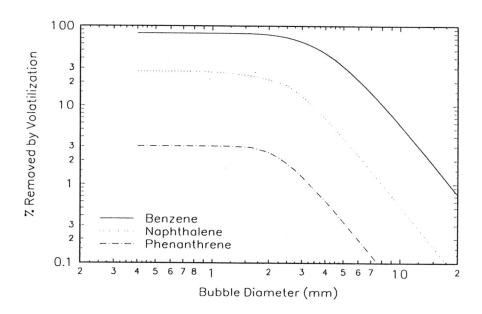

FIGURE 2. Effect of bubble diameter on predicted removal efficiency due to volatilization for benzene, naphthalene, and phenanthrene at a water flow-rate of 10 gpm (0.038 m³/min) and an air flowrate of 25 scfm (0.71 m³/min).

The results also suggest that the removal efficiency caused by volatilization is dependent on the air flowrate and bubble diameter. The dominating pathway of naphthalene removal could be volatilization, which will not be affected significantly by changes in water flowrate. Future studies should be conducted by using tracers having properties similar to those of the contaminants. Further

TABLE 2. Henry's law constant of pollutants and their removal efficiency in large-scale bioreactor.

No.	Pollutant	Henry's Law Constant (Dimensionless)	% Decrease in Degradation Removal Efficiency[a]
1	Naphthalene	2E-2	1
2	PCP	3E-6	10.6
3	Phenanthrene	6E-4	12.5

(a) The removal efficiency is the ratio of the concentration of the pollutant in the effluent to its concentration in the influent multiplied by 100. The decrease in removal efficiency is the difference in the degradation removal efficiency at 15 gpm (0.057 m³/min) and at 10 gpm (0.038 m³/min).

studies should also be conducted to establish the relationships among foaming, physicochemical properties of contaminants, residence time distribution, and rates of degradation of contaminants in a bioreactor.

REFERENCES

Hill, C. G. 1977. *Introduction to Chemical Engineering Kinetics and Reactor Design.* John Wiley & Sons, Inc., New York. 320 pp.

Shah, M. M. 1995. "Performance evaluation of the fixed film bioreactors at the Superfund site Libby, Montana." M.S. Thesis, Utah State University, Logan, UT.

Shah, M. M., D. K. Stevens, and H. C. Zhu. 1993. "Residence time distribution of Rhodamine B and KCl tracers in aerated packed bed reactors." *Joint CSCE-ASCE National Conference on Environmental Engineering,* Montreal, Vol. 2, pp. 1835-1842.

Stevens, D. K., M. M. Shah, and R. C. Sims. 1994. *Evaluation of the at-grade groundwater treatment system at Champion International, Libby,* Subcontract S-005-2950, Task Order Nos. 2150-401 and 2350-B13, Dynamac Corporation, Rockville, MD.

Valsaraj, K. J., X. Y. Lu, and L. J. Thibodeaux. 1991. "Continuous counter current solvent sublation for the removal of hydrophobic organics from water." *Wat. Res.* 25:1061-1072.

Biological Fluidized-Bed Treatment of Groundwater from a Manufactured Gas Plant Site

Gary M. Grey, O. Karl Scheible, Joy A. Maiello,
William J. Guarini, and Paul M. Sutton

ABSTRACT

Bench- and pilot-scale biological treatability studies were performed as part of a comprehensive study for developing an on-site treatment system for contaminated groundwater at a former manufactured gas plant site. The bench-scale work, which included evaluations of activated sludge and fluidized-bed biological processes, indicated that a carbon-based fluidized-bed process was most appropriate. The process was then demonstrated on a pilot level at the site. The bench and pilot studies demonstrated significant reductions of chemical oxygen demand (COD), and all target organics including polycyclic aromatic hydrocarbons (PAHs) and volatile organic compounds (VOCs).

INTRODUCTION

Fluidized-bed biological treatment was evaluated as part of a comprehensive bench- and pilot-scale groundwater treatability study for a manufactured gas plant (MGP) site located in Toms River, New Jersey (HydroQual 1994). The study was sponsored by Jersey Central Power and Light Company (JCP&L), New Jersey Natural Gas (NJNG), and the Electric Power Research Institute (EPRI), and was conducted over the period of April 1993 through August 1994.

The overall project focused on the evaluation of an integrated process sequence for treating contaminated groundwater. At the Toms River site, as well as most MGP sites, the groundwaters contain dense, nonaqueous-phase liquids (DNAPLs), tars, and emulsified oils. Physical/chemical processes were evaluated for removal of these constituents. The primary soluble contaminants are PAHs and VOCs, primarily benzene, toluene, ethylbenzene, and xylenes (BTEX), and styrene. Biological treatment, air stripping, and carbon adsorption were evaluated for their capacity to remove these constituents. Air stripping and carbon adsorption were found to be effective as alternatives to biological treatment for VOC and PAH removal. This paper presents the results of the bench- and pilot-scale evaluations of fluidized-bed biological treatment. The results of the compre-hensive study are presented in a companion paper (Scheible et al. 1995).

EXPERIMENTAL METHODS

The objective of the study was to evaluate candidate treatment technologies through bench-scale testing, followed by an evaluation of appropriate treatment trains through pilot testing. The bench-scale studies were conducted on ground-water samples collected at the site. The pilot studies were conducted on site with groundwater drawn directly to the pilot plant from selected wells. Both study phases included operation over a range of groundwater characteristics that might be encountered by the full-scale system over its operating life.

The bench studies determined that dissolved air flotation (DAF) was an appropriate pretreatment process producing a subnatant suitable for biological treatment. The bench-scale biological treatment evaluations included evaluations of activated sludge, sand-based (SFB), and carbon-based (CFB) fluidized-bed processes. The activated sludge unit consisted of a 3-L reactor and an external clarifier. The fluidized-bed units consisted of a 1-L reactor with a 0.5-L fluid-bed volume and an external recycle aeration cell. The activated sludge and sand-based fluidized-bed systems were initially seeded with an activated sludge from an oil refinery to hasten the acclimation period. The bench-scale CFB was started with carbon used on a previous MGP site treatability study. All three bench units were fed DAF subnatant from groundwater taken from a highly contaminated (high-strength) source area of the site. After this, only the CFB was operated and fed a DAF subnatant from more moderately contaminated (moderate-strength) site groundwaters.

The pilot study consisted of a series of three campaigns which encompassed a wide range of groundwater contaminant levels. The onsite pilot-scale biological treatment process was preceded by a DAF system when treating high-strength (Campaign 1) and moderate-strength (Campaign 2) groundwaters. The DAF pro-cess was not included when treating low-strength (Campaign 3) groundwater because free oils or tars were not present. The skid-mounted CFB unit was pro-vided by Envirogen, Inc., Lawrenceville, New Jersey. It consisted of a 1-ft (0.3-m)-diameter by 15-ft (4.6-m)-deep reactor with feed and recycle pumps, oxygenator, recycle tank, chemical feed equipment, and microprocessor controls.

BENCH-SCALE EVALUATION RESULTS

Table 1 presents a summary of the operations and performance data for the bench-scale systems. The organic loadings to the bench-scale systems were relatively low: approximately 0.2 gCOD/g-d for the suspended growth system and 1.8 to 2.0 gCOD/L-d for the fluidized-bed bioreactors. No attempt was made to optimize loadings; rather, the bench studies were conducted to assess

TABLE 1. Summary of bench-scale biological treatment performance.

		High Strength			Moderate-Strength Carbon-Based Fluidized-Bed
		Activated Sludge	Fluidized-Bed		
			Sand	Carbon	
Organic Loading		0.19[a]	1.8[b]	1.8[b]	2[b]
Detention Time	(h)	6.5	1.7	1.7	0.9
Influent					
COD-Total	(mg/L)	128	128	128	74
COD-Filtered	(mg/L)	83	83	83	62
BTEX	(µg/L)	11	11	11	6
PAH	(µg/L)	1,224	1,224	1,224	78
Effluent					
COD-Total	(mg/L)	63	74	57	44
COD-Filtered	(mg/L)	48	59	37	38
BTEX	(µg/L)	<5	<5	<5	<5
PAH	(µg/L)	24	45	<5	<5

(a) gCOD/gMLSS.
(b) gCOD/L-d.

the biodegradability of the dissolved organics and to determine which process configuration would be most appropriate.

All three biological treatment systems were found to provide a substantial reduction of COD and PAHs. The CFB provided the best treatment with a 71% reduction of COD on a total-in minus filtered-out basis and effluent PAH concentrations less than 5 µg/L. Based on this performance, the CFB process was selected for continued operation on the moderate-strength subnatant. The effluent produced from using the moderate-strength subnatant was of similar quality to that produced from treating the high-strength subnatant, with an effluent soluble COD of 38 mg/L and effluent PAHs less than 5 µg/L. The bench-scale studies demonstrated that biological treatment was viable and that the CFB was the most appropriate process. For these reasons, the CFB process was incorporated into the pilot-scale treatment system at the Toms River MGP site. The studies suggested that there was sufficient nitrogen, but that phosphorus was deficient for effective long-term biological treatment. Phosphorus addition thus was included as standard practice for the pilot studies.

PILOT STUDY RESULTS

The CFB process was initially charged with coconut-based granular activated carbon and operated at a hydraulic flux of approximately 530 m^3/m^2-d with tap water to wash out fines. High-strength DAF subnatant was then fed to the unit for approximately 3 weeks to saturate the carbon with organics before seeding. This startup procedure was used to minimize any subsequent biases to the organics removal that might be ascribed to the adsorptive capacity of the virgin carbon. Once saturated, the bed was seeded with activated sludge taken from a local oil refinery's treatment plant.

Figure 1 presents a chronological plot of CFB loadings and performance, denoted by campaign. During the three campaigns, COD loads ranged from 2 to 4.5 gCOD/L-d, progressively increasing during the program. Influent total and filtered COD decreased from one campaign to the next, reflecting the change in groundwater characteristics. Effluent quality, as measured by filtered COD, averaged 30 to 80 mg/L, responding slightly to influent COD concentration. Sand filtration was practiced to remove CFB effluent suspended solids. In a full-scale system, a clarifier or high-rate prefiltration may be more appropriate for effluent solids removal. Subsequent filtration would then be applied if carbon adsorption were to be used for final effluent polishing.

Figure 2 presents COD removal as a function of COD loading for all campaigns. The line shown through the data is a least-squares regression, indicating an average COD removal rate of 56%. COD removal dropped to approximately 42% during Campaign 3, although the effluent-filtered COD was the lowest of all campaigns, averaging 37 mg/L. The lower removal rate is an artifact of the low initial COD of the groundwater.

Table 2 presents a summary of influent and effluent quality with respect to the organic constituents for each campaign. During Campaign 1, the CFB process was capable of reducing total volatiles by 99% to a total effluent VOC concentration of 102 µg/L. PAHs were reduced by 95% to 92 µg/L. All of the residual PAH was adsorbed to effluent suspended solids; the PAH concentrations were less than 5 µg/L after filtration. The CFB feeds for Campaigns 2 and 3 were generally similar, with influent total VOCs ranging from 2,500 to 1,600 µg/L and PAHs ranging from 210 to 270 µg/L. During both campaigns, essentially complete removal of BTEX and PAH was observed, with effluent concentrations of less than 2 µg/L and less than 5 µg/L, respectively.

CFB off-gas flows were measured and samples taken to determine the significance of volatilization as a removal mechanism for VOCs. Essentially no off-gas flow was detected, and gas samples of the reactor headspace contained less than 1 ppm VOCs. The lack of any off-gas was attributed to the high oxygen transfer efficiencies of the oxygenator with a pure oxygen supply. Also, the CO_2 produced as a by-product of biochemical oxidation tended to stay in solution and depress the pH rather than strip out as an off-gas. A pH depression was observed in the pilot study to a point where the caustic feed system was activated to maintain pH at approximately 7. Caustic requirements averaged 30 to 40 mg/L as $CaCO_3$.

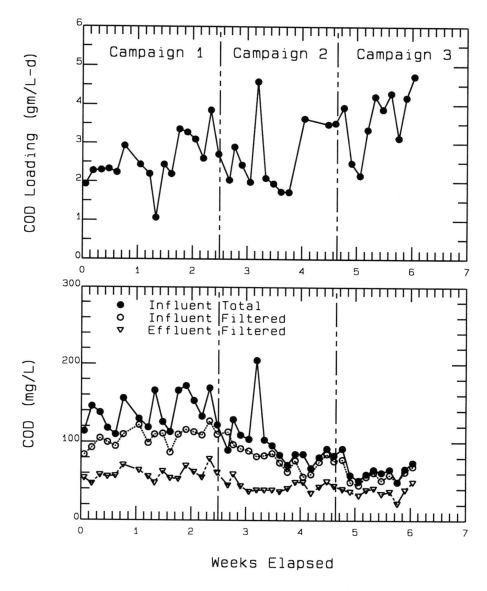

FIGURE 1. Chronological plot of COD loading and performance.

Analysis of fluidized-bed media indicated biomass concentrations of 5,000 to 10,000 mg/L. During Campaigns 1 and 2, net sludge production was estimated to be between 0.2 to 0.4 gTSS/g COD removed. TCLP analysis of the biological sludge indicated compliance with all Toxicity Characteristic Leaching Procedure (TCLP) regulatory limits.

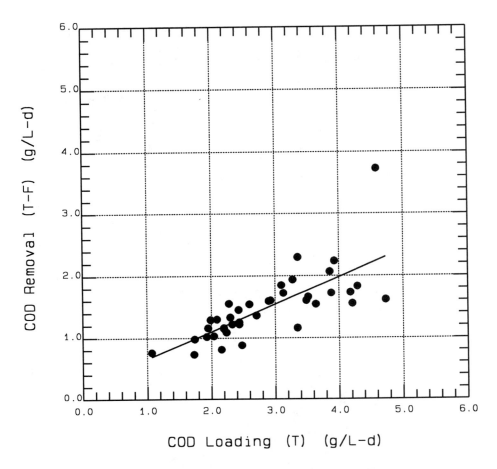

FIGURE 2. COD removal as a function of the total COD loading rate.

SUMMARY

The carbon-based fluidized-bed biological treatment system was successful in treating DAF subnatant generated from high- and moderate-strength ground-waters and for direct application to low-strength groundwaters. Substantial COD and complete BTEX and PAH removals were observed. Residual PAHs present in the CFB effluent were found to be adsorbed to effluent suspended solids, and removed by subsequent filtration. Biodegradation is believed to be the primary removal mechanism for BTEX; volatilization was found to be insignificant. Phosphorus addition was necessary to satisfy nutrient requirements. Alkalinity addition was also necessary for effective pH control. Oxygen requirements and sludge production estimates appear to be similar to those of an activated sludge system. The biosolids produced are nonhazardous based on TCLP criteria.

TABLE 2. Summary of CFB influent and effluent BTEX and PAH results.

	Campaign 1		Campaign 2		Campaign 3	
	Influent	Effluent	Influent	Effluent	Influent	Effluent
Volatiles (µg/L)						
Benzene	1,900	9.2	960	<2	540	<2
BTEX	5,200	98	2,500	<2	1,600	<2
Total VOCs	16,800	102	2,500	<2	1,600	<2
PAHs (µg/L)						
Naphthalene	610	<5	3.5	<5	48	<5
Total PAH	1,800	92	210	<5	270	<5

ACKNOWLEDGMENTS

Gary Grey, Karl Scheible, and Joy Maiello are with HydroQual, Inc., which conducted this study under subcontract to Foster Wheeler Environmental Corp, Lyndhurst, New Jersey. William Guarini is with Envirogen, Inc., Lawrenceville, New Jersey, which supplied the fluidized-bed system. Paul Sutton is with Paul M. Sutton Associates, which provided consultation to the project. The work was overseen by the EPRI Steering Committee, comprised of Dr. I. Murarka (EPRI), Mr. C. R. Sweeney (JCP&L), Mr. E. Sawicki (NJNG), and Mr. A. Jain (ERM, Inc.). Mr. Jain served as the Project Manager for the EPRI Steering Committee.

REFERENCES

HydroQual, Inc. 1994. *Bench- and Pilot-Scale Treatability Studies for Contaminated Ground Waters from the Toms River Former Manufactured Gas Plant Site.* Prepared under subcontract to Foster Wheeler Environmental Corporation, Lyndhurst, NJ, for the Electric Power Research Institute, Jersey Central Power and Light, and New Jersey Natural Gas.
Scheible, O. K, G. M. Grey, and J. A. Maiello. 1995. "Evaluation of an Integrated Treatment System for MGP Site Groundwaters." In R. E. Hinchee, R. E. Hoeppel, and D. B. Anderson (Eds.), *Bioremediation of Recalcitrant Organics*, pp. 63-70. Battelle Press, Columbus, OH.

Treatment of PAHs in Waters Using the GAC-FBR Process

Robert F. Hickey, April Sunday, Dan Wagner,
Veronica Groshko, Raj V. Rajan, A. Leuschner,
and Tom Hayes

ABSTRACT

Pilot studies were conducted to determine the utility of the granular activated carbon fluidized-bed reactor (GAC-FBR) process to treat groundwater from manufactured gas plant (MGP) sites containing polycyclic aromatic hydrocarbons (PAHs) and a process effluent water from a deep subsurface dense, nonaqueous-phase liquid (DNAPL) removal process at an MGP site. Removal of naphthalene exceeded 99.9%, and overall PAH removals of 99+% were observed at organic loading rates (OLRs) exceeding 4-kg chemical oxygen demand (COD)/m^3-d and a hydraulic retention time (HRT) of about 6 min. Analysis of PAHs accumulated on GAC and oxygen consumption clearly demonstrated that removal of 2- to 4-ring PAHs was due primarily to biological oxidation and not to adsorption. Analysis of influent and effluent samples using Microtox® indicated removal of toxicity. Full-scale application of the GAC-FBR process has begun at a Superfund site in Pennsylvania. The GAC-FBR is being used to treat a 15-gal per min (gpm) (57 L per min) process effluent flow from a subsurface DNAPL removal process (Crow™ process). Initial results confirm the ability of the process to treat PAHs at high OLRs and short HRTs. The GAC-FBR is scheduled for use at a second MGP site for treating groundwater contaminated with PAHs; benzene, toluene, ethylbenzene, and xylenes (BTEX); phenolics; and cyanide.

INTRODUCTION

Development of the biological FBR process dates back to the late 1960s (Jeris et al. 1974). Initial work focused on laboratory and pilot-scale denitrification of nitrified municipal wastewater (Jeris and Owens 1975; Jeris et al. 1977). Application of the process for aerobic degradation of organics, nitrification (Sutton et al. 1979), anaerobic treatment of high-strength wastes (Hickey and Owens 1981; Suidan et al. 1981) and sewage (Switzenbaum and Jewell 1980) followed.

In the mid-1980s, it was recognized that the technology had the potential of substantially reducing the cost of treating contaminated groundwater and process effluents containing low levels of toxics. The key features offered by the FBR process are:

- Large surface area for biomass attachment
- High biomass concentrations
- Ability to control and optimize biofilm thickness
- Minimal plugging, channeling, or gas holdup
- High mass transfer properties through maximum contact between biomass and substrate
- No off-gas produced; therefore, no air quality concerns
- Tailored biomass carrier to optimize the system for certain contaminant classes.

The key advantages to using GAC as the biomass carrier are:

- High removal efficiencies from time zero, because the GAC adsorbs contaminants during startup
- The system performs as a biological reactor once the biofilm is established
- Contaminants such as BTEX, PAHs, and solvents adsorbed onto GAC during startup are, to a large extent, desorbed and degraded (bioregeneration) once an active biofilm is established
- Dual biological and adsorptive removal mechanisms ensure robust performance during perturbations in contaminant concentrations, interruption in feed, etc.
- Slow-to-degrade contaminants are held in the system for a longer time, resulting in more complete degradation
- The adsorption mechanism is available for removing nonbiodegradable compounds.

The GAC-FBR process has been successfully used to treat groundwater containing BTEX and other petroleum hydrocarbons at a number of sites (Hickey et al. 1993, 1991). The utility of GAC-FBR technology to treat waters from MGP sites where PAHs are the primary contaminants of concern was examined using a 550-gpd (2,080 L per day) pilot-scale system. Based on results generated, the GAC-FBR process has been installed at a Superfund site to process a 15-gpm effluent water stream containing high levels of PAHs from a subsurface DNAPL removal process. Presented below are results from the pilot-scale system.

PILOT REACTOR

A pilot-scale, 550-gpd GAC-FBR constructed from glass and Teflon™ components was used for this study (Figure 1). The reactor was fed a synthetic

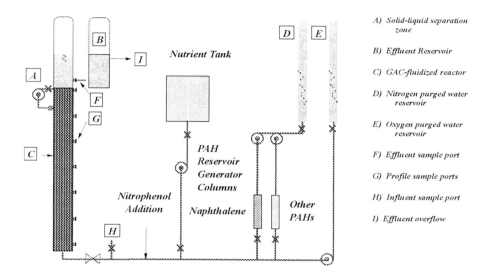

FIGURE 1. Schematic of pilot-scale GAC-FBR treating PAHs and nitrophenols.

wastestream containing 8 PAHs for over 1 year. The PAHs added were representative of the major components identified in samples of groundwater from an MGP site and a process effluent from pilot work associated with subsurface DNAPL removal at this site. The contaminated wastestream was amended with nitrogen, phosphorus (at a COD/N/P ratio of 100/5/1), trace nutrients, and oxygen. Influent feedwater was at room temperature (23°C). The HRT was maintained between 6 and 7 min. The reactor was operated as a one-pass system (no recycle used). The GAC-FBR was inoculated with a consortium able to degrade 2- to 4-ring PAHs obtained from soil samples from several MGP sites.

RESULTS

Startup

The GAC-FBR was biologically started in June 1992. An initial charge of virgin carbon was loaded into the FBR. The initial applied OLR to the system was approximately 1.6 kg COD/m³-d. More than 80% of this applied OLR was naphthalene. Complete removal of naphthalene was observed from system startup. Effluent naphthalene concentrations were near or below detection limits (3 µg/L) over the first 10 days, with removals of greater than 99%. Dissolved oxygen (DO) consumption through the reactor was initially much lower than required to oxidize the amount of naphthalene removed in the GAC-FBR. Removal during the initial startup was due primarily to adsorption. By day 16, the measured DO consumption was commensurate with oxygen required to completely oxidize the mass of PAHs being removed in the reactor, based on COD removal and an assumed biomass yield of 0.3 mg total suspended solids (TSS)/mg COD removed (Figure 2). Biofilms visible to the naked eye were easily observed on the GAC particles by day 17; the presence of fungi within the biofilm present on the GAC particles was observed shortly thereafter.

Reactor Performance

The OLR to the GAC-FBR was varied from 1.6 to 4.6 kg COD/m^3-d. Total PAH removal ranged from 97.9 to 99.7% during these steady-state periods. Only results from the highest applied OLR are detailed here.

A steady-state monitoring period at the maximum OLR was performed approximately 4 months into the experimental program. The OLR during this period was 4.6 kg COD/m^3-d. At this OLR, naphthalene was 90% of the OLR applied to the system. Naphthalene and acenaphthene effluent concentrations were consistently below detection limits (3 µg/L and 10 µg/L, respectively).

Removal of the remaining PAHs ranged from 98.7% for phenanthrene to 83.6% for benzo[a]pyrene (Table 1). Dissolved oxygen consumption was 13.6 mg/L. This oxygen consumption is approximately commensurate with the amount of PAHs removed.

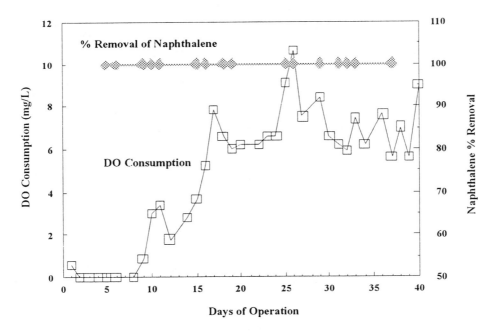

FIGURE 2. Performance data for startup period of a GAC-FBR treating PAHs; initial organic loading = 1.6 kg COD/m^3-d.

TABLE 1. Performance of a pilot GAC-FBR treating PAHs at an applied OLR of 4.6 kg COD/m^3-d.

Compound	Influent (µg/L)	Effluent (µg/L)	% Removal
Naphthalene	5,600 (±672)	<3.0	>99.9
Acenaphthene	413 (±169)	<10.0	>97.2
Phenanthrene	103 (±26.8)	1.4 (±1.3)	98.7
Fluoranthene	41.6 (±9.2)	1.1 (±0.8)	97.8
Pyrene	30.3 (±7.6)	0.98 (±0.73)	97.2
Benzo[b]fluoranthene	0.55 (±0.20)	0.11 (±0.13)	85.2
Benzo[a]pyrene	0.62 (±0.22)	0.12 (±0.11)	83.6
Total 3+4 ring	588	13.5	97.7
Total PAH	6,189	16.7	99.7

Oxygen Consumption and Solids Production

The ratio of DO consumed to COD removed ranged from 0.71 to 0.77 during the course of this work. The fraction of COD applied to the reactor that is not consumed for energy is incorporated into cell mass (assuming no significant adsorption). Control of the expansion of the bed height at a fixed level in the FBR resulted in the biomass that was produced being wasted in the effluent flow. Concentrations of total suspended solids (TSS) in the effluent ranged from 1.8 mg/L at an organic loading rate of 1.6 kg COD/m^3-d to 8.4 mg/L at an OLR of 4.6 kg COD/m^3-d. Most of the solids produced in the reactor were organic in nature; the volatile suspended solids (VSS)/TSS ratios averaged 0.87.

PAH Removal Through the Profile of the FBR

The GAC-FBR can be characterized hydraulically as a plug flow reactor. Some information on removal rates and capacity of the system can, therefore, be obtained by collecting data on the concentration profiles of PAHs and DO through the system. Samples were taken from the GAC-FBR while receiving an OLR of 4.6 kg COD/m^3-d. Superimposing the concentration of naphthalene in these samples and the concentration of dissolved oxygen consumed (Figure 3) illustrates the coinciding consumption of oxygen and naphthalene (PAH) removal. Naphthalene removal and oxygen consumption were rapid in the lower portion (first 66 cm) of the bed (75 s of empty bed retention time). At this point, 5,110 µg/L of naphthalene and 7.8 mg/L of oxygen had been consumed. Oxygen removal continued past 70 cm in the profile of the vertical axis, albeit at a much lower rate.

Reactor profile samples were also analyzed for 3- to 6-ring PAH concentrations. Most of these PAHs were removed in the first half of the bed. This

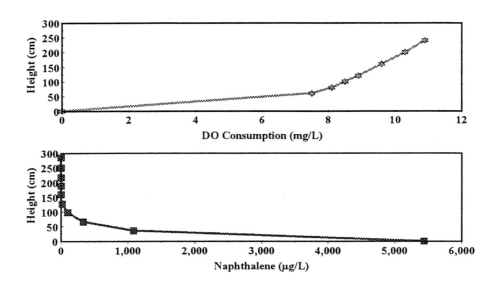

FIGURE 3. Profile of naphthalene and DO concentrations through the GAC-FBR operated at an OLR of 4.6 kg COD/m^3-d.

indicates that the system has the potential to handle considerably higher PAH concentrations and applied OLRs than tested during this work.

PAHs Adsorbed onto GAC

After 7 months of operation, GAC samples were removed from the GAC-FBR and the mass of PAHs were quantified via sequential extraction of samples taken through the vertical profile of the system. Results comparing the mass of PAHs applied versus the amount recovered from the sequential extractions with methylene chloride are presented in Table 2. An average of 1.2% of the PAHs applied was retained on the GAC. Further evidence supporting biological oxidation as the primary removal mechanism was obtained from batch assays conducted using biomass from the system. The seven PAHs (no naphthalene) supplied to the pilot-scale reactor were tested with abiotic and inactivated biomass controls. Results demonstrated that removal of the 3- and 4-ring PAHs was due to biological oxidation.

As a final test of the reversibility of adsorbed PAHs, the feed of naphthalene to the pilot reactor was stopped for 5 days. The mass of naphthalene adsorbed onto the GAC decreased from 8.70 to 3.67 mg/g GAC in the reactor. The standard deviation of the analysis was 14.3%.

Toxicity Assays

Influent and effluent samples were taken from the GAC-FBR and tested for toxicity using the Microtox® assay. Samples of influent and effluent were

TABLE 2. Mass of PAHs applied to laboratory-pilot GAC-FBR and recovered on the GAC after 7 months of operation.

Compound	Mass Applied (g)	Mass on GAC (g)	Percent on GAC
Naphthalene	1,144.0	7.86	0.7
Acenaphthene	130.3	6.89	5.3
Phenanthrene	26.0	0.86	3.3
Fluoranthene	10.3	0.17	1.7
Pyrene	8.7	0.13	1.5
Total PAHs	1,319.0	15.9	1.2

Note: Benzo[b]fluoranthene and benzo[a]pyrene were below quantitation limits.

taken when the system was operated at an OLR of 1.6 kg COD/m³-d. No toxicity was detected in the effluent from the FBR (Figure 4). The influent sample taken during the same period had an EC50 of 9.5% (luminescence from the organism was decreased by 50% with only 9.5% of the actual influent). This demonstrates that toxic or inhibitory intermediates of PAH degradation were not present to any significant extent (if at all) in the system effluent.

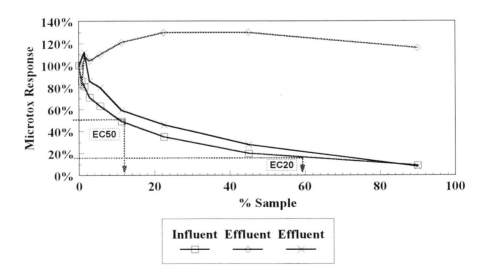

FIGURE 4. Microtox® toxicity assay performed on influent and effluent samples taken from a GAC-FBR treating PAHs; organic loading rate of 1.6 kg COD/m³-d.

SUMMARY

The ability of the GAC-FBR to treat groundwater containing PAHs was shown to be feasible, using a 550-gpd pilot. Removal efficiencies of 2- to 4-ring PAHs were 99% for all OLRs applied. Based on results from concentration profiles of PAHs and DO and batch tests using biomass from the system, this removal was due primarily to biological oxidation once a mature biofilm developed (less than 3 weeks). Only 1.2% of the PAHs fed to the GAC-FBR over a 7-month period was observed on the GAC. The removal of 5- and 6-ring PAHs was likely due, in large part, to adsorption onto the GAC. Full-scale application at a Superfund site in Pennsylvania is under way.

REFERENCES

Clesceri, L. S., A. E. Greenberg, and R. R. Trussel. 1989. *Standard Methods for the Examination of Water and Wastewater*, 17th ed. APHA, AWWA, and WPCF, Washington, DC, pp. 2-71 through 2-79.

Hickey, R. F., et al. 1993. "Application of the GAC-FBR for Treatment of Gas Industry Wastewaters." Presented at Institute of Gas Technologies 6th International Symposium on Gas, Oil and Environmental Biotechnology, Institute of Gas Technologies, Colorado Springs, CO, November.

Hickey, R. F., and R. W. Owens. 1981. "Methane Generation from High-Strength Industrial Wastes with the Anaerobic Biological Fluidized Bed." In *Biotechnology and Bioengineering Symposium No. 11*, pp. 399-413. John Wiley and Sons, New York, NY.

Hickey, R. F., D. Wagner, and G. Mazewski. 1991. "Treating Contaminated Groundwater Using a Fluidized Bed Reactor." *Remediation*, 2:447-460.

Jeris, J. S., C. Beer, and J. A. Mueller. 1974. "High Rate Biological Denitrification Using a Granular Fluidized Bed." *JWPCF*, 46:2118-2128.

Jeris, J. S., R. W. Owens, R. F. Hickey, and F. Flood. 1977. "Biological Fluidized Bed Treatment of BOD and Nitrogen Removal." *JWPCF*, 49:816-831.

Jeris, J. S., and R. W. Owens. 1975. "Pilot Plant, High-Rate Biological Denitrification." *JWPCF*, 47:2043-2057.

Suidan, M. T., W. H. Cross, M. Fong, and J. W. Calvert, Jr. 1981. "Anaerobic Carbon Filter for Degradation of Phenols." *Journal of the Environmental Engineering Division*, ASCE, 107, EE3: 563-579.

Sutton, P. M., W. K. Shiek, C. P. Woodcock, and R. W. Morton. 1979. "Oxitron System Fluidized Bed Wastewater Treatment Process: Development and Demonstration Studies." Presented at the Joint Annual Conference on the Air Pollution Control Association and the Pollution Control Association of Ontario, Toronto, Canada, April.

Switzenbaum, M. S. and W. J. Jewell. 1980. "Anaerobic Attached-Film Expanded Bed Reactor Treatment." *J. Water Pollution Control Federation*, 52:1953-1965.

Biological Activated Carbon Fluidized-Bed System to Treat Gasoline-Contaminated Groundwater

Thomas C. Voice, Xianda Zhao,
Jing Shi, and Robert F. Hickey

ABSTRACT

An integrated biological granular activated carbon fluidized-bed reactor (GAC-FBR) and a biological fluidized-bed reactor (FBR) charged with nonactivated carbon were evaluated for treating groundwater contaminated with the gasoline constituents benzene, toluene, and xylenes (BTX). The systems were studied under several conditions including startup, steady-state, and step-load increase conditions. Development of bioactivity in the GAC-FBR was faster than in the FBR using a nonactivated carbon biomass carrier. Under two steady-state conditions, organic loading rates of 3 and 6 kg-chemical oxygen demand (COD)/m³-day, BTX removal was similar in the two systems with more than 90% of applied BTX removed. The GAC-FBR produced superior effluent quality during step organic load rate (OLR) increases compared to the FBR. The results from an extremely high step OLR increase show the formation of partial oxidization products from the degradation of BTX. Significant adsorption capacity was still observed after the biofilm developed, although capacity gradually decreased over a 6-month period of operation to approximately 50% of its original value.

INTRODUCTION

The most widely used remediation techniques, liquid-phase adsorption using granular activated carbon (GAC) and air stripping, simply transfer contaminants from one phase to another (McCarty 1983; Voice 1989). Further treatment or disposal of the receiving phase normally is required. Biological treatment is a desirable alternative to such techniques because it has the potential to completely destroy the contaminant compounds and it is generally less expensive than physical-chemical treatment processes. Biological treatment has not been widely accepted for groundwater treatment, however, due to the widespread impression that biological systems are not sufficiently stable to consistently meet the stringent discharge limitations that are often required. In cases where biological treatment has been employed, typically it is followed by GAC adsorption for effluent polishing and to provide backup treatment in the event of failure of the biological system. A promising new approach, termed biological activated carbon (BAC), integrates biological removal and granular activated carbon adsorption into a single unit process (Weber et al. 1978; DiGiano 1981; Rice and Robson 1982; Suidan et al. 1987; Speitel et al. 1989). BAC has only recently been considered as a potential treatment process for groundwater contaminated with relatively low levels of toxic materials, such as petroleum hydrocarbons or chlorinated solvents (Hickey et al. 1990).

In this research, an integrated BAC system employing granular activated carbon as a biofilm support for a biological fluidized-bed reactor (GAC-FBR) was evaluated and compared to biological removal only (FBR) with a nonadsorbent biomass carrier for the remediation of groundwater contaminated with BTX.

MATERIALS AND METHODS

Fluidized-Bed Reactors and Experimental Phases

The laboratory-scale fluidized-bed reactors used in phases one and two of this study were glass columns with a 2.5 cm diameter and 92 cm height (Figure 1). All reactors were operated as one-pass systems with no recycle. The GAC in the GAC-FBR was Calgon Filtrasorb 400 (Calgon Co., Pittsburgh, Pennsylvania).

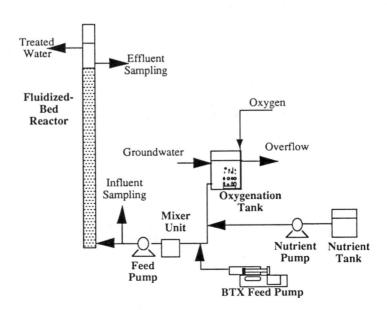

FIGURE 1. Schematic of the experimental FBR system.

The nonactivated carbon (termed "baker product" by the manufacturer) used in the FBR is the same material as GAC except that it has not undergone the activation step in the manufacturing process and therefore has little adsorption capacity. Three constituents — benzene, toluene, and p-xylene — were selected to represent the aromatic components in gasoline. The concentration of each BTX compound in the influents was approximately 1 mg/L. The GAC-FBR and FBR (biological only) systems were seeded with a mixed culture of microorganisms. This culture was taken from a pilot-scale fluidized-bed system that was originally seeded with activated sludge and was subsequently supplied with BTX as the sole carbon source. The reactors were provided with sufficient dissolved oxygen (DO) and essential nutrients to encourage biofilm formation and permit complete biological oxidation of the BTX. The groundwater (hardness 450 mg/L as $CaCO_3$, pH 7 to 8, and total organic carbon 2.5 mg/L) was pumped from a deep aquifer underlying the campus of Michigan State University. During the course of this experiment the water temperature averaged 15°C ± 1.0°C.

In phase one of the experiment, startup profiles, pseudo-steady-state operating data under two different organic load rates (OLRs: 3 and 6 kg-COD/m^3-day), and responses to 5-fold 8-hour step OLR increases in BTX were obtained for the two reactor types (GAC-FBR and FBR). Additional details of the experiment designs can be found in previous publications (Voice et al. 1992; Zhao 1994).

In phase two of the experiment, the stability of the systems when subjected to the extremely high OLR increases was investigated and the formation of the partial oxidization products was monitored. Following a period of pseudo-steady-state operation, the organic loading rates were increased by a factor of 20, 12 and 7 for benzene, toluene, and p-xylene, respectively. These step-OLR increases were maintained for 4 h. During the increased loading period, the feed concentrations of DO and essential nutrients were increased to ensure that these materials did not limit degradation.

In phase three, the adsorption capacity of biofilm-coated GAC was monitored in a pilot-scale fluidized-bed reactor, which was constructed using a 7.6-cm-diameter by 322-cm-high polyvinyl chloride (PVC) column to produce a system with a 14.7-L working volume. Carbon samples were collected from the middle portion of the reactor (150 cm from the bottom of the reactor) using a sampling bottle (50 mL). The carbon and water-filled bottles were sterilized by gamma irradiation from a cobalt-60 source located at the Phoenix Memorial Laboratory (Ann Arbor, Michigan) to stop all biological activity. The samples were sterilized for 1 h at a dose rate of 2 Mrad/h.

Analytical Methods

Aqueous BTX concentrations were determined using an automated headspace sampler (Perkin-Elmer, model HS-101) coupled to a gas chromatograph (Perkin-Elmer model 8700) equipped with a flame ionization detector (FID) using helium as a carrier gas. The detection limit of this method was 1 µg/L for each BTX component. The accuracy for measurements of a sample containing a concentration of 1.5 µg/L was ± 0.1 µg/L. Influent and effluent dissolved oxygen (DO) concentrations were analyzed using a polarographic electrode coupled to a digital pH/millivolt meter. The concentration of nonvolatile total organic carbon (nonvolatile TOC) was measured on a TOC analyzer (Shimazdu, Model TOC-500) equipped with an autosample injector (Shimazdu, ASI-502). Before the measurement of nonvolatile TOC, the water sample was acidified with hydrochloric acid and purged with oxygen for 4 min to remove inorganic carbon and volatile organic carbon. A water sample spiked with BTX and catechol (one of the expected intermediates) was tested to ensure the efficacy of the procedure. The results showed that the BTX and inorganic carbon were removed completely, and there was no significant removal of the nonvolatile organic carbon (catechol). The TOC detection limit was 0.6 mg/L and the accuracy for measurements of a standard solution of 1 mg/L was ± 0.2 mg/L.

RESULTS

Phase One: Breakthrough Profiles, Steady-State and OLR Increases

During the startup of the reactors, the concentration of each BTX component in the influents was maintained at approximately 1 mg/L. In the FBR system, where biodegradation was the only significant removal mechanism, breakthrough of BTX occurred soon after introduction of the influent. After approximately 400 h, significant DO consumption was observed in the FBR (nonactivated carbon) and the concentrations of BTX decreased in the effluent. Because of the adsorption capacity, the GAC-FBR performed much like an adsorption system during startup with complete removal of the added BTX. Significant DO consumption was observed after 140 h.

Under steady-state conditions, the applied OLRs were 3.8 and 2.9 kg COD/m^3-day for the GAC-FBR and FBR, respectively. More than 90% of the BTX (94% for benzene and toluene and 90% for xylene) was removed in the FBR. Despite the higher OLR applied to the GAC-FBR system, greater substrate removal was observed: 99% for benzene and toluene and 92% for xylene. With a total COD loading to both reactors of 6.0 kg COD/m^3-day, the two systems performed comparably. Removal rates averaged 94%, 90%, and 81% for benzene, toluene, and xylene, respectively.

After steady-state conditions were reached, the reactors were subjected to a five-fold step-load increase from a base OLR of 6.0 kg COD/m^3-day for 8 h. BTX removal and DO consumption in the GAC-FBR system were higher than in the FBR system. The higher DO consumption indicates enhanced bioactivity. Adsorption also accounted for some of the observed removal of BTX.

Phase Two: Extremely High OLRs and Partial Oxidized Product Formation

For the first extremely high OLR increase, the benzene concentration in the feed was increased from the steady-state value of 1 mg/L to 45 mg/L for 4 h.

After the reactors returned to steady-state conditions again, the concentration of toluene in the influent was increased from 0.9 mg/L to 27 mg/L for 4 h. For the last step-load increase experiment, the concentration of xylene in the influents was increased from 0.6 mg/L to 13 mg/L for 4 h. The DO concentration in the effluents of both reactors was maintained at greater than 10 mg/L during these three step increases. The removal rates for applied BTX in both reactors are shown in Table 1. Carbon balances for the experiments are presented in Table 2. DO consumption was similar in the two systems during the increases. BTX removals in the GAC-FBR were significantly higher than in the FBR. Production of nonvolatile TOC was considerably higher in the FBR system effluent throughout the step OLR increase compared to the GAC-FBR.

TABLE 1. BTX removal during extremely high OLR increases for each BTX compound.

Compound Used for the Increase	Reactor	Compound	Influent (mg/L)	Effluent (mg/L)	Removed (mg/L)	Removal (%)
		Benzene	45.31	16.52	28.81	64
	GAC-FBR	Toluene	0.85	0.15	0.71	85
Benzene		p-Xylene	0.60	0.04	0.56	93
		Benzene	44.88	33.69	11.17	25
	FBR	Toluene	0.83	0.38	0.46	55
		p-Xylene	0.58	0.06	0.52	89
		Benzene	1.98	0.40	1.58	80
	GAC-FBR	Toluene	27.92	4.23	23.69	85
Toluene		p-Xylene	2.42	0.65	1.77	73
		Benzene	1.98	1.38	0.60	30
	FBR	Toluene	27.77	18.40	9.38	34
		p-Xylene	2.40	2.10	0.29	12
		Benzene	1.06	0.00	1.06	99
	GAC-FBR	Toluene	1.23	0.02	1.21	98
p-Xylene		p-Xylene	13.48	1.17	12.31	91
		Benzene	1.04	0.04	1.00	96
	FBR	Toluene	1.19	0.08	1.10	93
		p-Xylene	13.29	5.63	7.67	58

TABLE 2. Carbon balances during extremely high OLR increases for each BTX compound (% of influent carbon).

				Removal by			
				Biodegradation			
		Total			By-Product		
Load		Removal		Biomass			
Increase	Reactor	of BTX	Adsorption	and CO_2	in Eff.	Adsorbed	Total
Benzene	GAC-FBR	64.2%	35.6%	18.6%	2.9%	7.1%	28.6%
	FBR	26.1%	N/A	17.0%	9.2%	N/A	26.1%
Toluene	GAC-FBR	83.9%	44.8%	20.9%	3.4%	14.8%	39.1%
	FBR	32.1%	N/A	17.2%	14.9%	N/A	32.1%
Xylene	GAC-FBR	93.0%	32.3%	43.2%	11.4%	6.1%	60.7%
	FBR	63.4%	N/A	45.2%	18.2%	N/A	63.4%

Phase Three: Adsorption Capacity of GAC in GAC-FBR

For this experiment, 3 mg/L of toluene was fed to a pilot-scale GAC-FBR. The adsorption capacity of GAC in the GAC-FBR was determined by a bottle point method at regular intervals (Voice et al. 1992). An extended Freundlich isotherm equation was fit to the data to provide a convenient mathematical description of the adsorption characteristics over the course of the experiment:

$$q_e + q_o = K_f C_e^{\frac{1}{n}}$$

where q_e is the additional amount of toluene adsorbed (mg/g), q_o is the initial amount of adsorbed toluene (mg/g) that can be determined by solvent (methanol) extraction, K_f is a Freundlich constant, C_e is the equilibrium concentration of toluene in water (mg/L), and $1/n$ is a exponent constant.

The measured isotherm parameters are presented in Table 3. The value of K_f decreased gradually over a period of 120 days. Adsorption capacity remained at approximately 50% of that found with virgin carbon after this time until the conclusion of the experiment at 187 days. Values of $1/n$ increased slightly throughout the experimental period.

DISCUSSION AND CONCLUSIONS

This study demonstrates that the use of activated carbon as a biomass carrier in fluidized-bed reactors produces a system in which both adsorption and biodegradation affect substrate removal. To a large extent, the performance of such biological activated carbon systems during nonsteady-state conditions is controlled

TABLE 3. Adsorption isotherm parameters for GAC collected mid-height from a GAC-FBR over a 6-month period.

Day Sampled	Data Points	Isotherm Parameters		
		q_o (mg/g)[a]	K_f (mg/g)[b]	$1/n$[c]
Clean GAC	70	—	79.3 ± 3.3[d]	0.37 ± 0.03
0	10	0.30	79.3 ± 5.9	0.43 ± 0.06
4	10	1.11	71.5 ± 1.4	0.38 ± 0.02
7	10	2.13	69.6 ± 1.4	0.37 ± 0.02
17	20	5.61	62.6 ± 1.7	0.44 ± 0.01
21	20	6.87	61.0 ± 1.1	0.42 ± 0.03
28	10	4.16	67.5 ± 1.7	0.44 ± 0.01
35	20	6.22	64.5 ± 3.0	0.35 ± 0.04
49	10	5.48	56.7 ± 3.9	0.43 ± 0.04
64	10	6.38	66.7 ± 2.8	0.37 ± 0.02
77	10	5.58	46.3 ± 1.2	0.42 ± 0.02
94	42	4.01	40.9 ± 2.5	0.48 ± 0.02
122	29	3.71	34.4 ± 1.2	0.54 ± 0.02
158	44	2.56	35.7 ± 0.9	0.51 ± 0.01
187	20	2.54	37.7 ± 1.2	0.51 ± 0.02

(a) q_o is the initial amount of adsorbed toluene (mg/g) which can be determined by solvent (methanol) extraction.
(b) K_f is the Freudlich constant.
(c) $1/n$ is the exponent constant.
(d) 95% confidence interval.

by the additive contributions of these two removal mechanisms. During the startup period, before a fully functional biomass has developed, substrate is removed primarily by adsorption. Biomass developed more quickly on the surface of activated carbon than it did on nonactivated carbon having a similar surface texture during this startup period. After the biomass is established and steady-state conditions are reached, system performance is dominated by biodegradation. The system retains adsorption capacity under steady-state conditions as evidenced by the response of the system to in-line concentration shocks. Lower effluent concentration levels were observed in the GAC-FBR systems during a step concentration increase compared to similar systems using a nonadsorbing biomass carrier. The data suggest that this results from adsorption of a portion of the increase.

The production of partial oxidization products was not observed under steady-state conditions, at an organic loading rate of 2.2 kg COD/m³-day. Under application of the OLR increases of 20, 12, and 7-fold by increasing the influent benzene, toluene, and *p*-xylene concentrations, however, significant concentrations of partial oxidization products were produced. A significant amount of BTX was biologically transformed to partial oxidization products (35%, 46%, and 29% for benzene, toluene, and xylene increases, respectively) in the FBR with a non-adsorbent biomass carrier. The use of GAC as a biofilm carrier in the GAC-FBR dramatically increased the total removal of BTX and significantly reduced the amount of partially oxidized products observed in the effluent.

The adsorption capacity of biocoated carbon decreased gradually over the 6-month period of study. The fact that a mature biofilm was present early in this period suggests that the biofilm itself does not significantly inhibit adsorption. After 6 months of operation, the remaining capacity was approximately 50% of the initial value, suggesting that the concentration-dampening capability of the system is significant over extended periods of operation.

The results demonstrate that the biological granular activated carbon system was able to deliver good effluent water quality and stable performance during nonsteady-state conditions. This suggests that such a system should be a viable treatment technology for groundwater remediation.

REFERENCES

DiGiano, F. A. 1981. "Influence of biological activity on GAC performance." Presented at Conference on Applications of Adsorption in Wastewater Treatment, Enviro Press, Nashville, TN.

Hickey, R. F., D. Wagner, and G. Mazewski. 1990. "Combined biological fluid bed — carbon adsorption system for BTEX contaminated groundwater remediation." Presented at the Fourth National Outdoor Action Conf. on Aquifer Restorating, Groundwater Monitoring and Geophysical Methods, Las Vegas, NV.

McCarty, P. L. 1983. "Removal of organic substances from water by air stripping." In B. Berger (Ed.), *Control of Organic Substances in Water and Wastewater*. EPA-600/8-83-011.

Rice, R. G., and C. M. Robson. 1982. *Biological Activated Carbon — Enhanced Aerobic Biological Activity in BAC Systems*. Ann Arbor Sci. Publ., Ann Arbor, MI.

Speitel, G. E., Jr., C. J. Lu, X. J. Zhu, and M. H. Turakhia. 1989. "Biodegradation and adsorption of a bisolute mixture in GAC columns." *Journal of Water Pollution Control Federal* 61(2): 221-229.

Suidan, M. T., P. Fox, and J. T. Pfeffer. 1987. "Anaerobic treatment of coal gasification waste-water." *Water Science Technology* 19: 229-236.

Voice, T. C. 1989. "Activated carbon adsorption." In H. Freeman (Ed.), *Standard Handbook for Hazardous Waste Treatment and Disposal*. McGraw-Hill, New York, NY.

Voice, T. C., D. Pak, X. Zhao, J. Shi, and R. F. Hickey. 1992. "Biological activated carbon in fluidized bed reactors for the treatment of groundwater contaminated with volatile aromatic hydrocarbons." *Water Research*, 26(10): 1389-1401.

Weber, W. J., Jr., M. Pirbazari, and G. L. Melson. 1978. "Biological growth on activated carbon: An investigation by scanning electron microscopy." *Environmental Science and Technology* 12(7): 817-819.

Zhao, X. 1994. "Biological activated carbon in fluidized bed reactors for the treatment of groundwater contaminated with volatile aromatic hydrocarbons." Ph.D. Dissertation, Michigan State University, East Lansing, MI.

Activated Carbon-Enhanced Biological Groundwater Treatment: Superfund Site Case Study

Lucy B. Pugh, Richard R. Rediske,
David F. Rogers, and David H. Peden

ABSTRACT

Two activated carbon-enhanced biological treatment processes, GAC/FB (granular activated carbon/fluidized bed) and two-stage PACT® (powdered activated carbon treatment) were evaluated for treatment of impacted groundwater at a former organic chemicals manufacturing facility, known as the Ott/Story/Cordova Superfund Site near Muskegon, Michigan. The groundwater contains an estimated 80 mg/L of ammonia-nitrogen and 1,500 mg/L of chemical oxygen demand (COD). Of the COD, 30% is comprised of a complex mixture of 50 targeted compounds including halogenated and aromatic organics. The remaining COD includes an unidentified group of process intermediates and degradation products related to historical production at the facility. GAC/FB and PACT® were initially evaluated using treatability-scale reactors. The GAC/FB reactor achieved consistent, high removals of all organic compounds, but consistent nitrification could not be maintained. The two-stage PACT® system achieved nearly complete removal of all organic compounds and ammonia. A 9.5-L/minute pilot system was subsequently operated at the site to provide additional performance information. Three of the most highly impacted site wells were used to supply groundwater to the two-stage PACT® system. Excellent removal of all known and unknown organic compounds as well as nearly complete nitrification were consistently achieved. As a result of the on-site pilot study, a two-stage PACT® system is being implemented at the site.

INTRODUCTION

Impacted groundwater at the Ott/Story/Cordova Superfund site near Muskegon, Michigan, contains a complex mixture of organic compounds and nitrogen. The remediation plan for the site calls for extracting and treating 4,600 m³ per day of groundwater, estimated to contain 1,500 mg/L of COD and

80 mg/L of ammonia-nitrogen. Approximately 30% of the COD is comprised of a mixture of 50 Appendix IX compounds, including halogenated and aromatic organics. (Appendix IX is a comprehensive list of compounds used in groundwater monitoring, and is part of the Resource Conservation and Recovery Act, 40 CFR Part 264.) The remaining COD is contributed by an unidentified group of process intermediates and degradation products related to historical production at this former chemical production facility.

Discharge of the treated groundwater will be to a local surface water body. In addition to requiring removal of known organic compounds and ammonia-nitrogen, concern was expressed regarding the potential presence of unidentified compounds in the treatment system effluent. The objective of this work was to evaluate the effectiveness of two activated carbon-enhanced biological treatment systems, GAC/FB and PACT®, for achieving the expected standards for surface water discharge.

TREATABILITY STUDIES

Laboratory-scale treatability studies were conducted to compare the relative effectiveness of the GAC/FB and two-stage PACT® technologies for treatment of the Ott/Story/Cordova site groundwater. Schematics of the GAC/FB and PACT® laboratory reactors are shown in Figures 1 and 2, respectively.

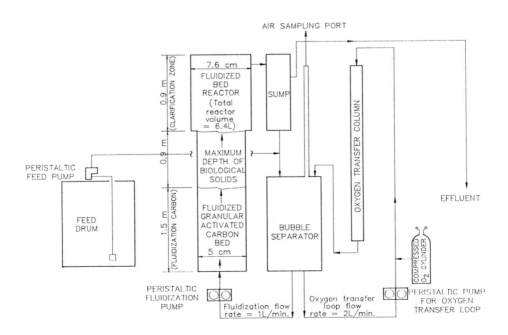

FIGURE 1. GAC/FB laboratory reactor schematic.

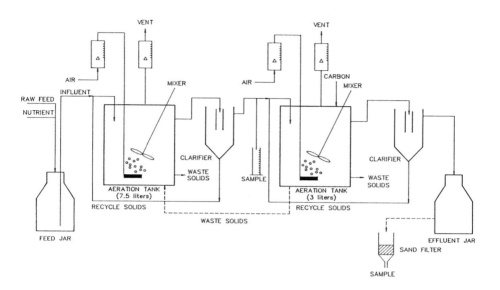

FIGURE 2. Two-stage PACT® laboratory reactor schematic.

Methods

The GAC/FB system was operated by EARTH TECH and consisted of a fluidized-bed reactor (manufactured by Envirex, Ltd.), which used granular activated carbon as a support medium for biological attachment (total bed volume of 6.4 L). The system received an initial charge of GAC; no additional GAC was added during the study. During operation, pure oxygen was injected into a recycle loop to maintain a residual concentration of dissolved oxygen (DO) in the effluent of the reactor above 2 mg/L. A portion of the effluent was recirculated through the system using a peristaltic pump to fluidize the carbon bed.

The two-stage PACT® system was assembled and operated by Zimpro Environmental, Inc., and consisted of a 7.5-L first-stage aeration tank, a first-stage clarifier, a 3-L second-stage aeration tank, a second-stage clarifier, and return sludge pumps for transfer of settled solids from each clarifier to its respective aeration tank. The aeration tanks each received an initial charge of powdered activated carbon (PAC). During operation, PAC was wasted from the system on a daily basis from the first-stage aeration tank, and was made up in the system through daily addition of virgin PAC to the second-stage aeration tank. Waste solids from the second-stage aeration tank were recycled to the first-stage aeration tank.

Both systems were initially seeded with return activated sludge from a local municipal wastewater treatment plant. A mix of site groundwater selected to represent the full-scale system influent quality was amended with phosphorus and pumped to each of the systems on a continuous basis for treatment. The systems were monitored on a regular basis over approximately a 5-month period for operating parameters (pH, DO, mixed liquor solids, etc.) and for influent/effluent concentrations of conventional pollutants and volatile/semivolatile organic compounds (U.S. EPA SW-846 Method 8260, 3rd ed., Update I, July 1992, and U.S. EPA SW-846 Method 8270, 3rd ed., 1986, Revision 0).

Results

At an organic loading rate of approximately 2 g COD/L-day, the GAC/FB reactor achieved 80 to 90% removal of COD, greater than 95% removal of biochemical oxygen demand (BOD_5), and greater than 95% removal of most individual organic compounds. By monitoring air emissions from the system, it was determined that less than 0.01% of the removal of volatile compounds was attributable to stripping. Nitrification in the GAC/FB reactor was inconsistent. The reactor was reseeded with nitrifying microorganisms (both cultured nitrifiers and nitrifying return activated sludge from the local municipal wastewater treatment plant) on several occasions in an attempt to maintain nitrification in the reactor. Although removal of ammonia-nitrogen was at times as high as 90 to 95%, nitrification could not be reliably maintained in the system. It is hypothesized that concentrations of some of the organic compounds adsorbed on the GAC in the system may have reached levels toxic to nitrifying microorganisms.

The two-stage PACT® system was loaded at a rate of 1.8 g COD/L-day in the first stage and 1.7 g COD/L-day in the second stage. The resultant removal of COD averaged 77%, and the removal of BOD_5 and ammonia-nitrogen exceeded 99%. Individual organic compounds were completely removed by the system. By comparing mass spectra of unidentified organic compounds for influent and effluent compounds, it was demonstrated that unidentified compounds were also completely removed by the two-stage PACT® system. In contrast to the GAC/FB system, through monitoring of air emissions from the reactors, it was determined that up to 1% of the overall removal of volatile compounds was due to stripping by aeration.

PILOT STUDY

Based on the success of the PACT® treatability study, a two-stage PACT® pilot system (manufactured by Zimpro Environmental, Inc.) was installed and operated at the site. A schematic of the pilot system is provided in Figure 3.

Methods

Each aeration tank received an initial charge of PAC (170 kg total), and was seeded with 11,400 L of return activated sludge from the local municipal wastewater treatment plant. Influent groundwater from the three most highly impacted wells at the site was pumped using submersible pumps through a header system to the first-stage aeration tank. A summary of the pilot system influent groundwater quality is provided in Table 1. Phosphorus was supplied on a continuous

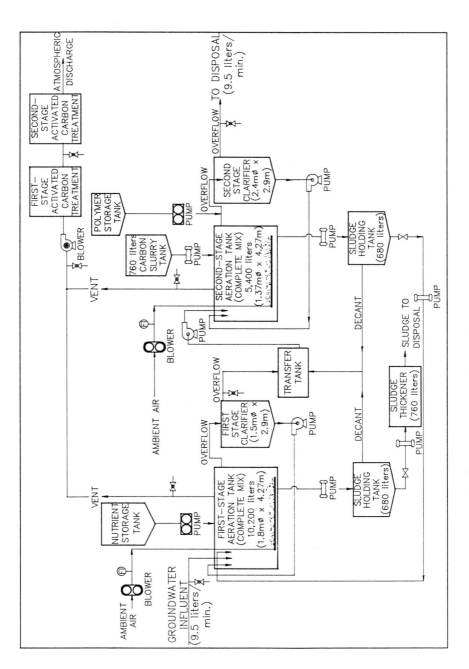

FIGURE 3. Two-stage PACT® pilot system schematic.

TABLE 1. Pilot system influent groundwater quality.

Target Parameters	Average Concentration	Target Parameters	Average Concentration	Target Parameters	Average Concentration
Conventional Parameters, mg/L		*VOCs (continued)*		*SVOCs (continued)*	
COD	1,795	4-Methyl-2-pentanone	417	4-Chloroaniline	1,670
BOD, 5-day	692	Methylene chloride	915	2-Chlorophenol	436
Nitrogen, ammonia	81	Tetrachloroethene	5,163	1,2-Dichlorobenzene	879
		Tetrahydrofuran	6,852	1,4-Dichlorobenzene	424
Volatile Organic Compounds (VOCs), µg/L		Toluene	5,156	N,N-Dimethylbenzenamine	3,570
Acetone	34,259	1,1,1-Trichloroethane	5,678	N-Ethylbenzenamine	21,630
Benzene	774	1,2-Trichloroethane	685	2-Methylnaphthalene	418
Carbon tetrachloride	685	Trichloroethylene	770	2-Methylphenol	417
Chlorobenzene	685	Vinyl chloride	7,574	3,4-Methylphenol	420
Chloroethane	685	Xylenes	2,337	N-Nitrosodiphenylamine	417
Chloroform	685			Nitrobenzene	417
1,1-Dichloroethane	2,000	*Semivolatile Organic Compounds (SVOCs), µg/L*		Di-*n*-octylphthalate	417
1,1-Dichloroethene	922	Aniline	3,770	1,1,3,3-Tetramethylurea	835
1,2-Dichloroethane	83,481	Benzoic acid	4,639	1,2,4-Trichlorobenzene	417
cis-1,2-Dichloroethene	689	Benzyl alcohol	4,174	2,4-Dimethylphenol	417
trans-1,2-Dichloroethene	685	*bis*(2-ethylhexyl)phthalate	417	2-Ethylbenzenamine	835
1,2-Dichloroethene (total)	1,374	Butyl benzyl phthalate	417	5-Methyl-2-Hexanone	1,252
Ethylbenzene	704	Di-*n*-butylphthalate	417	1,1-Dichloro-2,2-diethoxyethane	835
Isopropanol	34,259	Camphor	7,619		

Note: In cases where analytes were present at concentrations less than the detection limit, the average values were calculated using the detection limit. Therefore, in those cases, the true average is less than what is shown.

basis to the first-stage aeration tank. After the first-stage aeration tank, the mixed liquor overflowed on a continuous basis to the first-stage clarifier for gravity separation of solids. The first-stage effluent overflowed to a transfer tank, from which it was pumped to the second-stage aeration tank. The second-stage aeration tank and clarifier operated in a similar fashion to the first-stage tank. Both aeration tanks were covered and vented to a vapor-phase activated carbon adsorption system for control of air emissions. Polymer was added to the overflow from the second-stage aeration tank to aid the settling of solids in the second-stage clarifier. Second-stage effluent overflowed from the clarifier into a frac tank for storage and periodic off-site disposal.

The first-stage clarifier underflow was recycled to the first-stage aeration tank, and the second-stage clarifier underflow was recycled to the second-stage aeration tank. On a daily basis, sludge was wasted from the first and second stages, and PAC was added to the second stage. A summary of operating conditions maintained during the pilot study is provided in Table 2.

The PACT® pilot system was operated for 5 months, divided into "routine" and "performance" monitoring phases. The routine monitoring phase encompassed the initial 3 months, during which time the system was acclimated. In addition, steady operating conditions were maintained during a period of two solids residence times, to document stable performance prior to initiating the performance monitoring phase. During the 2-month performance monitoring phase, influent, first-stage effluent, and final effluent were sampled and analyzed at least three times per week for volatile organic compounds (VOCs), semivolatile organic compounds (SVOCs), and conventional pollutants. Effluent aquatic toxicity testing was performed during three sampling events (EPA/600/4-90/027, "Methods for Measuring the Acute Toxicity of Effluents and Receiving Waters to

TABLE 2. Average PACT® pilot system operating conditions.

Process Operating Parameters	First Stage	Second Stage
Influent flowrate, L/min	9.5	9.5
Residence time, hours	18	10
Solids residence time, days	20	15
Carbon dose, kg/day	—	7.5
Polymer dose, mg/L	—	10
pH	7.8	7.8
Dissolved oxygen, mg/L	4.4	8.6
Temperature, °C	16.6	15.6
Mixed liquor suspended solids, mg/L	19,400	22,900
Mixed liquor volatile suspended solids, mg/L	15,700	17,500

Freshwater and Marine Organisms" and EPA/600/4-89/001, "Short-Term Methods for Estimation of the Chronic Toxicity of Effluents and Receiving Waters to Freshwater Organisms"). Unidentified and tentatively identified organic compounds were quantified in influent and effluent samples during five sampling events. Waste sludge quality was evaluated by performing TCLP (Toxicity Characteristic Leaching Procedure, U.S. EPA Method 1311) analyses for VOCs and SVOCs.

Results

Treatment efficiencies for the two-stage PACT® system were very consistent throughout the performance monitoring phase. The removal of BOD_5 exceeded 99%, and the removal of COD exceeded 90%. The bulk of the removal of BOD_5 and COD occurred in the first stage of the PACT® system. The concentrations of alkalinity and sulfate increased significantly (180 mg/L as $CaCO_3$ and 72 mg/L, respectively) from the influent to the first-stage effluent as a result of the biological oxidation of carbon- and sulfur-containing compounds. Nearly complete nitrification also was consistently achieved, as shown in Figure 4. Decreases in ammonia-nitrogen concentrations occurred concurrent with stoichiometric increases in nitrate-nitrogen concentrations. In addition, the decrease in alkalinity matched well with the theoretical alkalinity requirement for nitrification (average 7.4 mg/L alkalinity as $CaCO_3$ per mg/L ammonia-nitrogen removed). The nitrification process took place primarily in the second stage of the PACT® system.

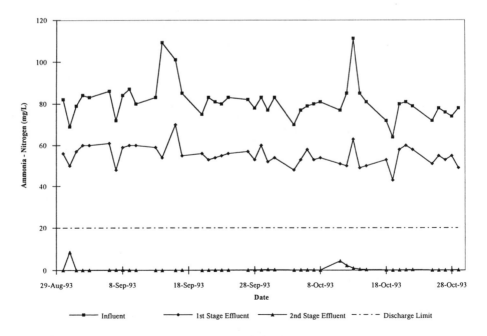

FIGURE 4. **Removal of ammonia-nitrogen in two-stage PACT® system.**

With the exception of 1,2-dichloroethane, complete removal of all Appendix IX VOCs and SVOCs was consistently achieved, as shown in Figure 5. Detection limits for most organic compounds in the effluent were 5 to 10 µg/L. The presence of 1,2-dichloroethane in the effluent can be attributed to its unusually high influent concentration, approximately five times higher than expected in the full-scale treatment system. Carbon isotherm studies conducted for the PACT® system effluent illustrated that additional removal of 1,2-dichloroethane could be achieved using higher PAC dosages.

Only one organic compound, 1,2-dichloroethane, was detected in TCLP extracts of waste sludge. The removal of 1,2-dichloroethane by wasting of sludge from the pilot system, estimated from the results of TCLP analyses, accounts for 8% of the overall removal of this compound observed during the study.

Only three unidentified compounds were present in both influent and effluent samples of the PACT® system; these were small-molecular-weight alcohols, esters, or hydrocarbons, which exhibited low potential for toxicity or bioaccumulation. Through cataloging and comparing mass spectra, complete removal of the other tentatively identified and unidentified compounds in the PACT® system influent was demonstrated.

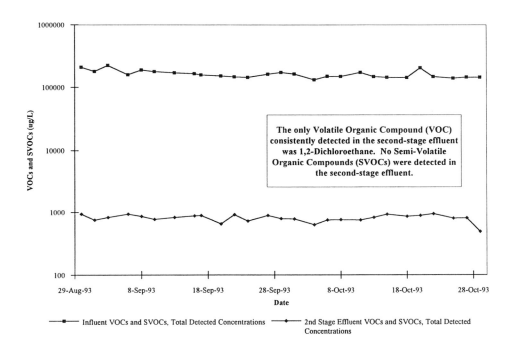

FIGURE 5. Removal of organic compounds in two-stage PACT® system. Note: Due to the differing magnitudes between influent and effluent data, concentrations are presented on a logarithmic scale.

No adverse effects were observed during acute toxicity evaluations using *Daphnia magna* or *Pimephales*. In addition, no adverse effects were observed during chronic toxicity evaluations using *Pimephales*; a slight chronic toxicity effect was observed for *Ceriodaphnia*.

CONCLUSIONS

Both the laboratory reactor and on-site two-stage PACT® pilot system demonstrated consistent performance, with effluent quality well within expected discharge limitations for all conventional pollutants, as well as identified and unidentified organic compounds. As a result of the pilot study work, a full-scale, 4,600 m³/day, two-stage PACT® system is being implemented at the site, with startup scheduled for early 1996.

Utilizing the Fluidized Bed to Initiate Water Treatment on Site

Hassan Ahmadvand, Gerry Germann,
John P. Gandee, and Verne T. Buehler

ABSTRACT

Escalating wastewater disposal costs coupled with enforcement of stricter regulations push industrial sites previously without water treatment to treat on site. These sites, inexperienced in water treatment, require a treatment technology that is easily installed, operated, and maintained. The aerobic granular activated carbon (GAC) fluidized bed incorporates biological and adsorptive technologies into a simple, cost-effective process capable of meeting strict effluent requirements. Two case studies at industrial sites illustrate the installation and operation of the fluidized bed and emphasize the ability to use the fluidized bed singularly or as an integral component of a treatment system capable of achieving treatment levels that allow surface discharge and reinjection.

INTRODUCTION

Case Histories

At a pipeline company tank farm terminal, tank bottom water is contaminated with benzene, toluene, ethylbenzene, and xylenes (BTEX) compounds and ammonia. A surface discharge permit requires total BTEX below 0.75 mg/L, benzene below 0.05 mg/L and ammonia below 5 mg/L. Previously, the tank bottom water was hauled off site for treatment. Currently, tank bottom water is pumped continuously to the fluidized bed. Effluent flows by gravity to a storage tank, and its contents are analyzed to verify permit compliance prior to surface discharge. An employee of the pipeline company, previously without water treatment experience, was trained in 2 days on the operation of the fluidized bed. Furthermore, the operator continues with her previous duties.

At a manufacturing facility, a remediation plan was developed to treat groundwater contaminated with BTEX compounds. Biological treatment with the fluidized bed meets permit requirements allowing reinjection: benzene <0.005 mg/L, toluene <1.0 mg/L, ethylbenzene <0.7 mg/L, and xylenes <10 mg/L.

The addition of hydrogen peroxide, nitrogen, and phosphorous to the fluidized bed effluent provides reinjection water for in situ treatment. The fluidized bed was installed followed by 1 week of startup and operator training.

At both sites a 4.5-ft-diameter (1.4-m-diameter) fluidized bed requiring 190 gpm (720 L/min) for fluidization was chosen to meet treatment objectives. Therefore, the discussion of system specifications, installation, startup, operation, monitoring, and maintenance applies to both sites unless otherwise stated.

Process Description

The GAC fluidized bed incorporates fixed-film biological and carbon adsorption into a single process to treat contaminated waters. Biological treatment destroys contaminants rather than transferring them from water to air. GAC treatment maintains a high-quality effluent via adsorption through varying influent quality. The GAC fluidized bed combines treatment methods and realizes the benefits of each (Edwards et al. 1994).

The GAC particles are porous solids with up to 900 m^2 of internal surface area per gram. The GAC particles provide large surface areas for microbial growth in contact with contaminated water. The pores and internal passages are properties of the GAC that facilitate microbial growth on the particles (Pirbazari et al. 1990). Adsorption and desorption on the GAC particle cause rapid microbial establishment on the GAC (Edwards et al. 1994). Microbial growth results in a continuous addition of solid mass to the fluidized bed, causing the bed to expand and rise. Narrow GAC interparticle spacing and high GAC interparticle velocities serve to maximize the contact between microbes and contaminants.

Figure 1 is a process flow diagram of the fluidized bed. Influent combines with a recycle stream of treated effluent forming the fluidization flow. The process hydraulics and valving cause the recycle flow to decrease as influent increases; therefore, the fluidization flow remains constant over varying influent flow rates. Oxygen is generated in a pressure swing adsorption (PSA) system supplied with pressurized air from the compressor. An oxygen holding tank smooths the delivery flow and pressure. The fluidization flow is injected with oxygen and the oxygen-enriched water enters a bubble trap. Undissolved oxygen collects in the top of the bubble trap and is reinjected upstream via an eductor. Oxygenated water exits the bubble trap and is monitored for dissolved oxygen.

Prior to entering the fluidized bed, nutrients are added to the fluidization flow. When the water enters the fluidized bed, it rises through the bed with a uniform velocity profile that separates the particles from one another and allows for particle mobility. When the bed nears the impeller of the growth control mixer, the induced shear force of the water breaks excess cell matter from the GAC particle. The separated cell mass leaves the reactor with the effluent and the bare GAC particles fall back through the active bed.

INSTALLATION AND STARTUP

The fluidized bed is prefabricated and shipped in two pieces. One piece is the skid which contains the piping and controls; the other piece is the bioreactor. During the first day the bioreactor is attached to the skid. Also, electrical power is brought to the control panel and the production of oxygen

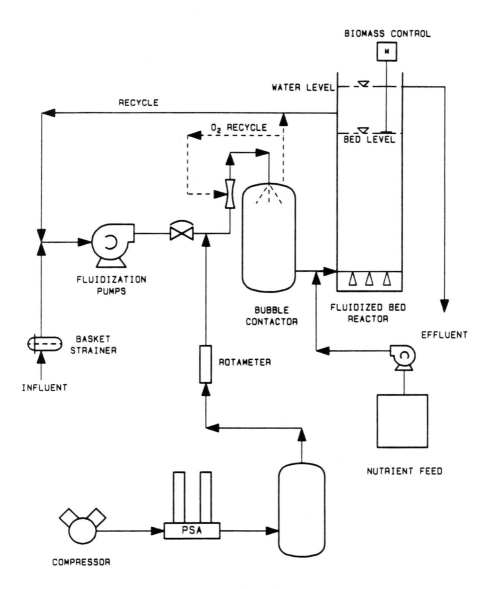

FIGURE 1. Fluidized-bed process flow diagram.

begins. Next, the bioreactor and system piping are filled with water to check for leaks and to verify the proper calibration of flow meters and dissolved oxygen probes.

On the second day, approximately 1,361 kg of GAC are added to the bioreactor and saturated with oxygen. On the following day, forward flow is initiated to load the carbon with organics to provide a food source for the microorganisms. Then the fluidized bed is put on recycle and inoculated with microorganisms, which are activated sludge from a local municipal treatment plant. The fluidized bed remains on recycle for approximately 12 hours to promote attachment of the microorganisms to the GAC. The adsorptive properties of carbon coupled with its surface characteristics have been shown to shorten biological startup times (Voice et al. 1992).

OPERATION AND MONITORING

Maintaining proper dissolved oxygen (DO) and nutrient levels is stressed when training the new operators. The oxygen uptake of a healthy fluidized-bed system is directly related to the organic loading placed on the system: the greater the organic loading, the greater the oxygen uptake. The oxygen uptake is readily determined by reading the influent and effluent dissolved oxygen values off the control panel and calculating the difference. The oxygen uptake is used to adjust the nutrient flowrate. Considering 12.4% by weight of new cell production is nitrogen and one-fifth of this value is phosphorous (Tchobanoglous & Burton 1991), an equation based on a chemical oxygen demand (COD):N:P of 100:5:1 was generated to facilitate adjusting the nutrient dosage. Also assumed is a nutrient solution that is 5.8% by weight nitrogen and 1.2% by weight phosphorous. The following equation results:

$$Q_n = Q_f \times [\text{Influent DO(mg/L)} - \text{Effluent DO(mg/L)}] \times 0.0012$$

where Q_n = nutrient flowrate (gpd)
       Q_f = fluidization flowrate (gpm)
      0.0012 = conversion factor

The calculated value for Q_n is used with Figure 2 to set the nutrient pumping rate. Monthly analysis showing that treatment objectives have been met verifies that adequate oxygen and nutrients have been added to the system. Other biweekly checks include verifying that the fluidized bed is running with no alarm lights illuminated, and noting the influent and effluent dissolved oxygen, influent flowrate, and nutrient consumption.

At the Case I site, maintaining an alkaline pH is important to enhance nitrification. Therefore, pH is analyzed daily during the week utilizing pH strips to make adjustments to the nutrient pumping rate. The nutrient solution here consists of trisodium phosphate and caustic soda.

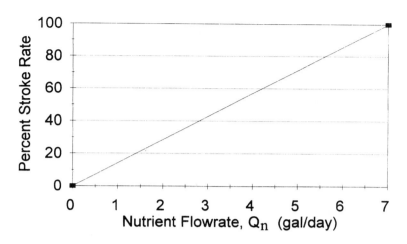

FIGURE 2. Nutrient pump curve.

MAINTENANCE

Weekly cleaning to remove any buildup from the influent dissolved oxygen probe and monthly verification of calibration of the self-cleaning dissolved oxygen probe are sufficient to maintain reliable dissolved oxygen readings. Approximately every other week, the 60-gal (227-L) nutrient tank is refilled. Oil in the air compressor is checked once a week. The oil and the oil filter are changed every 6 weeks. The oil in the biomass control mixer is changed every 6 to 8 weeks. To reduce maintenance, the fluidization pumps have been designed with permanently lubricated bearings.

TREATMENT RESULTS AND OPERATING COSTS

Case I

At a pipeline company terminal tank farm, bottom water from several storage tanks is stored in one tank that holds 3.4 million gal (12.8 million L). Previously the bottom water was hauled off site and treated at the local municipal treatment plant at a cost of $0.19/gal. Table 1 shows the monthly effluent requirements that had to be met for discharge to the surface and to avoid disposal costs. The operating cost of the fluidized bed is $0.01/gal with 1 gpm (3.78 L/min) forward flow and $0.10 kWh electric cost. Other assumptions include 7½-HP fluidization pump power cost = $13.42/day, 5-HP air compressor power cost to produce 44 lb (20 kg) of oxygen = $2.64/day, and $0.54/day for carbon replacement due to attrition.

Table 1 shows the average tank bottom water characteristics from October 1992 through October 1993. Table 1 also shows the averaged monthly effluent

TABLE 1. Analytical results and permit requirements.

	Permit Requirement (mg/L)	Averaged Influent to Fluidized Bed (mg/L)	Averaged Effluent from Fluidized Bed (mg/L)	Percent Removal	Analysis for Performance Guarantee
CASE I					
Total BTEX	<0.75	27.4	<0.015	>99	<0.005
Benzene	<0.05	7.4	<0.005	>99	<0.005
NH$_3$	<5	23.63	1.7	92.8	8.01
Ethylbenzene	—	0.66	<0.005	>99	<0.005
Toluene	—	14.03	<0.005	>99	<0.005
Xylenes	—	5.27	<0.015	>99	<0.005
COD	—	888.3	41.3	95.4	<15
					Analysis for Demonstration Test
CASE II					
Benzene	<0.005	<0.012	<0.005	—	<0.005
Toluene	<1.0	2.71	<0.008	>99	0.0021
Ethylbenzene	<0.7	0.55	<0.001	>99	<0.0033
Xylenes	<10	1.17	<0.0049	>99	0.023

analysis, which has consistently allowed for surface discharge of the fluidized bed effluent. The tank bottom water is pumped through an oil-water separator prior to delivery to a 6,000-gal (23,000-L) feed tank. Water is pumped from the feed tank at 1 gpm. Effluent, containing biosolids removed from the GAC, flows from the fluidized bed into a 6,000-gal storage tank. Every third day the fluidized bed is put on recycle for approximately 3 hours allowing the biosolids to settle in the storage tank. The treated water in the storage tank is analyzed for ammonia to verify the 5 mg/L permit requirement. Effluent from the storage tank is pumped through six 55-gal (208-L) carbon drums prior to surface discharge. The installation of these carbon drums facilitated acquisition of a discharge permit. The analysis taken for the performance guarantee prior to installation of the carbon drums demonstrated that removal of the organics occurs in the fluidized bed, see Table 1. Furthermore, the carbon in the drums has not been replaced since initiation of surface discharge in June 1993.

The solids that are collected in the treated water storage tank have been removed from this tank twice via a vacuum truck. The solids were disposed of at the local municipal treatment plant at no charge.

Case II

A groundwater recovery and reinjection system was implemented to remediate contaminated soils and groundwater resulting from an underground storage tank release. The remediation system had to fulfill three U.S. Environmental Protection Agency (EPA) mandates: (1) provide maximum reinjection of the recovered groundwater; (2) demonstrate phase control via groundwater monitoring and modeling; and (3) obtain maximum contaminant levels (MCLs) for drinking water on the treated groundwater.

A GAC fluidized bed was selected for treatment of the extracted groundwater because of its low operating cost compared with conventional activated carbon treatment (Perpich & Laubacher 1992). Furthermore, the fluidized bed offered high treatment capacity and fewer of the iron fouling problems associated with conventional systems.

Upon startup, the fluidized bed met MCL criteria during an initial 1-hour demonstration test. This demonstration test was required by the EPA and involved treating approximately 4,000 gal (15,000 L) of contaminated groundwater (see Table 1). The GAC fluidized bed has been in operation approximately 7 months. Approximately 25% of the effluent from the fluidized bed, 15 gpm, is filtered and amended with nutrients and hydrogen peroxide for reinjection to stimulate in situ bioremediation. To date, the solids concentrated from the filters in the backwash holding tank have not reached a level requiring disposal. The effluent has met the EPA requirement for obtaining MCLs as presented in Table 1. Carbon addition has not been required. The operating cost for the fluidized bed is less than $0.01/gal. Refer to Case I for the assumptions used to determine the operating cost. The remaining effluent from the fluidized bed, approximately 45 gpm (170 L/min), is sent to a publicly owned treatment works (POTW) at a cost of $0.22/gal.

REFERENCES

Edwards, D. E., M. A. Heitkamp, and W. J. Adams. 1994. "Laboratory-Scale Evaluation of Aerobic Fluidized Bed Reactors for the Biotreatment of a Synthetic, High-Strength Chemical Industry Waste Stream." *Water Environment Research* 66(1): 70-83.

Perpich, Jr., W., and R. Laubacher. 1992. "Implementation of GAC Fluidized Bed for Treatment of Petroleum Hydrocarbons in Groundwater at Two BP Oil Distribution Terminals, Pilot and Full Scale." Presented at the International Symposium on the Implementation of Biotechnology in Industrial Waste Treatment and Bioremediation, September 15-16, 1992. Grand Rapids, MI.

Pirbazari, M., T. C. Voice, and W. J. Weber. 1990. "Evaluation of Biofilm Development on Various Natural and Synthetic Media." *Hazardous Waste Hazardous Materials* 7: 239.

Tchobanoglous, G., and F. L. Burton (Eds.). 1991. "Design of Facilities for the Biological Treatment of Wastewater." In: Metcalf & Eddy, Inc., *Wastewater Engineering: Treatment, Disposal, and Reuse*, 3rd ed. pp. 536-537.

Voice, T. C., D. Pak, X. Zhao, J. Shi, and R. F. Hickey. 1992. "Biological Activated Carbon in Fluidized Bed Reactors for the Treatment of Groundwater Contaminated With Volatile Aromatic Hydrocarbons." *Waste Resources* 26(10): 1389-1401.

Biological Pretreatment to Reduce Off-Gas Treatment Costs for Air Strippers

Charles J. Gantzer

ABSTRACT

Biological pretreatment of petroleum-contaminated groundwater prior to air stripping was evaluated as a means of reducing the cost of off-gas treatment. The goal was to reduce benzene emissions for the entire site below the regulatory threshold for installing off-gas treatment equipment (0.87 lb benzene/day). The operating costs for two 2,000-gal (7,571-L) fixed-bed bioreactors receiving a 15 gal/min (57 L/min) flow with 20 mg/L of benzene were compared using existing process models. The first bioreactor was air sparged and consequently had air emissions. The second bioreactor was emission-free, because its oxygen source was a bubble-free membrane oxygenator. The estimated granular activated carbon (GAC) replacement cost for the existing air stripper was $54/day, the estimated minimum GAC cost for the hybrid system using a sparged bioreactor was $30/day, and the hybrid system with the emission-free bioreactor had no GAC replacement costs. Capital costs for the emission-free bioreactor could be recovered in about 90 days, based on the operating cost savings.

INTRODUCTION

Air stripping is a common pump-and-treat technology for groundwater contaminated with volatile organic compounds (VOCs), because of its low capital and operating costs. However, when the total emissions for the site exceed the regulatory threshold for installing air pollution control equipment, then air stripping can become expensive. For example, our client was recovering gasoline-contaminated groundwater at 15 gal/min (57 L/min) that contained initially 20 mg/L of benzene. This benzene concentration did not drop after 3 weeks of pumping. The existing air stripper was emitting about 3.6 lb/day (1.6 kg/day) of benzene, which was above the regulatory threshold of 0.87 lb/day (0.39 kg/day). Treatment of the exhausted air with GAC canisters would have had a GAC replacement cost of $54/day, assuming a 50:1 air to water flowrate in the air stripper, the VentSorb™ (Calgon Corporation) isotherm for benzene, and a GAC replacement cost of $3.00/lb (0.45 kg) GAC (including labor). A hybrid treatment system consisting of a 2,000-gal (7,571-L) bioreactor (made from existing tankage on site) and the existing air stripper was proposed. If the biological pretreatment could reduce benzene concentrations from 20 to 4.8 mg/L, then the existing air stripper could be used without air pollution control equipment.

Two possible bioreactor configurations were evaluated. The first consisted of packing the tank with structured media and sparging air through diffusers at the tank bottom as an oxygen source. Sparging would make the bioreactor behave as a completely mixed reactor (Perry & Chilton 1973). This bioreactor option would have benzene emissions. The second option consisted of packing the tank with structured media and installing membrane oxygen-dissolution modules in an effluent recycle loop as the oxygen source. Because the membrane modules dissolve oxygen without bubbles, the second bioreactor option would have no benzene emissions.

REACTOR MODELING

The performance of the sparged and emission-free fixed-bed bioreactors was evaluated, assuming that benzene was the only electron donor present in the water. The flux of benzene and oxygen into a biofilm was modeled as a function of water-phase mass transfer, molecular diffusion within the biofilm, and Monod kinetics within the biofilm. Rather than attempting to model dual-substrate limitation in the biofilms, the present study used an algorithm to calculate oxygen and benzene fluxes by (1) determining the water-phase substrate concentrations that define the boundaries between dual- and single-substrate limitation for both substrates, (2) defining an appropriate boundary-defining oxygen or benzene limitation, and (3) assuming that all points within the simulated bioreactor are either benzene or oxygen limited (Gantzer 1995). The benzene and oxygen concentrations that define the boundaries between single- and dual-limitation biofilm kinetics were determined by the algorithms developed by Rittmann and Dovantzis (1983). Flux into biofilms limited by a single substrate were calculated by a pseudoanalytical model (Sáez & Rittmann 1988).

Sparged Bioreactor Evaluation

The sparged 2,000-gal (7,571-L) fixed-bed bioreactor was modeled as a completely mixed biofilm reactor (Rittmann 1982) using the volatilization model of Roberts et al. (1984)

$$V\frac{dS}{dt} = Q\,S_o - QS - AJ - GH_c\,[1 - \exp(-\phi)]\,S \qquad (1)$$

in which V is the volume of the bioreactor (m^3), t is time (h), Q is influent water flowrate (m^3/h), S_o is the influent benzene concentration (g/m^3), S is the effluent benzene concentration (g/m^3), A is the biofilm surface area (m^2), J is the flux of benzene into the biofilms $(g/m^2 \cdot h)$, G is the air flowrate (m^3/h), H_c is the Henry's law constant for benzene $(m^3 \text{ water}/m^3 \text{ air})$, and ϕ is the stripping factor (unitless) defined by

$$\phi = \frac{K_L a\ V}{G\ H_c} \qquad (2)$$

in which $K_L a$ is the overall mass transfer rate coefficient for the movement of benzene from the water phase to the inside of the air bubbles $(1/h)$. The model used the benzene Henry's law coefficient reported by Freedman et al. (1993), the $K_L a$ correlations provided in Weber (1972), the benzene and oxygen kinetic coefficients obtained from several sources (McCarty 1972, Goldsmith & Balderson 1988), and established mass transfer correlations (Perry & Chilton 1973).

Figure 1 illustrates the predicted steady-state effluent benzene concentrations and benzene emissions from the sparged 2,000-gal (7,571-L) fixed-bed bioreactor as a function of air flowrate. The model results suggest that above an air flowrate of 5 standard cubic feet per minute (SCFM), the reactor's biofilms are no longer oxygen limited. At 5 SCFM, the sparged bioreactor has the predicted ability to remove only 62% of the influent benzene mass (45% by biological oxidation and 17% by volatilization) and does not reduce benzene concentrations to the desired 4.8 mg/L. Increasing the air flowrate above 8 SCFM results in a greater total removal of benzene, but air emissions exceed the 0.87 lb/day (0.39 kg/day) threshold for the site, thus requiring GAC canisters for both the bioreactor and

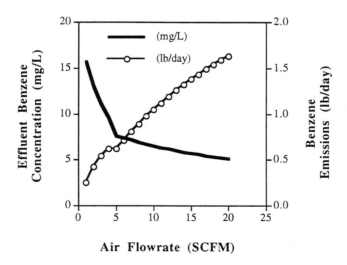

Air Flowrate (SCFM)

FIGURE 1. **Predicted effluent benzene concentrations and benzene emissions for the sparged bioreactor as a function of air flowrate, assuming a feed flow of 15 gpm, a feed benzene concentration of 20 mg/L, and a temperature of 10°C, where lb/day is 0.45 kg/day.**

the air stripper. The greatest reduction in GAC usage for the site would occur when the extent of biological oxidization of the benzene is maximized and total emissions are minimized, i.e., at an air flowrate of 5 SCFM.

Emission-Free Bioreactor Evaluation

The hybrid treatment process consisting of the emission-free bioreactor and the air stripper is illustrated in Figure 2. The emission-free bioreactor was modeled as a fixed-bed biofilm reactor with an effluent recycle loop (Rittmann 1982),

$$\frac{dS}{dt} = D_h \frac{d^2S}{dx^2} - u \frac{dS}{dx} - J\,a \tag{3}$$

in which D_h is the hydraulic dispersion coefficient (m^2/h), x is distance from the inlet (m), u is the superficial water velocity (m/h), and a is the specific surface area of the structured media $(1/m)$. The benzene and oxygen flux values were modeled in the same manner as for the sparged bioreactor. The benzene and oxygen mass balance equations were solved via an iterative approach using the quasilinear formulation of Odencrantz et al. (1990) with the resulting tridiagonal matrix solved by the Thomas algorithm (Pinder & Gray 1977). The modeling

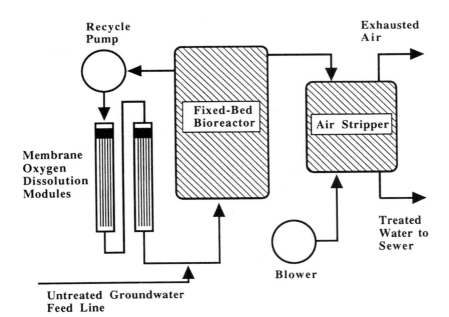

FIGURE 2. Schematic diagram of the evaluated hybrid system that uses an emission-free bioreactor as a pretreatment step prior to an air stripper. The 2,000-gal (7,571-L) emission-free bioreactor has the predicted ability to reduce benzene concentrations below 5 mg/L, allowing the air stripper to operate without air pollution control equipment.

assumed that two Membran BMA-31-1 oxygen dissolution modules (Semmens & Gantzer 1993) were operated in series in the recycle line. A schematic diagram of a Membran gas-dissolution module is provided in Figure 3.

The predicted steady-state benzene and oxygen concentrations for the emission-free bioreactor are provided in Figure 4. The predicted results indicate that the 2,000-gal (7,571-L) emission-free fixed-bed bioreactor is able to reduce the benzene concentrations below the goal of 4.8 mg/L, i.e., 75% of the influent benzene is biologically oxidized. Because the bioreactor has no emissions and because the emissions for the subsequent air stripper are less than 0.87 lb/day (0.39 kg/day), there is no need for installing air pollution control equipment on the site.

COST COMPARISON

The projected capital costs for the major components of the two hybrid treatment systems and the projected operating costs for the three treatment alternatives (existing air stripper, hybrid system with sparged bioreactor, and hybrid system with emission-free bioreactor) are provided in Table 1. The listed capital costs for modifying the existing tankage into bioreactors are low, because not all required components are included (e.g., nutrient addition equipment and control systems) and because fabrication costs are largely ignored. The listed operating costs assumed that electricity costs $0.50/kW·h, oxygen costs $0.40/lb (0.45 kg), and

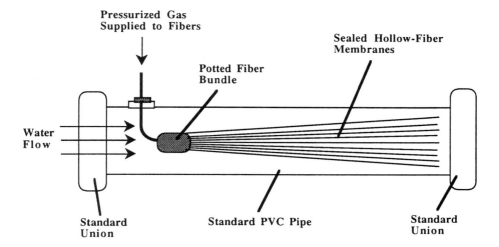

FIGURE 3. Schematic diagram of a Membran bubbleless gas-dissolution membrane module. Pressurized oxygen is applied to the lumen of microporous hollow-fiber membranes. The distal ends of the fibers are sealed. Oxygen diffuses across the wall of the microporous fibers and into the water flowing past the fibers.

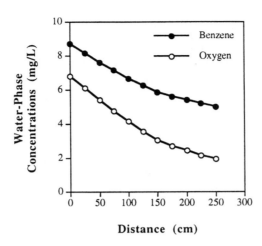

Distance (cm)

FIGURE 4. Predicted steady-state benzene and oxygen concentrations as a function of distance from the inlet of the 2,000-gal (7,571-L) emission-free fixed-bed bioreactor. The above curves assume a feed flowrate of 15 gal/min (57 L/min), a feed benzene concentration of 20 mg/L, a recycle ratio of 3, a water temperature of 10°C, and an oxygen pressure of 45 psig applied to the two Membran BMA-31-1 oxygen dissolution modules.

TABLE 1. Comparison of projected capital costs for major bioreactor components and of operating costs for the three treatment alternatives.

Treatment Alternative	Capital Cost for Modifications (U.S. $)	Operating Cost (U.S. $ per day)
Existing Air Stripper		
5 HP blower		$ 4.50
GAC replacement		53.85
Totals		**58.35**
Sparged Bioreactor with Air Stripper		
5 HP blower bioreactor	$ 1,500.00	4.50
structured media	2,500.00	
GAC replacement bioreactor		9.30
5 HP blower air stripper		4.50
GAC replacement stripper		20.60
Totals	**4,000.00**	**38.90**
Emission-Free Bioreactor with Air Stripper		
1.5 HP recycle pump	500.00	1.30
oxygen cost		2.00
oxygen dissolution system	2,000.00	
structured media	2,500.00	
5 HP blower air stripper		4.50
Totals	**$ 5,000.00**	**$ 7.80**

GAC has a replacement cost of \$3.00/lb (0.45 kg). The present value of the three treatment alternatives as a function of years of operation is provided in Figure 5. The present value calculations assumed a 6% discount rate, an annual replacement of the hollow-fiber membranes in the emission-free bioreactor (\$1,000/y), and no change in influent benzene concentrations. Figure 5 suggests that the hybrid treatment system with the emission-free bioreactor will have substantially reduced costs compared to the other two treatment alternatives. The present value analysis also indicates that the hybrid system with the emission-free bioreactor would be less expensive after 1 year of operation than the continued use of the existing air stripper with GAC, even if the total capital cost for converting the existing 2,000-gal (7,571-L) tank into an emission-free bioreactor was \$16,000.

SUMMARY

Biological pretreatment of contaminated groundwater prior to air stripping can reduce off-gas treatment costs. The maximum potential cost savings may not be realized with an air-sparged bioreactor, because the off-gas from the bioreactor must be included in the total VOC emissions. An emission-free bioreactor that uses a bubbleless membrane oxygenator does not require off-gas treatment, which allows the cost savings potential for biological pretreatment to be realized.

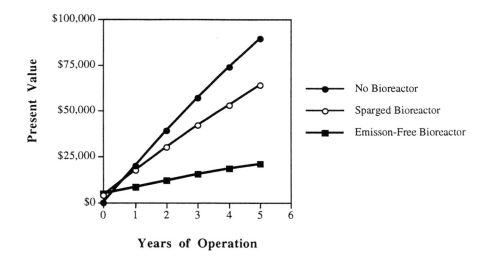

FIGURE 5. Present value comparison for the three treatment alternatives as a function of years of operation. The plotted curves use the values listed on Table 1, assume that the hollow-fiber membranes are replaced at the end of each year, and assume that the influent benzene concentration remains 20 mg/L.

REFERENCES

Freedman, D. L., P. Chennupati, and B. J. Kim. 1993. "Temperature variability of Henry's Constants and the effect of VOC emissions from wastewater treatment plants." *66th Annual Conference of the Water Environment Federation* (Anaheim, CA), Research Symposium. pp. 95-106.

Gantzer, C. J. 1995. *Sparged Bioreactor for Macintosh: User's Guide and Reference.* Gantzer Environmental Software and Services, Llc., Minneapolis, MN.

Goldsmith, B. D., and R. K. Balderson. 1988. "Biodegradation and growth kinetics of enrichment isolates on benzene, toluene and xylene." *Water Science and Technology 20*: 505-507.

McCarty, P. L. 1972. "Stoichiometry of biological reactions." Presented at International Conference, Towards a Unified Concept of Biological Treatment Design, Atlanta, GA.

Odencrantz, J. E., B. Wookeun, A. J. Valocchi, and B. E. Rittmann. 1990. "Stimulation of biologically active zones (BAZ's) in porous media by electron-acceptor injection." *Journal of Contaminant Hydrology 6*: 37-52.

Perry, R. H., and C. H. Chilton. 1973. *Chemical Engineers' Handbook.* McGraw-Hill, New York, NY.

Pinder, G. F., and W. G. Gray. 1977. *Finite Element Simulation in Surface and Subsurface Hydrology.* Academic Press, New York, NY.

Rittmann, B. E. 1982. "Comparative performance of biofilm reactor types." *Biotechnology and Bioengineering 24*: 1341-1370.

Rittmann, B. E., and K. Dovantzis. 1983. "Dual limitation of biofilm kinetics." *Water Research 17*: 1727-1734.

Roberts, P. V., C. Munz, P. Dändliker, and C. Matter-Müller. 1984. *Volatilization of Organic Pollutants in Wastewater Treatment: Model Studies.* EPA-600-2-84-047. Report prepared for the Municipal Environmental Research Laboratory, Office of Research and Development, U.S. Environmental Protection Agency, Cincinnati, OH.

Sáez, P. B., and B. E. Rittmann. 1988. "Improved pseudoanalytical solution for steady-state biofilm kinetics." *Biotechnology and Bioengineering 32*: 379-385.

Semmens, M. J. and C. J. Gantzer. 1993. "Gas transfer using hollow fiber membranes." *66th Annual Conference of the Water Environment Federation* (Anaheim, CA), Research Symposium. pp. 395-406.

Weber, W. J. 1972. *Physicochemical Processes for Water Quality Control.* Wiley-Interscience, New York, NY.

Pilot Study: Fixed-Film Bioreactor to Enhance Carbon Adsorption

Gregory J. Gromicko, Mark Smock,
Arthur D. Wong, and Bill Sheridan

ABSTRACT

A pilot study was performed to evaluate a novel bioreactor for treatment of groundwater at a former wood-preserving facility. Groundwater, impacted with wood-treating preservatives, is currently being recovered and treated using chemical/physical pretreatment followed by granular activated carbon (GAC) adsorption. The bioreactor was evaluated to reduce GAC usage. The study tested AlliedSignal's Immobilized Cell Bioreactor (ICB) on a sidestream from the groundwater treatment system. The ICB technology is a fixed-film, fixed-bed bioreactor that utilizes a dual microbe support media system consisting of (1) a patented, carbon-coated, polyurethane foam packing; and (2) conventional, random, plastic packing. The ICB was tested at four discrete operating conditions representing hydraulic retention times of 16.6, 8.3, 4.2, and 2.1 h. Influent and effluent samples collected at each condition were analyzed for chemical oxygen demand (COD), phenolics, polycylic aromatic hydrocarbons (PAHs), and benzene, toluene, ethylbenzene, and xylenes (BTEX). Results indicated that a maximum removal of 86.9% for COD, 96.3% for total phenols, 98.9% for total PAHs, and 97.2% for BTEX. Based on these data, a full-scale ICB as a pretreatment step would decrease GAC usage by a factor of 3.5.

INTRODUCTION

The biological pilot study was conducted using groundwater containing polycyclic aromatic hydrocarbons (PAHs); phenolics; and benzene, toluene, ethylbenzene, and xylenes (BTEX). Pentachlorophenol, chromium, and arsenic are not present in the site groundwater being recovered. The recovered groundwater is treated using a treatment system that includes solids and oil removal by settling and filtration and removal of organic constituents using granular activated carbon (GAC) prior to National Pollutant Discharge Elimination System (NPDES) discharge. Due to the elevated concentration of organic constituents in the groundwater, AlliedSignal's ICB technology was tested to reduce organic loading to the GAC, thereby reducing carbon usage. A schematic of the ICB unit is shown in Figure 1 (Mashayekhi 1993).

The ICB technology utilizes dual support media. The first is a patented, polyurethane foam packing coated with activated carbon (Sanyal et al. 1993). The packing provides a large specific surface area for biomass growth and very limited adsorption capacity for organics. The second is conventional, random, plastic packing to provide structural rigidity and continuous redistribution of the fixed bed. The two media are mixed at approximately a 1:1 ratio.

OPERATION

The unit was installed indoors and connected to treat groundwater following the solids and oil separator, prior to filtration and GAC. The effluent from the ICB was directed to a small carbon canister to obtain on-line carbon usage measurements before being polished by the GAC and discharged.

To inoculate the ICB, biomass from another ICB installation treating PAHs was used. No microbial tests were performed on the biomass, but the inoculum was assumed to contain microorganisms acclimated to PAHs and was believed

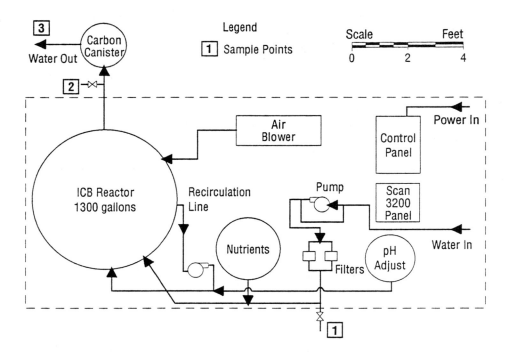

FIGURE 1. Schematic of ICB unit (where 1 ft = 0.3 m).

to be capable of creating a culture able to degrade the organics found at this site. The ICB was operated for 3 months. Four different influent flowrates were tested, representing hydraulic retention times (HRTs) of 16.6, 8.3, 4.2 and 2.1 h.

Each condition was tested for 2 weeks to allow acclimation of the microorganisms and for steady-state conditions to be achieved. Operating parameters were monitored closely during each condition and included controlling the influent flowrate, maintaining excess dissolved oxygen in the effluent, and maintaining residual ammonia nitrogen and orthophosphorous concentrations in the effluent.

SAMPLING AND ANALYSES

Field tests for COD and total phenols were performed on the influent and effluent to determine the instantaneous performance of the ICB (Chemetrics Inc. 1990/1991 and Hach Company 1993). Test kits were also used to monitor the residual ammonia nitrogen and orthophosphorous in the reactor. The influent and effluent from the ICB was sampled for laboratory analyses twice during each operating condition after an approximate 1- to 1½-week acclimation period. In addition, effluent from the carbon canister was analyzed for COD to estimate carbon usage for comparison to the existing treatment system. All analytical methods used are presented in Table 1.

RESULTS

Chemical Oxygen Demand (COD)

The influent and effluent COD was measured in both the field and laboratory. The average influent COD concentrations were 336 mg/L for the field and 418 mg/L for the laboratory. The effluent COD concentrations were also similar for both the field and laboratory measurements. This comparison verified that the field COD analyses could be used as a treatment indicator parameter. The

TABLE 1. Analytical methods.

Parameter	Methods
Chemical oxygen demand (COD)	Test Kit (Field) EPA 410.3
Total phenols	Test Kit (Field) EPA 9065/420.1 (4-AAP)
Phenolics	EPA 8040
Polycyclic aromatic hydrocarbons (PAHs)	EPA 8310
Benzene, toluene, ethylbenzene, xylenes (BTEX)	EPA 8020

laboratory average COD concentrations are presented in Figure 2. Results indicate a maximum COD removal of 86.9% and a trend of decreasing removal at lower HRTs. This does not necessarily imply that removal of all target constituents also decreased at the same rate. The COD measurement was a broad test used as an indicator parameter and measured other organics in addition to the target constituents. In addition, biodegradation of the soluble organics may produce intermediates which can be detected as COD but are not detected in the specific constituent analyses. Therefore, even if incomplete COD removal is observed, target constituents such as phenolics or PAHs may still be effectively removed.

Total Phenols

The influent and effluent total phenols were also measured in both the field and laboratory. The total phenols removal efficiencies calculated using both the field and analytical data compared favorably indicating that the data correlated relatively well, confirming the use of the test kits to monitor this constituent. The laboratory average total phenols concentrations are presented in Figure 3. The results show high removal rates at each condition, with a maximum removal of 96.3%, that decreased only slightly at the lower HRTs.

Phenolics

The average influent and effluent phenolics concentrations from the laboratory data are presented in Figure 4. The influent and effluent phenolics concentrations varied widely between samples ranging from 41 µg/L to 14,405 µg/L in the influent and varied from nondetectable to 1,248 µg/L in the effluent. Although high removals were observed at each operating condition, the influent and effluent phenolics concentrations measured during the first two conditions were very low.

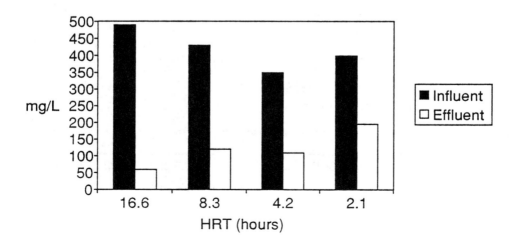

FIGURE 2. Average COD concentrations.

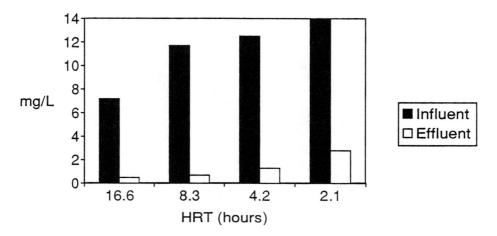

FIGURE 3. Average phenol concentrations.

Polycyclic Aromatic Hydrocarbons (PAHs)

The average influent and effluent PAH concentrations from the laboratory data are presented in Figure 5. The influent and effluent PAH concentrations varied from 2,733 µg/L to 9,833 µg/L in the influent and from 31 µg/L to 1,758 µg/L in the effluent. High removal efficiencies can be observed at all conditions with a maximum removal of 98.9%. Even at condition #4, using an HRT of 2.8 h which is a very low HRT for biodegradation of PAHs, 88.0% removal was observed (U.S. EPA 1987).

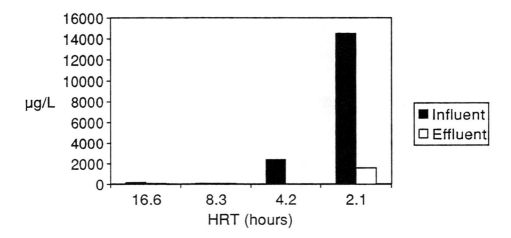

FIGURE 4. Average phenolics concentrations.

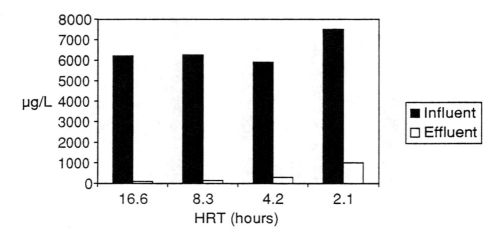

FIGURE 5. Average PAH concentrations.

Benzene, Toluene, Ethylbenzene, and Xylenes (BTEX)

The average influent and effluent BTEX concentrations from the laboratory data are presented in Figure 6. The influent and effluent BTEX concentrations varied from 93 µg/L to 231 µg/L in the influent and from nondetect to 11.2 µg/L in the effluent. High removal rates of BTEX were observed at every condition, with a maximum removal of 98.4%.

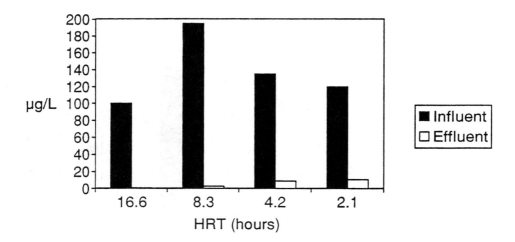

FIGURE 6. Average BTEX concentrations.

CONCLUSIONS

ICB treatment has been proven to be effective for removal of organics from the groundwater prior to treatment with GAC at this site. It can greatly reduce the organic loading to the GAC, resulting in reduced carbon usage rates. Currently, the carbon usage rate of the GAC system is estimated at 5.4 lb per 1,000 gal of groundwater treated. Data collected from the carbon canister following the ICB reactor suggest that the carbon usage rate would decrease to approximately 1.55 lb per 1,000 gal with ICB pretreatment. This is a decrease in carbon usage of approximately 3.5 times. Additionally, GAC may adsorb nonbiodegradable organics that may pass through the ICB, and would provide an effective treatment combination, although effluent solids issues were not examined during this study. The HRT selected for a full-scale ICB system for removal of these constituents to satisfactory levels would be approximately 4 h. A system of this size would require little space and minimal operator attention, as proven by a similar full-scale system in operation since 1990 (Lupton and Detroy 1991).

REFERENCES

Chemetrics, Inc. 1990/1991. *Advanced Systems for Water Analysis.*

Hach Company. 1993. *Hach COD System for Wastewater Testing.*

Lupton, F. S., and R. W. Detroy. 1991. "Immobilized Microbial Bioreactors for Petrochemical Wastewater Remediation." Presented at AIChE Summer National Meeting. Pittsburgh, PA, August 18-21.

Mashayekhi, M. 1993. "Full-Scale Experiences with the Fixed-Film Immobilized Cell Bioreactor." *Applied Bioremediation 93.* Fairfield, NJ, October 25-26.

Sanyal, S., T. P. Love, and L. J. DeFilippi. 1993. *Process and Apparatus for Removal of Organic Pollutants from Waste Water.* United States Patent Number 5,217,616.

U.S. Environmental Protection Agency. 1987. *Estimating Releases and Waste Treatment Efficiencies for the Toxic Chemical Release Inventory Form.* PB88-210380.

Performance of Prototype Bioreactors for Biodegradation of Trichloroethylene by *Methylocystis* sp. M

Fusako Okada, Tatsuo Shimomura,
Hiroo Uchiyama, and Osami Yagi

ABSTRACT

Two prototype bioreactors were tested for continuous biodegradation of trichloroethylene (TCE) by *Methylocystis* sp. M, a methane-utilizing bacterium. One bioreactor was a fluidized-bed bioreactor (FBB) containing alginate-immobilized cells and the other a membrane bioreactor (MB) in which hollow fiber membrane modules were employed for recovering free cells from treated water. TCE degradation efficiency was 80% to 90% with FBB, which was operated at an influent TCE concentration of about 1 mg/L and a hydraulic retention time (HRT) of 2.6 hours. MB, which was operated at an influent TCE concentration of about 0.25 mg/L and HRT of 3.3 hours, yielded a TCE degradation efficiency of 90 to 99%. In this study, MB was found to be more practical than FBB.

INTRODUCTION

Because contamination of soil and groundwater by chlorinated organic compounds is grave social concern in Japan, the authors have studied a variety of approaches for reclamation of TCE-contaminated groundwater by *Methylocystis* sp. M (strain M), a type II methanotroph (Uchiyama et al. 1989). In this paper, as strain M was hardly employable in biofilm form, a fluidized-bed bioreactor (FBB), which contained alginate-immobilized strain M cells, and a membrane bioreactor (MB), which was equipped with hollow fiber membrane modules for recovering free cells from the treated water, were compared for continuous treatment of TCE-contaminated groundwater.

MATERIALS AND METHODS

Microorganism and Immobilization

Methylocystis sp. M (strain M, Uchiyama et al. 1989) was used throughout this study. For FBB, strain M was cultured to the late log phase of growth at

30°C in a 10-L jar fermenter containing 7 L of mineral salts medium supplemented with methanol and methane as carbon sources (Shimomura et al. 1994). The composition of the mineral salts medium was as follows (mg/L): KH_2PO_4 450, K_2HPO_4 1170, $NH_4 \cdot Cl$ 2140, $Ca(NO_3)_2 \cdot 4H_2O$ 4.8, $MgSO_4 \cdot 7H_2O$ 121, $FeSO_4 \cdot 7H_2O$ 28, $MnSO_4 \cdot 4\text{-}5H_2O$ 0.6, H_3BO_3 0.05, $ZnSO_4 \cdot 7H_2O$ 0.1, $CuSO_4 \cdot 5H_2O$ 0.06, $Na_2MoO_4 \cdot 2H_2O$ 0.01, $Co(NO_3)_2 \cdot 6H_2O$ 0.6, $NiSO_4 \cdot 7H_2O$ 0.06, and H_2SeO_4 0.04, pH 7.2. Cells were harvested by centrifugation, resuspended in activated carbon (AC) treated tap water, and fixed with 2% calcium alginate to give gel beads (cell density of 13 g/L, average bead diameter 2.8 mm). For MB, strain M was cultured under the same conditions except that only methane was used as carbon source.

FBB and MB Bioreactors

Figures 1 and 2 schematically show the FBB and MB systems. The FBB and MB vessels were made of glass and had capacities of 20 L and 1.1 L, respectively.

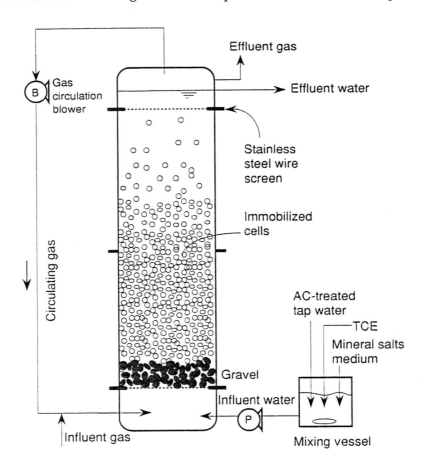

FIGURE 1. **Schematic diagram of the FBB system.**

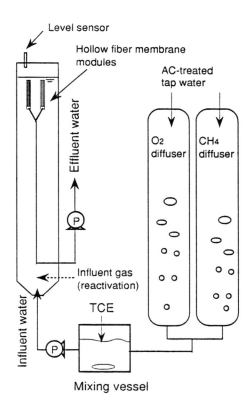

FIGURE 2. Schematic diagram of the MB system.

Eight polyethylene hollow fiber membrane modules STN0124T (pore diameter 0.1 micron, effective surface area 0.014 m²/module; Mitsubishi Rayon Engineering Co., Ltd.) was employed for MB. Table 1 summarizes the operational conditions of the two bioreactors employed in this study. The following nutrients were dissolved in AC treated tap water and fed for reactivation of the TCE degradation activity of strain M. FBB (mg/L): on days 1 and 2 $Ca(NO_3)_2 \cdot 4H_2O$ 96, KH_2PO_4 1.7, K_2HPO_4 2.7 and $FeSO_4 \cdot 4H_2SO_4$ 3.5; on days 3 to 37 $Ca(NO_3)_2 \cdot 4H_2O$ 288, KH_2PO_4 7.0, K_2HPO_4 10.9, and $FeSO_4 \cdot 4H_2SO_4$ 10.7; and on days 38 to 44 $Ca(NO_3)_2 \cdot 4H_2O$ 288, KH_2PO_4 7.0, and K_2HPO_4 10.9 in ¼ strength mineral salts medium; in batch mode, under daily replacement to a fresh solution. MB(mg/L): on days 8, 20, and 30 $Na(NO_3)_2 \cdot 4H_2O$ 340, KH_2PO_4 14, K_2HPO_4 22; and on days 24, 35, and 43 $Na(NO_3)_2 \cdot 4H_2O$ 510, KH_2PO_4 21, and K_2HPO_4 33; in ¼ strength mineral salts medium.

TCE Analysis

Headspace gas chromatography (Shimomura et al. 1994) was used to analyze TCE in the influent water, the effluent water of FBB and MB, and the effluent

TABLE 1. Operational conditions of the bioreactors.

Bioreactor		Influent Water (mL/min)	Hydraulic Retention Time (h)	Influent TCE Concentration (mL/min)	Influent Methane (mL/min)	Influent Air (mL/min)	Circulation Gas Flowrate (mL/min)	Influent Dissolved Methane (mg/L)	Influent Dissolved Oxygen (mg/L)	Initial Cell Density (mg/L)	Gel Volume (L)	Temperature (°C)
FBB	Degradation	130	2.6	0.9-1.6	10	90	3,000	ND	ND	1,500	2.2	20
	Reactivation[a]	260[c]	1.3[c]	0.0	100	400	4,000	ND	ND			28
MB	Degradation	5.5	3.3[d]	0.14-0.38	0[f]	0[f]	0[f]	3.5	15	1,500	—	20
	Reactivation[b]	0	—[e]	0.0[e]	30	150	0	—[e]	—[e]			28

(a) In FBB, each reactivation was carried out as shown in Figure 3.
(b) In MB, each reactivation was carried out for 24 hours on the indicated day in Figure 4.
(c) In FBB, the reactivation during the period from days 1 to 4 was carried out in batch mode (see the text).
(d) In MB, the membrane module was backwashed with treated water for 5 seconds every 14 minutes to prevent membrane fouling.
(e) Batch operation.
(f) In MB, methane and oxygen were supplied in dissolved form using separate oxygen and methane diffuser columns.
ND = not determined.

gas of FBB. Samples of strain M were periodically collected from the bioreactors to measure the specific TCE degradation activity (constant k_1 in L/g/h, Speitel et al. 1993) by shaking the cells (cell density 1,000 mg/L) at 28°C and 120 rpm in 33-mL serum bottles containing 5 mL of medium plus 1 mg/L TCE.

RESULTS AND DISCUSSION

Figure 3 elucidates the changes in the TCE degradation activity of FBB. Specific TCE degradation activity (k_1=2.44 L/g/h) of FBB rapidly dropped to approximately 50% of the initial activity after only 1 day of operation. Reactivation was repeated in batch mode on days 1 and 2, but without significant recovery. On day 3, higher concentrations of nutrients were added (see Materials and Methods section) without effect on degradation activity. On day 5, the mode of reactivation was changed to continuous flow, with the same composition of nutrients. After 2 days of reactivation in continuous-flow mode, the specific TCE degradation activity unexpectedly rose to about 250% of the starting activity (k_1 = 6.0 L/g/hr; *cf.* k_1 of MB below). Although the specific TCE degradation activity of immobilized strain M was markedly recovered by reactivation, TCE degradation efficiency of FBB was 80% to 90% at an influent TCE concentration of about 1 mg/L and a hydraulic retention time (HRT) of 2.6 h. Compared to MB, the specific TCE degradation of FBB seems more unstable. Figure 4 shows the results of performance analysis of MB. TCE degradation efficiency of MB was 90% to 99%, at an influent TCE concentration of about 0.25 mg/L and HRT of 3.3 h. At the beginning of operation, the first-order reaction constant (k_1)

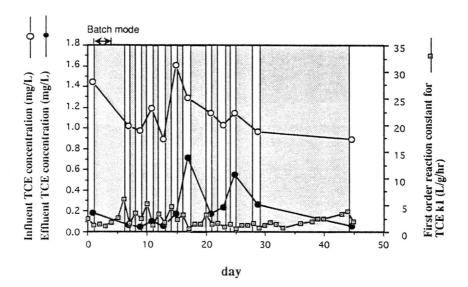

FIGURE 3. Continuous TCE degradation by FBB.

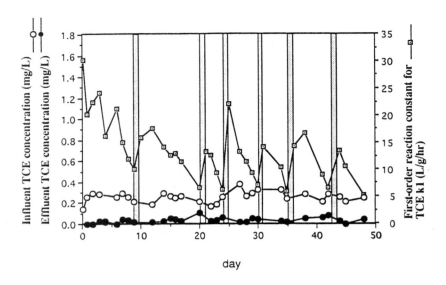

FIGURE 4. Continuous TCE degradation by MB.

of MB is 30.3 L/g/h (FBB 3.89 L/g/h), meaning MB has a 8-fold higher k_1 value than FBB.

    Critical comparison of FFB and MB is not yet possible, because important conditions, such as the influent TCE concentration and HRT, differ unpermissibly. It seems probable, however, that MB is superior to FBB for degradation of TCE in water, because the specific TCE degradation activity of MB is 8-fold higher than that of FBB (low specific TCE degradation activity of FBB is presumably due to immobilization). In addition, FBB, which requires a far longer period of reactivation than MB and involves an extra cost and labor for immobilization, allows no plug-type flow, which is said to be suitable for first-order reactions such as degradation of relatively low concentrations of halogenated compounds. Practical utility of strain M for TCE degradation using intact and immobilized cells will be further studied under long-term operation.

REFERENCES

Shimomura, T., F. Okada, K. Mishima, H. Uchiyama, and O. Yagi. 1994. "Change in Trichloroethylene Decomposition Activity of *Methylocystis* sp. M During Batch Culture." In R. E. Hinchee (Ed.), *Bioremediation of Chlorinated and Polycyclic Aromatic Hydrocarbon Compounds.* Lewis Publishers, Ann Arbor, MI. pp. 298-302.

Speitel, Jr., G. E., R. C. Thompson, and D. Weissman. 1993. "Biodegradation Kinetics of *Methylosinus trichosporium* OB3b at Low Concentrations of Chloroform in the Presence and Absence of Enzyme Competition by Methane." *Wat. Res.* 27: 15-24.

Uchiyama, H., T. Nakajima, O. Yagi, and T. Tabuchi. 1989. "Aerobic Degradation of Trichloroethylene by a New Type II Methane-Utilizing Bacterium, Strain M." *Agric. Biol. Chem.* 53: 2903-2907.

Two-Stage Bioreactor to Destroy Chlorinated and Nonchlorinated Organic Groundwater Contaminants

Brian R. Folsom, Andrew K. Bohner,
Thomas Burick, and William J. Guarini

ABSTRACT

Both chlorinated and nonchlorinated volatile organic compounds are found as common contaminants of groundwater across the nation. Two field-pilot bioreactors successfully treated contaminated groundwater at Robins Air Force Base (AFB). The fluidized-bed bioreactor (FBR) effectively removed >97% of the 1,2-dichlorobenzene (DCB) and >95% of the benzene, toluene, ethylbenzene, and xylene(s) (BTEX) from more than 210,000 gal (794,934 L) of contaminated groundwater. The FBR removed 84% of the trichloroethylene (TCE), also found in this groundwater, based on a total mass balance beyond carbon adsorption. Enhanced operational stability was demonstrated for the gas-phase reactor (GPR) with 10 months of continuous operation in the laboratory and 2 months in the field. TCE concentrations in contaminated air entering the pilot GPR were reduced by 75% on average. Capital and operating costs for the FBR system were compared to other treatment options including ultraviolet (UV)-peroxidation, air stripping with carbon adsorption, and wet carbon adsorption. GPR economics were compared to carbon adsorption at two TCE concentrations. These bioreactor systems provide economical, destructive technologies for treating either contaminated water or contaminated air originating from air stripping, air sparging, or soil vapor extraction operations and will be effective remedial options at many sites.

INTRODUCTION

Through a combination of accidental discharges and previously accepted disposal practices, chemical contaminants have been introduced to soils and surface water at sites across the United States. Benzene, a constituent of jet aviation fuel, is one contaminant commonly found at defense sites. Benzene and related aromatic hydrocarbons tend to be readily biodegradable. TCE is one of the most prevalent organic chemical contaminants found in soil and groundwater. At many sites, these and other chemicals threaten groundwater supplies.

TCE contamination is aggravated by its relatively high solubility, high density, and recalcitrance. Characterization of anaerobic and aerobic biotransformation of TCE and related volatile organic compounds (VOCs) has been ongoing for more than a decade (Ensley 1991). Under anoxic conditions, TCE undergoes reductive dehalogenation to dichloroethylene, vinyl chloride, and finally to ethylene. Under aerobic conditions, a broad variety of microorganisms cometabolize TCE following growth with methane, propane, ammonia, phenol, or toluene. Although these bacteria apparently cannot grow with, or derive energy from, the transformation of TCE, they can oxidize it to innocuous products. TCE represents a class of recalcitrant chemicals that often require a unique set of conditions for biodegradation to occur.

A variety of bioreactor configurations has been studied to establish feasibility and to overcome limitations associated with cometabolic degradation of TCE and related VOCs (Folsom and Chapman 1991; Coyle et al. 1993; Landa et al. 1994; Ensley and Kurisko 1994). Envirogen has developed a GPR system for destroying TCE, overcoming a variety of microbial and engineering limitations. Both laboratory-scale and pilot-scale systems were designed to balance the mass transfer of TCE from air to water with biodegradative capacity.

The major objective of this project was to demonstrate the efficacy of a bioreactor system for treating groundwater contaminated with hydrocarbon-based fuels and solvents commonly found at Air Force installations. Initial screening experiments demonstrated that (1) TCE-degradative microorganisms were not capable of degrading benzene unless they had been induced following growth with toluene and/or phenol, (2) TCE did not inhibit benzene degradation over the range of concentrations tested, and (3) benzene-degradative organisms did not degrade TCE (Folsom 1993). The number of chemicals to be treated during the field demonstration was expanded to include BTEX, DCBs, and TCE (Folsom 1995).

METHODOLOGY

The base industrial area (OT20 site) at Robins AFB was selected as the demonstration site. The test site was next to a fuel tank storage facility and the base industrial area. Principal contaminants in the groundwater included benzene (46 µg/L), TCE (1,445 µg/L), toluene (40 µg/L), ethylbenzene (23 µg/L), xylene(s) (50 µg/L), and DCBs (2,014 µg/L), primarily the 1,2-DCB isomer, although all isomers were present.

The laboratory-scale FBR systems was constructed of glass with Teflon™ tubing and polypropylene fittings and had a total liquid volume of 4 liters and an empty bed volume of 800 mL. Contaminated water was fed at 15 to 20 mL/min, which resulted in an empty-bed hydraulic retention time (HRT) of approximately 40 to 50 min. Dissolved oxygen (DO) levels were maintained above 2 ppm, pH was automatically controlled between 6.8 and 7.0, and temperature was maintained between 22 and 30°C. Liquid samples were analyzed using GC/FID purge-and-trap methods.

The laboratory GPR system, liquid volume of 2 L, was constructed from glass, Teflon™, and stainless steel. A proprietary nutrients mixture was added at a rate of about 0.7 mL/min which gave a hydraulic retention time of about 10 days. Liquid pH was automatically controlled between 6.8 and 7.0, and temperature was maintained at 28°C. Air contaminated with TCE entered through a ⅛-inch tube at the bottom of the vessel and exited after passing through the liquid column with suspended bacteria. An automated gas sampling system was connected to both inlet and outlet gas streams, and TCE concentrations were monitored by an automated gas chromatograph/electron capture detector (GC/ECD).

The field pilot FBR system was constructed of stainless steel, had approximate dimensions of 12.5 ft long by 5.5 ft wide by 15 ft high (3.8 m long by 1.7 m wide by 4.6 m high), with an empty-bed volume of about 66 gal (250 L). All process controls and equipment are weatherproof, and used a computer control and data logging system. DO levels were maintained above 2 ppm and pH was maintained between 6.8 and 7.0. Approximately 210 lb (95 kg) of GAC was used as the bed support. Bed height level was monitored and logged. The system was operated with a 2-gpm (7.6-L/min) feed flow with a 4.5-gpm (17-L/min) upflow through the reactor yielding a 25% bed expansion. An automated purge-and-trap GC system was used to automatically monitor feed and effluent water streams.

A stirred-tank field-pilot GPR system had approximate dimensions of 8 ft long by 8 ft wide by 11 ft high (2.4 m long by 2.4 m wide by 3.4 m high) and held approximately 750 gal (2,840 L) of liquid. System pH was controlled automatically and water temperature was controlled at 88°F (31°C) and operated with a hydraulic retention time of about 10 days. Average air flow to the reactor was 7 cfm with contamination levels in feed and effluent air streams automatically monitored using a GC/ECD system.

Standard methods for water analysis were used to monitor a variety of operating parameters including, pH, biomass density, and nutrient concentrations. VOC concentrations were automatically monitored using GC systems. Two GC systems were employed, one for monitoring VOC concentrations in water and one for monitoring TCE concentrations in air. Both systems were routinely calibrated. Concentrations were determined by an external standard method (Folsom 1993).

RESULTS

Laboratory Study

Results from operation of a laboratory FBR system using both site water and a surrogate water at the concentrations listed above demonstrated efficient removal of BTEX, DCB, and TCE. The granular-activated carbon (GAC) used as a biofilm support in the FBR was saturated with chemicals before inoculation with degradative bacteria. Following inoculation and adaptation, effluent concentrations for all of the chemicals decreased. Once steady-state operation was achieved after 3 weeks of operation, greater than 90% of all chemicals and 80 to 90% of the TCE were degraded in the FBR (Folsom 1995). Three major points were concluded from this test. First, chemical removal was biologically mediated and not merely a physical process. Second, there was an apparent adaptation of bacteria responsible of degrading toluene, ethylbenzene, and TCE. Third, no apparent toxic or inhibitory effects were detected following long-term operation.

Laboratory efforts focused on GPR operational stability, which was enhanced following a change to continuous operation. Typically, between 200 and 600 µg TCE/L air entered the reactor at an air flowrate of 70 mL/min with effluent concentrations close to, or below, detection limits (Folsom 1995). Performance was extended to beyond 10 months of continuous operation using laboratory systems. Overall TCE removal efficiencies exceeded 95% though there were several minor operational upsets which included plugged feed lines and interruptions in electrical service. Following each event, biological activity in the reactor recovered without amendments to, or replacement of, the bacteria. The bacterial composition of the reactor became a mixed culture after the first week of operation. Modifications made in GPR operation led to stable and reliable performance for extended time periods.

Pilot Demonstration

The field demonstration system included an FBR for primary treatment of groundwater, an air stripper for transferring residual volatile organic compounds (VOCs) to an air stream, and a GPR for final treatment of those residual VOCs, primarily TCE. Due to the enhanced performance of the FBR towards TCE, the demonstration essentially became a test of two independent pilot systems, one for treating contaminated water in the FBR and the second for treating contaminated air in the GPR.

The FBR pilot was assembled and filled with GAC. Contaminated groundwater was pumped through the reactor for 4 weeks prior to adding the bacterial inoculum. The system was operated at a flowrate of 2 gpm with an empty-bed HRT of about 30 min, pH 6.7, 24°C, and 4.6 mg/L dissolved oxygen. The FBR demonstrated greater than 95% removal of most targeted contaminants and greater than 85% removal of TCE from groundwater (Table 1). More than 210,000 gal (794,934 L) of contaminated groundwater were treated in the pilot FBR with effluent quality close to drinking water standards (Folsom 1995). Pilot FBR performance was fully consistent with the laboratory studies.

The bed material removed from the FBR was also subjected to methanol extraction to quantify the amount of key organic chemicals bound to the GAC. A methanol extraction method was able to extract >95% of the bound organic chemicals from virgin GAC. Chemical recoveries from the GAC were consistent with the isotherms generated using virgin carbon. The amounts of chemicals bound to the carbon were greatest at the bottom of the reactor (Table 2). This was consistent with the pseudo-plug flow operation of the FBR system in which

TABLE 1. Feed and effluent chemical concentrations during steady-state operations of the pilot FBR.

Chemical	Feed (µg/L)	Effluent (µg/L)	% Degraded
benzene	46 ± 30	<10 ± 4	>78[a]
TCE	1,445 ± 173	206 ± 142	86
toluene	40 ± 46	<11 ± 5	>73[a]
ethylbenzene	23 ± 17	<12 ± 8	>46[a]
xylenes	50 ± 29	<20 ± 7	>40[a]
1,3-DCB	123 ± 10	10 ± 2	92
1,4-DCB	227 ± 20	9 ± 2	96
1,2-DCB	1,664 ± 134	13 ± 5	99

(a) Degradation based on detection limit for compounds giving a conservative estimate of performance.

the highest concentration of chemicals in the water was greatest at the bottom of the reactor. The total organic load to the carbon was less that the maximal holding capacity due to isotherm effects at the chemical concentrations in the feed and effluent streams. Maximal loading capacities are achieved only when the carbon can be saturated, which clearly was not achievable under these operating conditions. A total mass balance was determined for operation of the pilot FBR system (Table 3). Overall, 83% of the TCE, 93% of the 1,2-DCB, and 67% of the BTEX were destroyed during FBR operation using the conservative estimate of bound chemicals. These results clearly demonstrated a significant loss of chemical as a result of biodegradative activity beyond the binding capacity of the GAC in the reactor.

TABLE 2. Methanol extractable organic chemicals removed from pilot FBR[a].

Chemical	Bottom Sample (mg/g GAC)	Middle Sample (mg/g GAC)	Top Sample (mg/g GAC)
TCE	9.61	<0.01	<0.01
1,2-DCB	1.95	0.20	0.31
BTEX	5.16	0.84	0.69

(a) A 1-g GAC sample was extracted with 9 mL of methanol. The methanol phase was injected onto a GC and the amount of chemical was quantified using an external standard. The bottom sample represents the average of four separate extractions and analyses.

TABLE 3. Chemical mass balance for key contaminants during operation of the pilot FBR[a].

Chemical	Total Input in Feed (g)	Total Output in Effluent (g)	Net Load to Reactor (g)	Total Bound to GAC (g)	Net Destroyed (g)
TCE	6,041	518	5,523	912	4,611
1,2-DCB	2,830	92	2,738	185	2,553
BTEX	1,536	65	1,471	490	981

(a) Loading calculations were based on weekly averages for feed and effluent chemical concentration and totaled liquid flow to system. The loading calculations also included the chemicals added to enhance breakthrough adjusted for losses to the effluent. GAC loading was based on 95 kg of activated carbon in reactor and the amount of organic sorbed to carbon at the bottom of the reactor from Table 2.

The GPR was then prepared for normal operation using an inoculum of TCE degradative bacteria grown with phenol as the sole carbon source. The system was operated at a 10-day HRT once biomass concentrations reached 0.35 mg/mL. A chemical addition system was installed to deliver TCE into the air entering the GPR. Over the course of steady-state operation, average feed and effluent concentrations for TCE were 371 and 80 µg/L air, respectively. A pH excursion during the sixth week of operation resulted in a 75% loss of biomass which significantly lowered the volumetric performance of the reactor. The pH control issue was rectified and the reactor was temporarily switched to batch operation, and within 24 hours biomass levels doubled from 0.4 to 0.8 mg/mL protein, at which time the TCE feed was re-initiated and normal operation resumed. This system demonstrated stable operation for 2 months before it was shut down. TCE was reduced by an average of 75% in the GPR. This removal rate can easily be increased to over 90% by increasing biomass concentrations in the reactor. Pilot operation and stability were consistent with laboratory studies, although overall performance was lower than normal during the demonstration.

DISCUSSION

An economic evaluation was performed based on key parameters expected to represent those typically found at contaminated sites. The standard contaminated groundwater characteristics used for analysis were a feed flow rate of 100 gpm, containing 15 ppm BTEX (ratio of 7.5:4:1.5:2) with 1 ppm TCE. Capital and operating costs were developed for an FBR system, wet carbon adsorption, air stripping/dry carbon adsorption, and UV-peroxidation using this set of flow rates and concentrations (Figure 1). All cost estimates include installation of

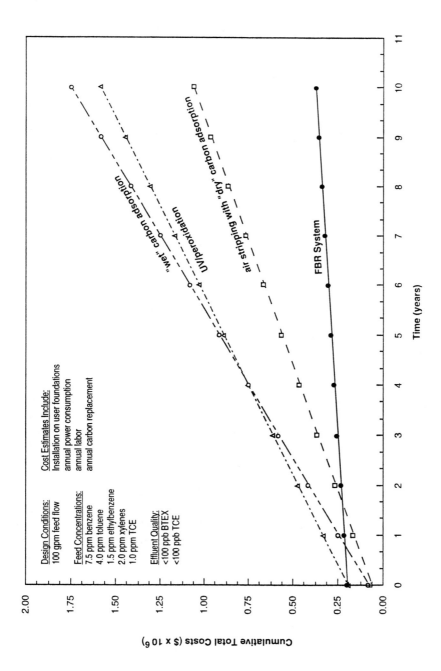

FIGURE 1. Life-cycle cost comparison for FBR versus wet carbon adsorption, UV-peroxidation, or air stripping followed by carbon adsorption.

the complete system on customer-supplied foundations and exclude (1) routing of groundwater to the system; (2) routing treated effluent from the system; (3) startup; (4) field supervision; (5) equipment freight; (6) taxes; and (7) additional, site-specific pretreatment or posttreatment equipment requirements. Power calculations were based on a rate of $0.07/kWh and a labor rate of $50/h was used. Carbon replacement costs were set at $2.00/lb (0.45 kg), which included replacement carbon and extras such as vacuuming, shipping, removal, and disposal of spent carbon. FBR carbon attrition rate was assumed to be 5% (approximately 320 lb [145 kg]/year). Estimates of carbon usage for the air stripping with carbon adsorption, obtained from two vendors, were 112 and 136 lb (51 and 62 kg)/day. For the FBR system, the desired effluent quality can be met using a 5-ft (1.5-m)-diameter by 11-ft (3.3-m)-tall fluidized bed bioreactor with a system capital cost of $200,000 and an operating cost of $17,520/year including power, nutrients, carbon replacement due to attrition, manpower, and maintenance costs. As shown in Figure 1, the break-even point for the FBR is 1.6 years compared to the best alternative technology, i.e., air stripping/carbon adsorption. This payback reflects an $82,000 savings in yearly operating and maintenance costs for the FBR system compared to air stripping/carbon adsorption. The cumulative total cost savings (operating, maintenance, and capital) for a 10-year project would be $690,000.

An economic evaluation was performed for the GPR system based on key parameters expected to represent those typically found at contaminated sites during soil vapor extraction (SVE) operations. Assuming an air flowrate of 300 cfm and TCE concentrations in the air of either 100 or 300 ppmv, an 11-ft (3.4-m)-diameter, 7,500-gal (28,390-L) GPR is required to achieve the desired treatment level (Figure 2). System costs were estimated at $125,000 ± 15% installed on customer-supplied foundations. Operating costs were estimated at $25,000/year including power (@ $0.07/kWh), nutrients, manpower, and maintenance costs. Comparable capital and operating costs were developed for carbon adsorption using the same set of flowrates and concentrations. Carbon consumption was based on theoretical isotherm data and changed significantly for the two concentrations of TCE used (Figure 2). Capital costs were estimated at $10,000 ± 15%. System costs included complete carbon adsorption system installed on customer-supplied foundations. Costs excluded installation of SVE system, routing air to and from the carbon adsorbers, and any startup and field supervision services. The annual operating costs were estimated to range from $50,000 to $250,000 depending on TCE concentrations. Carbon replacement service costs may vary depending on site location.

Due to the high performance of the FBR, the vapor entering the GPR for treatment had to be spiked with TCE. Operational stability was successfully demonstrated with 10 months of continuous operation using the laboratory system and 2 months using the field-pilot system. Commercial GPR systems have been designed to treat 50 to 300 cfm of contaminated air, typical for soil vapor extraction operations. The economic analysis generated as part of these projects indicates typical savings in operating costs of 70 to 80% using the biological treatment system compared to using carbon adsorption (Figure 2).

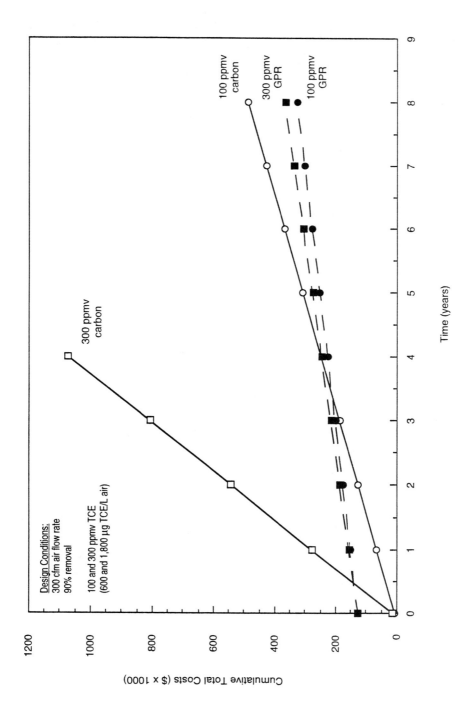

FIGURE 2. Life-cycle cost comparison for GPR versus carbon adsorption.

In essence, two independent field demonstrations were successfully performed. Due to the high performance of the FBR, the vapor entering the GPR for treatment had to be spiked with TCE. All hazardous chemicals were treated to concentrations near or below drinking water standards. An economic evaluation of these innovative technologies to conventional carbon adsorption suggests a significant cost savings. If chemical concentrations were higher than the assumptions used in the cost estimates, operating costs for carbon adsorption would increase, whereas FBR and GPR operating costs would not change significantly. Biological treatment provides an economical, destructive technology for remediating contaminated air or water.

ACKNOWLEDGMENTS

The project was performed under a contract (FO8635-91-C-0198) sponsored and administered through HQ AFCESA/RAVW Tyndall Air Force Base, Florida, 32542-5000. The authors wish to acknowledge the support provided by the Environmental Restoration Division at Robins AFB, Georgia.

REFERENCES

Coyle, C. G., G. F. Parkin, and D. T. Gibson. 1993. "Aerobic, phenol-induced TCE degradation in completely mixed, continuous-culture reactors." *Biodegradation* 4:59-60.

Ensley, B. D. 1991. "Biochemical Diversity of Trichloroethylene Metabolism." *Annu. Rev. Microbiol.* 45:283-299.

Ensley, B. D., and P. R. Kurisko. 1994. "A gas lift bioreactor for removal of contaminants from the vapor phase." *Appl. Environ. Microbiol.* 60:285-290.

Folsom, B. R. 1993. *Liquid-Phase Bioreactor for Degradation of Trichloroethylene and Benzene.* ESL-TR-92-02, Engineering and Services Laboratory, Headquarters Air Force Engineering and Services Center, Tyndall AFB, FL.

Folsom, B. R. 1995. *Pilot Demonstration of Bioreactor System for Degradation of Chlorinated Solvents and Hydrocarbons.* Armstrong Laboratory Technical Report Al/EQ-TR-1995-0001, Dual Phase Bioreactor for the Degradation of TCE & Benzene, Engineering and Services Laboratory, Headquarters Air Force Engineering and Services Center, Tyndall AFB, FL.

Folsom, B. R., and P. J. Chapman. 1991. "Performance characterization of a model bioreactor for the biodegradation of trichloroethylene by *Pseudomonas cepacia* Strain G4." *Appl. Env. Microbiol.* 57:1602-1608.

Landa, A. S., E. M. Sipkema, J. Weijma, A.A.C.M. Beenackers, J. Dolfing, and D. B. Janssen. 1994. "Cometabolic degradation of trichloroethylene by *Pseudomonas cepacia* G4 in a chemostat with toluene as the primary substrate." *Appl. Env. Microbiol.* 60:3368-3374.

Groundwater Treatment in a Field Pilot Methanotrophic Rotating Biological Contactor

David M. Belcher, Alex Vira,
Maureen A. Dooley, and Jaret C. Johnson

ABSTRACT

A pilot-scale rotating biological contactor (RBC) was operated under field conditions for approximately 1 month to remove chlorinated and nonchlorinated organic compounds from groundwater. Methanotrophic (methane-oxidizing) conditions were successfully established and maintained in the RBC during the field program. Results of the pilot program indicated that low concentrations of *cis*-1,2-dichloroethene (*cis*-DCE) and vinyl chloride (less than 1,000 and 50 µg/L, respectively) could be treated to below the maximum contaminant levels (MCLs) of 70 and 2 µg/L, respectively. Maximum removal rates for *cis*-DCE and vinyl chloride measured during the pilot study were 2.14 µg *cis*-DCE/ft^2 disc media-minute (952 µg *cis*-DCE/mg volatile solids [VS]-day) and 0.3 µg vinyl chloride/ft^2-minute (143 µg vinyl chloride/mg VS-day), respectively. Chlorinated ethene removal efficiencies decreased after the first 2 weeks of operation. Low concentrations of toluene, ethylbenzene, and total xylenes (TEX) (less than 500 µg/L) were effectively removed from groundwater throughout the course of the pilot study. The maximum observed TEX removal rate was 3.0 µg TEX/ft^2-minute (1,105 µg TEX/mg VS-day).

INTRODUCTION

Methanotrophic cometabolism of chlorinated aliphatics such as trichloroethene (TCE), dichloroethene (DCE), and vinyl chloride has been documented by several researchers (Anderson and McCarty 1994; Speitel and Leonard 1992; Strand et al. 1991; Vira et al. 1991). In essence, methanotrophic cometabolism of chlorinated aliphatics consists of the enzyme-catalyzed oxidation of the chlorinated aliphatic molecule by the methane monooxygenase (MMO) enzyme, which is induced in certain methanotrophic bacteria in the presence of methane and oxygen.

The pilot-scale RBC study discussed herein was conducted as part of an interim measure (IM) program at a landfill in the southeast United States. The specific objectives of the pilot-scale RBC program were to (1) provide sufficient data to confirm that target compounds can be treated biologically, and (2) gather information for evaluation of full-scale site groundwater remediation.

METHODS AND MATERIALS

A schematic of the pilot-scale RBC system is presented in Figure 1. The primary components of the system included the RBC unit, an influent pump, a water heater, nutrient and caustic injection systems, and monitoring instrumentation. The totally enclosed pilot-scale RBC unit was constructed of aluminum and was approximately 108 gal (409 L) in total volume. Groundwater volume was maintained at approximately 60 gal (227 L). The RBC (water and headspace) was divided into four separate compartments (chambers) by aluminum and polyethylene dividers. Groundwater passed through each chamber through the media shaft orifice holes (located beneath the water surface). The RBC media discs were 18 in. (46 cm) in diameter and were made of polyethylene (effective surface area of 306 ft²). Prior to the introduction of groundwater, an active methanotrophic consortium (as determined from methane and oxygen consumption rates) was established upon the media disks of the RBC reactor.

Experimental parameters that were monitored during the pilot-scale RBC study included influent and effluent target compound concentrations, headspace target compound concentrations, methane and oxygen headspace concentrations, groundwater pH, temperature, and nutrient concentrations. RBC influent and effluent water samples as well as headspace air samples were analyzed using a gas chromatograph (modified EPA method 8010/8020). A gas analyzer system was used to continuously monitor, record, and control methane and oxygen concentrations in the RBC headspace.

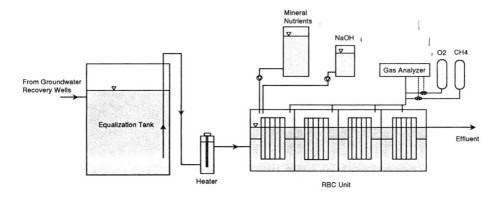

FIGURE 1. RBC system schematic.

Methane and oxygen were added to the RBC headspace from outside cylinders using gas analyzer-activated solenoid valves. Reactor pH was maintained near neutral using a pH controller and injection system. Reactor water temperature was monitored using a temperature gauge and was maintained at 25°C ± 5°C (natural groundwater temperature was approximately 15°C). Mineral nutrient (nitrate and phosphorus) concentrations were measured using colorimetric field methods.

The pilot-scale RBC study was divided into three tests:

- Experiment 1: 72 h, two of five extraction wells on, influent flowrate of 0.5 gal/min (1.9 L/min) (hydraulic residence time [HRT] = 2 h)
- Experiment 2: 19 days, five of five extraction wells on, influent flowrate of 0.5 gal/min (1.9 L/min) (HRT = 2 h)
- Experiment 3: 9 days, five of five extraction wells on, influent flowrate of 1.0 gal/min (3.8 L/min) (HRT = 1 h).

RESULTS

Experiment 1

Methane and oxygen removal rates indicated that an active methanotrophic environment existed in the RBC during Experiment 1. The removal efficiency for vinyl chloride was greater than 99% during Experiment 1, and the maximum contamination level (MCL) of 2 µg/L was achieved for all sampling points. The *cis*-DCE, although not present in the influent above its MCL (70 µg/L) during Experiment 1, was further reduced in concentration (57 to >70%) in the RBC during Experiment 1, indicating that low concentrations of *cis*-DCE can be treated biologically. TEX compounds, which were also present in groundwater influent below their respective MCLs (1,000, 700, and 10,000 µg/L, respectively), were also effectively reduced in the RBC during Experiment 1 (85 to >91%).

Experiment 2

During Experiment 2, the RBC received groundwater from all five extraction wells, providing higher concentrations of TEX and *cis*-DCE than in Experiment 1. Operating conditions, such as flowrate and temperature, remained unchanged from Experiment 1. Analysis of methane and oxygen utilization data indicated that methanotrophic conditions were maintained during Experiment 2 and that influent groundwater methane concentrations were increasing. Although influent characteristics were slightly different during Experiment 2 compared to Experiment 1, effective removal of the target compounds still occurred during most of Experiment 2. The removal efficiency for vinyl chloride was greater than 99% and the MCL of 2 µg/L was achieved for all but 2 sampling points during Experiment 2. According to concentration versus reactor location data (Figure 2), removal of vinyl chloride to below the 2 µg/L MCL occurred within the first 2 stages of the RBC during Experiment 2.

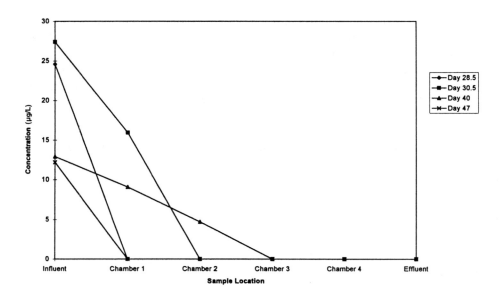

FIGURE 2. Vinyl chloride concentration vs. sample location, Experiment 2.

Removal efficiencies of *cis*-DCE ranged from 20 to greater than 99%. Several abrupt changes in *cis*-DCE removal patterns were observed during Experiment 2, such as the increase and decrease in effluent concentrations observed on days 35 through 37 in Figure 3. The changes appear to be the result of operational and

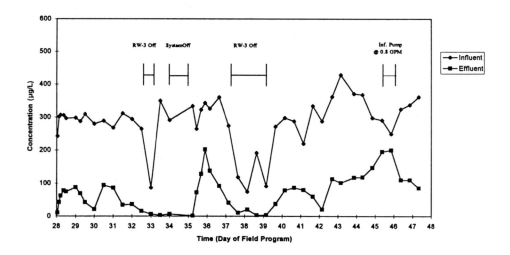

FIGURE 3. System operation changes and *cis*-DCE concentration vs. time, Experiment 2.

environmental system changes, such as disruption in the groundwater recovery system operation. According to concentration versus reactor location data (see Figure 4), removal of *cis*-DCE occurred steadily in groundwater as it passed through the RBC. TEX compounds, although present in the influent in concentrations less than respective MCLs, were further reduced in concentration during Experiment 2 (77 to >99% removal). According to concentration versus sample location data (Figure 5), TEX was removed primarily in the first two chambers of the RBC during Experiment 2 (days 28 through 47).

Experiment 3

The influent flowrate to the RBC was increased from 0.5 gal/min (1.9 L/min) to 1.0 gal/min (3.8 L/min) during Experiment 3. Analysis of methane and oxygen utilization data indicated that oxygen utilization continued to occur (and increase) during Experiment 3. Experiment 3 methane utilization data could not be computed due to high methane levels in site groundwater. In general, poor removal of *cis*-DCE and vinyl chloride occurred during the course of Experiment 3. Vinyl chloride removal typically was below 30% and *cis*-DCE removal ranged from 19 to 42%. The decrease in the rate and extent of *cis*-DCE removal during Experiment 3, compared to Experiment 2, can be observed in Figure 4 (day 51). Although effluent TEX concentrations increased during Experiment 3, the influent concentrations also increased and the extent of TEX removal remained relatively unchanged during Experiment 3. In addition, the majority of TEX removal shifted from Chambers 1 and 2 of the RBC (Experiment 2, days 28 through 47) to stage 3 during Experiment 3 (day 51), when the flowrate was increased (see Figure 5).

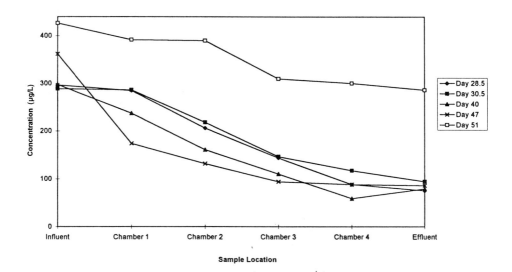

FIGURE 4. *cis*-DCE concentration vs. sample location, Experiments 2 and 3.

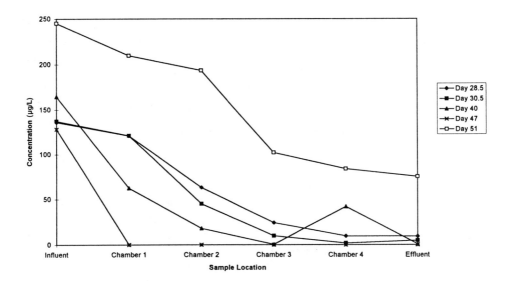

FIGURE 5. TEX concentration vs. sample location, Experiments 2 and 3.

DISCUSSION

Biological degradation was determined to be a major removal mechanism for target compounds during the RBC study based on the following observations: (1) the role of volatilization as a compound removal mechanism, as determined by headspace gas analysis, was limited (headspace concentrations were below theoretical equilibrium partitioning concentrations); (2) a methanotrophic environment capable of promoting methanotrophic cometabolism of chlorinated ethenes was successfully established and maintained in the RBC throughout the study; and (3) cis-DCE epoxide, a compound formed only from the microbiological cometabolism of cis-DCE, was detected in significant quantity in the reactor water during the RBC study. Although it is presumed that chlorinated aliphatic cometabolism was primarily due to methanotrophic activity, chlorinated aliphatic cometabolism due to toluene degradation (which occurred in the RBC throughout the course of the field program) cannot be discounted.

The overall decline in removal efficiency of target volatile chlorinated compounds during Experiment 3 was most likely a result of (1) the increase in groundwater flow through the reactor (decrease in HRT), and/or (2) the change in groundwater characteristics that occurred over time. An increase in reactor flow and/or the change in groundwater characteristics may have resulted in microbiological changes or decreased MMO enzyme availability due to toxicity, competitive inhibition, or microbiological selection. Additional testing would be required to determine which parameter(s) caused the decrease in cis-DCE and vinyl chloride removal during Experiment 3.

In summary, the following test objectives of the pilot-scale RBC program were accomplished: (1) sufficient data were obtained to confirm that vinyl chloride, *cis*-DCE, and volatile aromatic compounds present in site groundwater, can be treated in a methanotrophic RBC; and (2) a conceptual design, including sizing and cost estimates, could be outlined based on biodegradation rates observed in the pilot-scale study.

REFERENCES

Anderson, J. E., and P. L. McCarty. 1994. "Model for Treatment of Trichloroethylene by Methanotrophic Biofilms." *Journal of Environmental Engineering* 120(2):379-397.

Speitel, G. E., and J. M. Leonard. 1992. "A Sequencing Biofilm Reactor for the Treatment of Chlorinated Solvents Using Methanotrophs." *Water Environment Research* 64(5):712-719.

Strand, S. E., J. V. Wodrich, and D. H. Stensel. 1991. "Biodegradation of Chlorinated Solvents in a Sparged, Methanotrophic Biofilm Reactor." *Research Journal, WPCF* 63(6):859-864.

Vira, A., A. Moore, and R. Venterea. 1991. "Applications of Bioreactors for Treatment of Contaminated Groundwaters." *Hazardous Material Control/Superfund '91*. pp. 215-218. Hazardous Materials Control Research Institute, Greenbelt, MD.

Adsorption/Fermentation of Groundwater Contaminants in a Coal-Based Bioactive Filter

Felix Kudrjawizki, Sabrina Feldt, and Manfred Ringpfeil

ABSTRACT

A model of a bioactive filter for aerobic groundwater remediation is presented. It is based on a mixture of activated carbon and externally bred microorganisms, providing for a coupled process of adsorption and fermentation. Its function was demonstrated in a laboratory experiment. The filter consisted of a selected mixed culture of polycyclic aromatic hydrocarbon (PAH)-degrading bacteria and an activated brown coal coke. An aqueous extract from tar oil, rich in PAHs, was used as contaminant for tap water. The application of the adsorption/fermentation process resulted in stable degrees of PAH degradation despite experimentally produced differences in PAH load.

INTRODUCTION

Several varieties of combining adsorption and fermentation to liberate water from organic contaminants have been applied in technical practice. The oldest and most widely known application is the coke-based trickling filtration to purify sewage (Oehme 1984). Other applications for sewage purification are the fluidized-bed reactors using activated carbon as a carrier for microorganisms, for example the Biocarbon, Biokop, and Katox processes (Hartmann 1990). To purify potable water, fixed-bed reactors are used based on activated carbon as a carrier for the microorganisms (Foerstner 1992). In situ purification of groundwater is another application that should be considered.

DESCRIPTION OF THE SYSTEM

Activated carbon is mixed with a slurry of externally bred microorganisms so that the latter are adsorbed equally on the carbon surface. This mixture is used to form a vertical layer in the soil perpendicular to the flow direction of the groundwater that stretches from the upper level of the groundwater to the natural water-impermeable clay layer. The width of the barrier depends on the groundwater flowrate ($\dot{V}_L$). The activated carbon/microorganisms layer is artificially aerated (Figure 1). The system is composed of several phases:

- Outer water phase (OWP)
- Air phase dispersed in the outer water phase (DAP)
- Solid adsorption phase (SAP)
- Inner water phase (IWP)
- (Quasi)solid microbial phase (SMP).

The groundwater (OWP) is the continuous phase of the system, and there are two independent dispersed phases: the air bubbles (DAP) and a conglomerate consisting of the activated carbon (SAP), with the adhering microorganisms (SMP) surrounded by adhering water (IWP) (Figure 2).

From the biochemical point of view the contaminants (S) of the groundwater and the oxygen (C) in the air bubbles are the sources, and the microorganisms (X) are the sink of the system. Moreover, if the interaction of cell growth (Eq. 1) and cell degradation (Eq. 2) leads to a microcosmic behavior of the system. The net reaction approaches a combustion reaction (Eq. 3).

$$X + S + C + N,P.... \rightarrow (X + \Delta X) \qquad + CO_2 + H_2O + Q \qquad (1)$$

$$X \quad + C \qquad \rightarrow (X + \Delta X) + N,P... + CO_2 + H_2O + Q \qquad (2)$$

$$\overline{}$$

$$S + C \qquad \rightarrow \qquad\qquad\qquad CO_2 + H_2O + Q \qquad (3)$$

where X = microbial cells
 S = contaminants = carbon source = substrate
 C = molecular oxygen
N,P... = nutrients
 Q = reaction heat.

From a physical point of view the activated carbon (SAP) serves as an interim sink for the contaminants (S):

$$S \quad \underset{\leftarrow}{\overset{\rightarrow}{\rightleftharpoons}} \quad S_{ads} \qquad (4)$$

At the beginning of the process, the free adsorption capacity of the activated carbon (SAP) is high and the quick adsorption step (Eq. 4) will dominate over the slower microbial conversion step (Eq. 1).

When the adsorption capacity is going to be exhausted, desorption will increase. It is directed into the inner water phase (IWP) due to the metabolic activity of the microorganisms, which will decrease continuously the actual concentration of the contaminants in the inner water phase. Thus, the activated

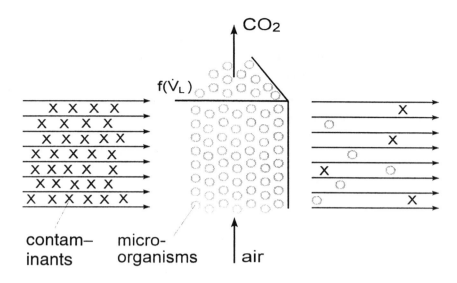

FIGURE 1. Bioactive barrier in a contaminated groundwater stream.

carbon (SAP) stabilizes the process in the case of fluctuating concentrations of contaminants in the groundwater. An increase of concentration will result in increasing adsorption, followed by increasing desorption due to the exhaustion of adsorption capacity balanced by increasing microbial conversion. A decrease

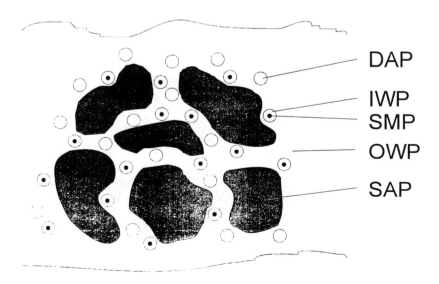

FIGURE 2. Phase composition of the aerated bioactive barrier.

in concentration will result in desorption because the decreasing concentration of contaminants in the inner water phase maintains microbial activity. These processes proceed properly if the oxygen transfer from the dispersed air phase (DAP) into the outer water phase (OWP) is not limiting.

From a kinetic point of view, the system is qualified as an open one due to the flow of the groundwater through the microbial layer and the flow of the air bubbles from the sparger through the microbial layer to the free atmosphere. Mass transfer processes provide the connection between the sources S and C and the sinks X and SAP:

S: $OWP \rightarrow SAP \rightarrow IWP \rightarrow SMP$

and

C: $OWP \rightarrow IWP \rightarrow SMP$

Transport and conversion of S in the system can be described as follows:

$$\dot{s}_{in} - \dot{s}_{out} = \dot{s}_{OWP \rightarrow IWP} - \dot{s}_{OWP \rightarrow SAP} + \dot{s}_{SAP \rightarrow IWP} = \dot{s}_{consumption} =$$

$$(\beta / \alpha) \dot{x}_{growth} = (1 / \beta) \dot{c}_{consumption} \qquad (5)$$

where s = concentration of S
$\dot{s}$ = rate of conversion or transport, respectively
α, β = conversion factors
$\dot{x}$ = rate of cell growth
c = rate of oxygen consumption.

Correspondingly, transport and reaction of C in the system can be described as follows:

$$\dot{c}_{in} - \dot{c}_{out} = \dot{c}_{DAP \rightarrow OWP} = \dot{c}_{OWP \rightarrow IWP} = \dot{c}_{consumption} =$$

$$\beta \dot{s}_{consumption} = (1 / \alpha) \dot{x}_{growth} \qquad (6)$$

The application of the filter pursues the goal to receive a constant low concentration value of the contaminants in the groundwater stream after passing the filter: $\dot{s}_{opt}$. This value depends on several properties of the system:

$$\dot{s}_{opt} = f(s_G, \dot{V}_L, d, \dot{x}, \dot{c}_{DAP \rightarrow OWP}) \qquad (7)$$

where s_G = concentration of contaminants in the groundwater
$\dot{V}_L$ = flowrate of the groundwater
d = width of the filter.

The constancy of this value depends on the flexibility of the system, which is determined by the interaction of adsorption and fermentation as described above.

EXPERIMENTAL EXAMPLE

To verify the assumptions about the function of the system, a bench-scale experiment was performed over a period of 11 months (Feldt 1994).

Materials

An activated brown coal coke (NOA 1 — 4 mm) was used as the solid adsorbent phase (SAP). A bacterial mixed culture containing *Pseudomonas putida* (DSM 7248) and *Pseudomonas* sp. (DSM 7249) was applied. A solution of PAH, including some unidentified substances, was used to simulate contaminated groundwater. The solution was made by shaking tar oil (TEWE GmbH, Erkner, Germany) with tap water, and the purified water phase was fed back from the fermentor system. The concentration of dissolved PAH ranged between 1 and 2 mg/L. Chemical oxygen demand (COD) values were measured up to 27 mg/L.

Apparatus

The microbial layer was constructed as a fixed-bed fermentor with inner circulation (Figure 3). In the flask (1), the PAH solution is prepared continuously.

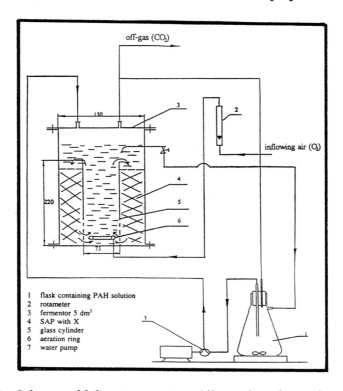

FIGURE 3. Scheme of laboratory system (dimensions in mm).

The pump (7) contains a filter to hold back undissolved drops of tar oil. The fermentor vessel (3) is divided by a vertical glass cylinder (5). The outer part is filled with the microbial layer (4). The cylinder contains holes in its lower part to allow the flow of the solution through the microbial layer. In the inner part of the cylinder, an air sparger (6) is mounted to provide the solution with dissolved oxygen and to cause an airlift effect, which sucks the solution through the fixed bed. The air volume is measured by a rotameter (2). The mixing time in the vessel is regulated by the airflow and determined in the range of minutes. Thus, the whole system could be characterized as an ideally mixed vessel. The concentration of PAH in the water phase of the system is determined as the out-flowing concentration. The solution is fed back in part to the flask (1). The addition of tap water to this flask sets the retention time in the fermentor system.

Methods

PAHs were determined by high-performance liquid chromatography (HPLC) (Shimadzu, Japan). Extraction from the water phase was performed by *n*-hexane, drying the extract and dissolving it in acetonitrile (Bundt et al. 1990). Recovery from the solid phase was performed by Soxhlet extraction with toluene (Krengel-Rothensee 1992). The number of microorganisms was determined by most probable number counting on selective tar oil agar plates. Cells were liberated by ultrasonic treatment (50 W, 2 min) (Svjaginzev 1973).

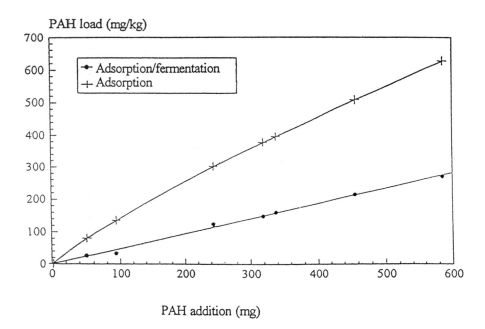

FIGURE 4. Comparison between calculated and determined PAH level.

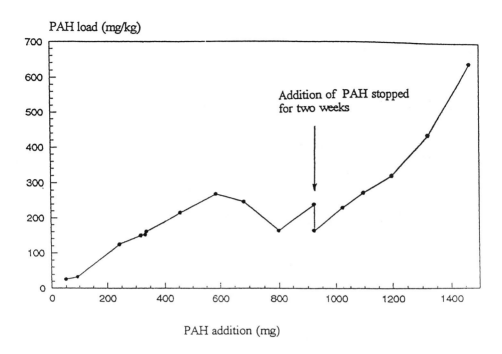

FIGURE 5. PAH load over lengthy PAH addition to the system.

RESULTS

Pure adsorption of PAH at the activated coke system follows the equation

$$q = 2,971 \ c^{0,84} \tag{8}$$

corresponding to the top curve (+ + +) in Figure 4. Measurement of the adsorbed PAH showed significantly lower adsorption values corresponding to the bottom curve (. . .), indicating that the difference in PAH was metabolized by the microorganisms. Carbon dioxide measurements in the off-gas ranging from 20 to 120 mg/h confirmed this conclusion. The number of microorganisms at the coke remained relatively constant at $N = 10^4/g$ dry matter even at residence times of 1.9 h. The number of microorganisms in the outflow ranged to $N < 10/mL$.

A lengthy experiment showed the characteristic course of adsorption and fermentation. Stopping addition of PAH at 900 mg for 2 weeks resulted in a significant desorption of adsorbed PAH due to microbial action (Figure 5). Application of different residence times changed the relation between adsorption and fermentation. Fermentation increased with rising residence time (Figure 6). Final concentrations of PAH depend on residence times. Adsorption/fermentation leads to 90% degradation when exceeding residence times of 4 h (Figure 7).

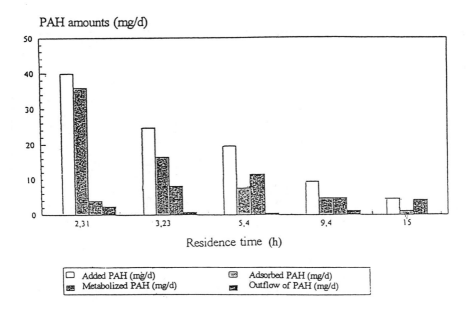

FIGURE 6. Balance of PAH based on residence time.

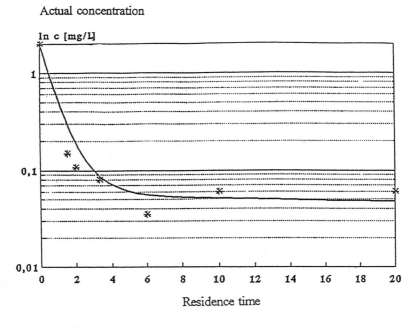

FIGURE 7. Residual concentration of PAH in the outflow of the system based on residence time.

DISCUSSION

All values observed in the experiments are in good compliance with the theory. The system is stable at the small concentrations of contaminants that are usually observed in groundwater. Experimental optimization and scaling up may lead to practical approaches that open new directions in groundwater cleaning. The method will allow a dynamic containment of hazardous deposits of organic origin by converting the outflowing components.

REFERENCES

Bundt, J., W. Herbel, and H. Steinhart. 1990. "Mineraloelanalytik von Bodenkontaminationen unter besonderer Beruecksichtigung von aromatischen Kohlenwasserstoffen." *Reinigung kontaminierter Boeden,*" Hamburger Berichte 3, Economica Verlag, Bonn, Germany.

Feldt, S. 1994. *Modelluntersuchungen an bioaktiven Sperrschichten für kontaminiertes Grundwasser.* Berlin, Germany.

Foerstner, U. 1992. *Umwelttechnik.* Springer Verlag, Heidelberg, Germany.

Hartmann, L. 1990. *Untersuchungen zum Phenolabbau mit Trichosporon beigelli, immobilisiert an Aktivkohle.* Berlin, Germany.

Krengel-Rothensee, K. 1992. "Erfahrungen in der Bodenanalytik." *Mikrobielle Reinigung von Boeden,* Beitraege des 9. DECHEMA Fachgespraechs Umweltschutz, Frankfurt/Main, Germany.

Oehme, C. 1984. *Traegerbiologien in der Abwassertechnik.* Chemie Ingenieur Technik, Weinheim, Germany.

Svjaginzev, D. 1973. *Adsorption von Mikroorganismen an festen Oberflaechen.* Verlag Moskauer Universitaeten, Moscow, Russia.

Economics of a Commercial Slurry-Phase Biological Treatment Process

Douglas E. Jerger and Patrick M. Woodhull

ABSTRACT

Slurry-phase bioremediation is an engineered process for treating contaminated soils or sludges in a slurry of water at 10 to 40% (weight/weight) solids. Slurry-phase treatment relies on the mobilization of contaminants to the aqueous phase, where they are susceptible to microbial degradation. The mobilization or dissolution step can be the result of either microbial or physical/chemical action. Operating parameters (nutrient concentrations, dissolved oxygen, pH, and temperature) are maintained to provide optimal conditions for biological treatment of the contaminants. OHM Remediation Services Corp. (OHM) has completed the full-scale slurry-phase biological treatment of 10,500 yd^3 (9,600 m^3) of Resource Conservation and Recovery Act (RCRA)-K001 wastes at the Southeastern Wood Preserving Superfund site in Canton, Mississippi. The polycyclic aromatic hydrocarbon (PAH) concentrations in the untreated material ranged from 8,000 mg/kg to 12,000 mg/kg dryweight. A 95% reduction in total PAH concentration was achieved following treatment.

FULL-SCALE REMEDIATION OF WOOD-TREATING WASTES

The full-scale slurry-phase biological remediation of creosote-contaminated wastes (RCRA-listed as K001) at the Southeastern Wood Preserving Superfund site in Canton, Mississippi, has been completed. A more detailed description of the laboratory-, pilot-, and full-scale treatment system has been published previously (Jerger et al. 1994, Woodhull & Jerger 1994).

PAH concentrations in the stockpiled soil ranged from 8,000 to 15,000 mg/kg dry-weight for total PAHs and from 1,000 to 2,500 mg/kg dry-weight for carcinogenic PAHs. Treatment efficiencies approaching 95% were achieved for total PAHs in the slurry-phase biological reactors, with the majority of the PAH biodegradation occurring in the initial 5 to 10 days of treatment (Jerger & Woodhull 1994). Risk-based treatment criteria were 950 mg/kg dry-weight for total PAHs and 180 mg/kg dry-weight for benzo(a)pyrene-equivalent carcinogenic PAHs.

The stockpiled material at the site was power-screened to prepare the material for treatment. The power screening separated the material into a large-debris (greater than 6-in. [15 cm]) pile, a power-screen-debris (greater than 0.5-in. [1.3 cm]) pile, and the material requiring further treatment (less than 0.5-in. [1.3 cm]). Approximately 10,500 yd^3 (14,140 tons) of material was screened and prepared for treatment (Table 1).

The slurry preparation system consisted of a slurry mix tank with a shaker screen and a hydrocyclone. The screened material was loaded into the slurry mix tank, and water was added to achieve the operating slurry solids concentration. Agitators mixed and homogenized the slurry, and a slurry pump transferred the slurry to the shaker screen. The shaker screen debris was stockpiled on site, and the slurry was pumped to a hydrocyclone that removed and stockpiled the remaining washed debris. An agitator mixed and blended the slurry, nutrients, and slurry-conditioning chemicals. The cost for the nutrients and slurry-conditioning chemicals was $20 per ton. The total cost for slurry preparation in U.S. dollars was $50 to $60 per ton (Table 2).

The prepared soil slurry was pumped into one of four biological slurry-phase reactors with an operating capacity of 180,000 gal (680,000 L). Aeration of the slurry was provided by diffusers and a blower. This equipment was sized to maintain a dissolved oxygen concentration of >2.0 ppm in the slurry. A mixer maintained the slurry in suspension. The reactors were monitored daily for pH, temperature, dissolved oxygen, and other biological monitoring parameters. The slurry temperature in the reactor ranged from 20 to 37°C during the operating period. An engineering-economic analysis compared the cost of heating the reactors during periods of low-ambient temperatures vs. extending operating time to achieve the cleanup criteria (Woodhull & Jerger 1995). Discrete slurry samples were collected and composited for PAH analysis to monitor biological treatment. The reactors were operated in a batch mode, with each batch typically requiring 8 to 12 days to achieve the treatment criteria. At the completion of

TABLE 1. Southeastern Wood Preserving Superfund Site contaminated material size fractions and quantities.

		Quantity	
Description	**Size Fraction**	**Cubic Yards**	**Tons**
Large debris	+6-inch	150	165
Power screen rejects	−6-inch + ½-inch	300	330
Shaker screen rejects	−½-inch + 12-mesh	1,500	1,825
Hydrocyclone rejects	−12-mesh + 200 mesh	1,500	1,825
Material to bioreactors	−200-mesh	7,050	9,995
	TOTAL	10,500	14,140

TABLE 2. Cost per ton breakdown for the full-scale slurry-phase biological remediation of RCRA K001 wastes.

Cost Category	Slurry Preparation	Slurry Biotreatment	Slurry Dewatering
Labor and equipment	$30 to $35	$10 to $15	$5 to $10
Supplies, materials, utilities	$20 to $25	$25 to $30	$15 to $20
Analytical	<$5	$5 to $10	<$5
TOTALS	$50 to $60	$40 to $55	$20 to $30

the biological treatment, the slurry was transferred to the slurry-dewatering unit. The cost to treat the slurry on a per ton basis was $40 to $55 (Table 2).

The slurry-dewatering unit was a 425-ft-long, 160-ft-wide, and 6-ft-deep (130 m × 50 m × 1.8 m) high-density polyethylene (HDPE)-lined cell. A water recovery system facilitated removal of the excess water, which was pumped to the 350,000-gal (1.3 × 10⁶-L) water management tank for reuse in the slurry preparation system. The soil in the dewatering unit was worked to further dewater and dry the treated material. The cost to dewater the treated slurry was $20 to $30 per ton (Table 2).

The cost for the full-scale slurry-phase biological treatment portion (slurry preparation, slurry treatment, and slurry dewatering) of K001 wastes to the risk-based treatment criteria was $110 to $145 per ton of material treated (Table 2). In addition to the costs directly associated with the treatment of the contaminated material, additional project costs included design engineering, treatability and pilot-scale testing, site closure, and project administration and support. The total project cost for the full-scale slurry-phase biological treatment of 14,140 tons of K001 waste material was approximately $2,800,000, or $190 to $200 per ton of material treated (Table 3).

FACTORS AFFECTING PROCESS ECONOMICS

Three variables have a significant impact on the total project cost of the treatment system. Treatment costs are affected by the slurry-phase reactor solids concentration, residence time in the reactors, and the percentage of material removed in the slurry preparation/soil washing system.

By increasing the slurry solids concentration in the reactors, the amount of soil treated per batch increases, decreasing the cost per ton of material treated (Figure 1). Higher reactor solids concentrations correspond to increased organic loading rates which typically result in higher degradation rates. During this project, first-order PAH degradation rates increased up to 20% solids concentration and remained constant from 20 to 30% solids concentrations. The maximum solids concentration which could be effectively treated using the reactor design

TABLE 3. Total project cost for the full-scale slurry-phase biological treatment of RCRA K001 wastes.

Task	Subtotal[a]
Treatability and pilot-scale testing	$200,000
Design engineering	$100,000
Soil screening and slurry preparation	$800,000
Slurry treatment	$700,000
Slurry dewatering	$400,000
Site preparation and closure	$500,000
Project administration and support	$200,000
Total project cost per ton of material treated:	$190 to $200

(a) Subtotal figures have been rounded up to the nearest $100,000.

implemented at this site was approximately 30%. Reactor modifications would be necessary to treat higher solids concentrations effectively. In general, for a given reactor configuration, the greater the slurry solids concentration, the lower the unit cost for the contaminated material.

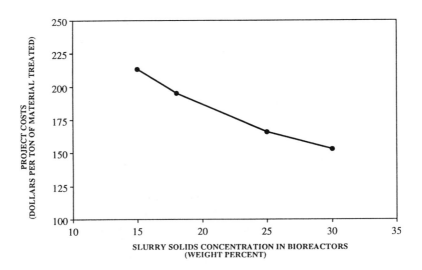

FIGURE 1. Effect of solids concentration in slurry-phase biological reactors on project costs. From P. M. Woodhull and D. E. Jerger, "Bioremediation Using a Commercial Slurry-Phase Biological Treatment System: Site-Specific Application and Costs," *Remediation* 4(3), pp. 353-362. Copyright© 1994. Reprinted by permission of John Wiley & Sons, Inc.

As the batch treatment time (residence time) for the reactors increases, the unit cost of material also increases (Figure 2). An increased residence time reduces the throughput of the system, which requires additional labor and equipment charges to treat the same amount of material per batch.

As the percent of material treated and removed in the slurry preparation/soil washing system increases, the overall unit cost for treatment decreases (Figure 3). The greater the quantity of material removed in the soil washing process, the less material remained requiring treatment and dewatering.

CONCLUSIONS

Slurry-phase biological treatment can be applied to a variety of contaminated wastes amenable to microbial degradation, including manufactured gas plant wastes, wood-treating wastes, refinery wastes, select halogenated compounds (including certain polychlorinated biphenyl [PCB] congeners), and petroleum hydrocarbons (Jerger et al. 1994; Liu et al. 1993; U.S. EPA 1993).

Treatment costs for comparable on-site treatment technologies vary widely. The following costs are typical of on-site remediation for PAH-contaminated soil (U.S. EPA 1993):

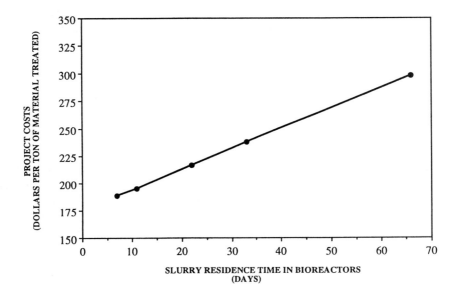

FIGURE 2. Effect of residence time in slurry-phase biological reactors on project costs. From P. M. Woodhull and D. E. Jerger, "Bioremediation Using a Commercial Slurry-Phase Biological Treatment System: Site-Specific Application and Costs," *Remediation* 4(3), pp. 353-362. Copyright© 1994. Reprinted by permission of John Wiley & Sons, Inc.

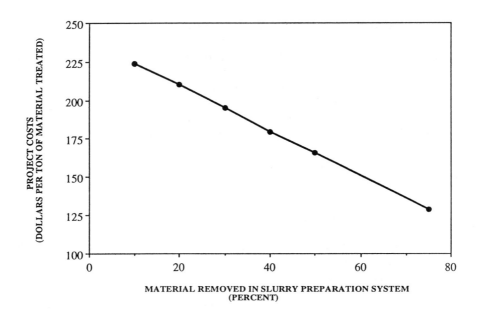

FIGURE 3. Effect of percent of material removed in slurry preparation system on project costs. From P. M. Woodhull and D. E. Jerger, "Bioremediation Using a Commercial Slurry-Phase Biological Treatment System: Site-Specific Application and Costs," *Remediation* 4(3), pp. 353-362. Copyright© 1994. Reprinted by permission of John Wiley & Sons, Inc.

Incineration	$400 to $1,000 per ton
Chemical oxidation	$200 to $500 per ton
Stabilization	$120 to $520 per ton
Solvent extraction	$100 to $500 per ton

Slurry-phase biological treatment costs range from approximately $190 to $200, as discussed in the case history. These costs were collected from a commercial treatment system and are comparable to or lower than costs for other destruction technologies (e.g., incineration or chemical oxidation). To our knowledge, this project provides the first well-documented cost estimates for slurry-phase treatment of PAHs.

Bioremediation offers two advantages over other remediation alternatives: on-site destruction of the contaminant(s) and lower total remedial costs. Slurry-phase biological treatment offers several advantages over conventional bioremediation technologies (e.g., land treatment). Slurry-phase biological treatment can achieve increased degradation rates, higher treatment efficiencies, greater control of environmental and operating conditions, and smaller treatment area requirements.

REFERENCES

Jerger, D. E., D. J. Cady, and J. H. Exner. 1994. "Full-Scale Slurry-Phase Biological Treatment of Wood Preserving Wastes." In R. E. Hinchee et al. (Eds.), *Bioremediation of Chlorinated and Polycyclic Aromatic Compounds*, pp. 480-483. Lewis Publishers, Chelsea, MI.

Jerger, D. E., and P. M. Woodhull. 1994. "Full-Scale Bioslurry Reactor Treatment of Wood Preserving Wastes at a Superfund Site." *Emerging Technologies in Hazardous Waste Management VI*, American Chemical Society. Vol. II, pp. 1370-1373.

Liu, B. Y., V. J. Srivastava, J. R. Paterek, S. P. Pradhan, J. R. Pope, T. D. Hayes, D. G. Linz, and D. E. Jerger. 1993. "MGP Soil Remediation in a Slurry-Phase System: A Pilot-Scale Test." Presented at the IGT Symposium on Gas, Oil, and Environmental Biotechnology. Colorado Springs, CO.

U.S. EPA. 1993. *Engineering Bulletin: Current Treatment and Site Remediation Technologies.* Office of Emergency and Environmental Response, Washington, DC; and Office of Research and Development, Cincinnati, OH.

Woodhull, P. M., and D. E. Jerger. 1994. "Bioremediation Using a Commercial Slurry-Phase Biological Treatment System: Site-Specific Application and Costs." *Remediation* 4(3): 353-362.

Woodhull, P. M., and D. E. Jerger. 1995. "Temperature Effects on the Kinetics and Economics of Slurry-Phase Biological Treatment." In R. E. Hinchee, C. M. Vogel, and F. J. Brockman (Eds.), *Microbial Processes for Bioremediation*, pp. 289-295. Battelle Press, Columbus, OH.

Chemical and Biological Oxidation of Slurry-Phase Polycyclic Aromatic Hydrocarbons

Kandi L. Brown, Brunilda Davila, John Sanseverino,
Mark Thomas, Craig Lang, Keith Hague, and Tony Smith

ABSTRACT

Bioslurry treatment of polycyclic aromatic hydrocarbon (PAH)-impacted soils was demonstrated under the Superfund Innovative Technologies Evaluation — Emerging Technologies Program (SITE ETP) as an extension of research previously funded by IT Corporation (IT) (Brown and Sanseverino 1993) and additional investigations supported by the U.S. Environmental Protection Agency (U.S. EPA) (Davila et al. 1994). During the demonstration, IT operated two 60-L TEKNO Associates bioslurry reactors (Salt Lake City, Utah) and a 10-L fermentation unit in semi-continuous, plug-flow mode for a 6-month period. The first 60-L reactor received fresh feed daily and supplements of salicylate and succinate to enhance PAH biodegradation. Effluent from the first reactor was fed to the second 10-L reactor in series, where Fenton's reagent ($Fe^{++}+H_2O_2$) was added to accelerate oxidation of 4- to 6-ring PAHs. The third reactor in series was used as a polishing reactor to biologically oxidize contaminants remaining following the addition of Fenton's reagent. R3 received no additions of salicylate and succinate. The reactor was aerated, nutrient-amended, and pH-adjusted only. During operation, the reactor system demonstrated average total PAH and carcinogenic PAH (CPAH) removals of 85 and 66%, respectively.

INTRODUCTION

The Comprehensive Environmental Response, Compensation, and Liability Act (CERCLA) mandates the U.S. EPA to select remedies that "utilize permanent solutions and alternative treatment technologies or resource recovery technologies to the maximum extent practicable." CERCLA also prefers remedial actions in which treatment that "permanently and significantly reduces the volume, toxicity or mobility of the hazardous substances, pollutants, and contaminants is a principal element." During this demonstration, innovative methods for the oxidation of PAH-impacted soils were evaluated to determine their compliance with the mandates of CERCLA.

The primary objective of reactor 1 (R1) operation was to increase the biological removal of organic carbon. Salicylate was used to induce the naphthalene degradation operon on NAH plasmids. This system has been shown to degrade phenanthrene and anthracene (Sanseverino et al. 1993). The naphthalene pathway may also play a role in CPAH metabolism. Succinate, a by-product of naphthalene metabolism, served as a general carbon source in R1 which removed easily degradable carbon and increased biological activity against more recalcitrant PAHs (i.e., 4-ring compounds and higher).

Effluent from R1 was fed to reactor 2 (R2) where Fenton's reagent was continuously introduced, resulting in chemical oxidation being the primary mechanism for PAH removal in this reactor. The pH in R2 was adjusted to 2.0 following the addition of R1 feed. Fenton's reagent (hydrogen peroxide in the presence of reduced iron salts) produces free radicals, which have been shown effective in extensively oxidizing multi-ring aromatic hydrocarbons in both soil and water systems (Gauger et al. 1991; Kelley et al. 1991; Elizardo 1991). The objective of Fenton's reagent addition was not PAH mineralization, but the hydroxylation of PAH, because hydroxylation of high-molecular-weight PAHs generally is the rate-limiting step in biological oxidation.

Reactor 3 (R3) was used for biological polishing of R2 effluent. R3 received no additions of salicylate and succinate. The reactor was aerated, nutrient-amended, and pH-adjusted following the introduction of R2 feed.

The primary objective of the investigation was to demonstrate increased CPAH and PAH removal using combined biological and chemical oxidation.

MATERIALS AND METHODS

PAH and CPAH-impacted soils were collected from a southeastern wood-treating facility. The soils, a blend of clayey and sandy material, were wet-sieved on site through a 30-mesh screen and submitted to IT's Biotechnology Applications Center (BAC) located in Knoxville, Tennessee, for testing. Oversized material was disposed of on site. Blended slurry PAH and CPAH concentrations averaged 4,380 and 300 mg/kg, respectively. Wet sieving the soils increased the uniformity of the slurry, thereby reducing the potential for sampling variability.

All reactors were loaded to establish 40% total solids (TS) slurry of the screened soils and operated as batch bioreactors for 2 weeks prior to the initiation of semicontinuous flow. The 40% solids loading was required to suspend the sandy slurry. During the course of reactor operation the initial 40% solids loading was decreased to a set point of 30%. This was accomplished due to the increased clay content of the influent feed. The clay content was increased by additional blending of clean, clayey soils with highly impacted screened material.

No microbial cultures were added to the reactors during the demonstration. All reactors were operated at approximately 20°C. During batch treatment, several operational difficulties were encountered, including significant foaming and tar ball formation.

During semicontinuous flow, 6 L/day of influent slurry was manually introduced to R1, resulting in a hydraulic retention time (HRT) of 10 days. To induce biological activity for PAH biodegradation, salicylate and succinate were added weekly to achieve final reactor concentrations of 86 and 4.4 mg/L, respectively. This addition rate was increased to 3 times per week after 4 weeks of operation to increase PAH removal in R1.

The second reactor in the series received dilute concentrations of Fenton's reagent to accelerate oxidation of PAH. Fenton's reagent was prepared by mixing a 1:1 volumetric ratio of 30% hydrogen peroxide and 8.4 millimolar (mM) iron sulfate solution. The iron sulfate solution and the hydrogen peroxide were individually applied to R2 below the slurry surface at a combined rate of 2.0 L/day. The influent feed rate to R2 was 6 L/day, resulting in a substrate HRT of 1 day. The reactor pH was maintained at 2 to enhance the effectiveness of the addition of Fenton's reagent.

The last reactor (R3) in the series was used as a polishing reactor for the removal of any partially oxidized contaminants remaining following addition of Fenton's reagent. The system received 8.0 L/day of influent feed from R2, resulting in an HRT of 7.5 days. All operational parameters were maintained similar to R1; however, no salicylate and succinate additions were made. The pH was buffered in the reactor due to the addition of R2 contents. The process flow diagram (PFD) for the pilot-scale system is presented in Figure 1.

Following 4 months of operation, the HRT in R1 and R3 was increased to 20 days. R2 HRT was increased to 2 days. This change was initiated to decrease the amount of total carbon introduced to the reactors.

Reactor pH, dissolved oxygen, ammoniacal nitrogen, and orthophosphate concentrations were monitored routinely during system operation. During operation, slurry in R1, R2, and R3 was monitored weekly for TS, volatile solids (VS) concentrations, and slurry density. Microbial enumeration of total heterotrophs and naphthalene degraders was also conducted on the same schedule.

Aqueous- and solid-phase PAH concentrations in R1, R2, and R3 were monitored once per week. Volatilization of constituents was monitored monthly in R1 off-gas. All PAH concentrations were determined using EPA Method 8310.

RESULTS AND DISCUSSION

The bioslurry reactor system demonstrated 85 and 66% removal of PAH and CPAH, respectively. Table 1 illustrates the reduction in concentration of all PAH compounds. Overall, the biologically active reactors (R1 and R3) illustrated a decreasing effectiveness in PAH removal as a function of compound molecular weight. R1 demonstrated 33% removal of CPAH, with approximately 66% removal of PAH. R2 demonstrated equal removals of PAH and CPAH, as expected during chemical oxidation. R3 CPAH and PAH removals averaged

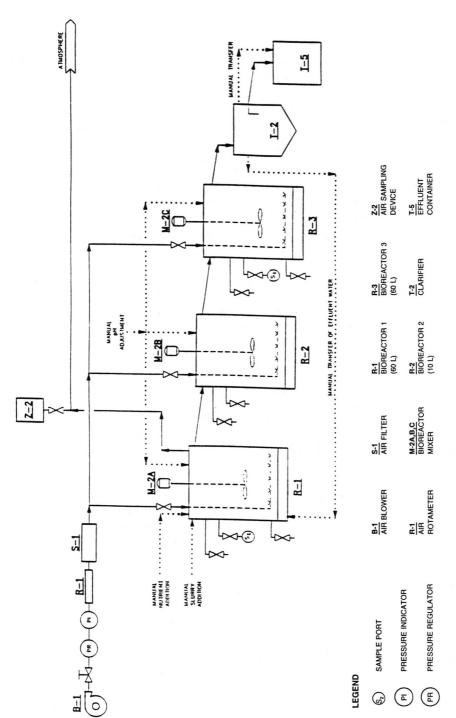

FIGURE 1. System process flow diagram.

TABLE 1. Bioslurry system PAH concentrations. (*text continues on page 126*)

OCTOBER 10, 1994

COMPOUND	INF	CONCENTRATION			% REDUCTION			OVERALL
		R1	R2	R3	R1	R2	R3	% REDUCTION
NAPHTHALENE	328	5.88U	5.88U	5.88U	99	0	0	99.1
ACENAPHTHYLENE	127	2.23U	2.23U	7.78	99	0	0	93.9
ACENAPHTHENE	142	67.7	78.1	60.2	52	0	22.9	57.6
FLUORENE	256	12.3	96.9	52.5	95	0	45.8	79.5
PHENANTHRENE	700	9.6	260	101	99	0	61.2	85.6
ANTHRACENE	129	13.6	80.6	48.1	89	0	40.3	62.7
FLUORANTHENE	866	218	442	321	75	0	27.4	62.9
PYRENE	748	129	409	288	83	0	29.6	61.5
BENZ(A)ANTHRACENE*	50.2	20.1	33.1	20.4	60	0	38.4	59.4
CHRYSENE*	62	25.1	38.6	27.3	60	0	29.3	56
BENZO(B)FLUORANTHENE*	22.3	9.83	17.2	8.09	56	0	53	63.7
BENZO(K)FLUORANTHENE*	14.5	6.64	10.3	3.41	54	0	66.9	76.5
BENZ(A)PYRENE*	20.5	7.58	10.9	2.92U	63	0	86.6	92.9
DIBENZ(A,H)ANTHRACENE*	10.9	4.11	6.19	4.44	62	0	28.3	59.3
BENZO(G,H,I)PERYLENE*	7.65	2.83	4.39	2.25	63	0	48.7	70.6
INDENOPYRENE*	6.47	2.23	3.28	2.14	66	0	34.8	66.9
TOTAL PAH	3489	533	1495	955	85	0	36.1	72.6
TOTAL CPAH	194	78.4	124	73.3	60	0	40.9	62.2

OCTOBER 19, 1994

COMPOUND	INF	CONCENTRATION			% REDUCTION			OVERALL
		R1	R2	R3	R1	R2	R3	% REDUCTION
NAPHTHALENE	5.88U	5.88U	5.88U	5.88U	0	0	0	0
ACENAPHTHYLENE	4.74	9.94	2.23U	2.23U	0	88.8	0	76.5
ACENAPHTHENE	81.2	178	34.2	6.60U	0	80.8	90	95.9
FLUORENE	324	64.4	11.8	1.29	80.1	81.7	89	99.6
PHENANTHRENE	1101	78.4	42.4	3.42	92.9	45.9	92	99.7
ANTHRACENE	252	68.4	7.68	1.69	72.9	88.8	78	99.3
FLUORANTHENE	1417	617	315	91.2	56.5	48.9	71	93.6
PYRENE	1358	454	200	83.8	66.6	55.9	58	93.8
BENZ(A)ANTHRACENE*	88.8	58.6	27.6	5.74	34	52.9	79	93.5
CHRYSENE*	111	74.2	36.2	10.6	33.2	51.2	71	90.5
BENZO(B)FLUORANTHENE*	37.6	26.2	12	11.3	30.3	54.2	5.8	69.9
BENZO(K)FLUORANTHENE*	20.7	17.3	5.44	5.92	16.4	68.6	0	71.4
BENZ(A)PYRENE*	30.3	24.9	2.92U	2.92U	17.8	94.1	0	95.2
DIBENZ(A,H)ANTHRACENE*	14	11.5	5.06	4.26	17.9	56	16	69.6
BENZO(G,H,I)PERYLENE*	7.45	8	3.82	3.54	0	52.3	7.3	52.5
INDENOPYRENE*	6.25	6.58	3.01	2.63	0	54.3	13	57.9
TOTAL PAH	4856	1700	710	234	65	58.2	67	95.2
TOTAL CPAH	316	227	97.8	49.6	28.2	56.9	49	84.3

U = Number is 1/2 MDL. The new MDL for naphthalene is based on the average relative
    SD and highest dilution factor used.
(+) = Area counts are over the highest calibration standard.
  * = Carcinogenic PAHs.

TABLE 1. (page 2 of 9)

OCTOBER 26, 1994

| COMPOUND | INF | CONCENTRATION | | | % REDUCTION | | | OVERALL |
		R1	R2	R3	R1	R2	R3	% REDUCTION
NAPHTHALENE	38.6	5.88U	5.88U	5.88U	92	0	0	92.4
ACENAPHTHYLENE	16.4	15.1	38.2	11.7	7.9	0	69.4	28.7
ACENAPHTHENE	112	147	75.1	14.7	0	48.9	80.4	86.9
FLUORENE	276	117	48	10.6	58	59	77.9	96.2
PHENANTHRENE	853	296	123	35.2	65	58.4	71.4	95.9
ANTHRACENE	181	131	34.2	8.27	28	73.9	75.8	95.4
FLUORANTHENE	1166	919	570	246	21	38	56.8	78.9
PYRENE	982	817	458	178	17	43.9	61.1	81.9
BENZ(A)ANTHRACENE*	123	112	62	34.9	8.9	44.6	43.7	71.6
CHRYSENE*	67.7	95.1	54.2	29.5	0	43	45.6	56.4
BENZO(B)FLUORANTHENE*	4.25	31.4	18	13.3	0	42.7	26.1	0
BENZO(K)FLUORANTHENE*	18.5	23.3	8.72	10.5	0	62.6	0	43.2
BENZ(A)PYRENE*	30.1	34.1	7.46	10.3	0	78.1	0	65.8
DIBENZ(A,H)ANTHRACENE*	18.6	11	6.6	4.73	41	40	28.3	74.6
BENZO(G,H,I)PERYLENE*	9.29	10.2	5.91	4.16	0	42.1	29.6	55.2
INDENOPYRENE*	9.43	8.47	4.89	3.39	10	42.3	30.7	64.1
TOTAL PAH	3906	2771	1518	618	29	45.2	59.3	84.2
TOTAL CPAH	281	326	168	111	0	48.5	33.9	60.5

NOVEMBER 2, 1994

| COMPOUND | INF | CONCENTRATION | | | % REDUCTION | | | OVERALL |
		R1	R2	R3	R1	R2	R3	% REDUCTION
NAPHTHALENE	38.6	9.67	5.88U	5.88U	74.9	69.8	0	92.4
ACENAPHTHYLENE	16.4	16.8	12.1	2.23U	0	28	91	93.2
ACENAPHTHENE	112	270	91.5	60.7	0	66.1	34	45.8
FLUORENE	276	17.2	15.3	2.34	93.8	11	85	99.2
PHENANTHRENE	853	16.4	30.7	5.76	98.1	0	81	99.3
ANTHRACENE	181	18.5	7.87	1.63	89.8	57.5	79	99.1
FLUORANTHENE	1166	690	471	257	40.8	31.7	45	78
PYRENE	982	425	287	177	56.7	32.5	38	82
BENZ(A)ANTHRACENE*	123	64.5	41.6	25.4	47.6	35.5	39	79.3
CHRYSENE*	67.7	98.5	65.7	41	0	33.3	38	39.4
BENZO(B)FLUORANTHENE*	4.25	33.6	20.3	18.5	0	39.6	8.9	0
BENZO(K)FLUORANTHENE*	18.5	16.9	9.27	8.8	8.65	45.1	5.1	52.4
BENZ(A)PYRENE*	30.1	25.8	7.13	8.45	14.3	72.4	0	71.9
DIBENZ(A,H)ANTHRACENE*	18.6	14.6	7.95	7.52	21.5	45.5	5.4	59.6
BENZO(G,H,I)PERYLENE*	9.29	11.8	6.64	6.4	0	43.7	3.6	31.1
INDENOPYRENE*	9.43	8.6	4.62	4.72	8.8	46.3	0	49.9
TOTAL PAH	3906	1738	1082	629	55.5	37.7	42	83.9
TOTAL CPAH	281	274	163	121	2.49	40.5	26	56.9

U = Number is 1/2 MDL. The new MDL for naphthalene is based on the average relative SD and highest dilution factor used.
(+) = Area counts are over the highest calibration standard.
* = Carcinogenic PAHs.

TABLE 1. (page 3 of 9)

NOVEMBER 9, 1994

COMPOUND	INF	CONCENTRATION			% REDUCTION			OVERALL
		R1	R2	R3	R1	R2	R3	% REDUCTION
NAPHTHALENE	38.6	96	56.9	35.9	0	40.7	36.9	6.99
ACENAPHTHYLENE	16.4	31.6	26.3	7	0	16.8	73.4	57.3
ACENAPHTHENE	112	194	86.9	38.1	0	55.2	56.2	66
FLUORENE	276	15	10.1	2.28	94.57	32.7	77.4	99.2
PHENANTHRENE	853	21.7	32.1	5.06	97.46	0	84.2	99.4
ANTHRACENE	181	14.4	6.73	1.43	92.04	53.3	78.8	99.2
FLUORANTHENE	1166	609	488	256	47.77	19.9	47.5	78
PYRENE	982	327	243	178	66.7	25.7	26.7	81.9
BENZ(A)ANTHRACENE*	123	79.4	60.9	30.2	35.45	23.3	50.4	75.4
CHRYSENE*	67.7	92.1	76.1	44.5	0	17.4	41.5	34.3
BENZO(B)FLUORANTHENE*	4.25	34	26.6	16.3	0	21.8	38.7	0
BENZO(K)FLUORANTHENE*	18.5	19.7	14.8	6.88	0	24.9	53.5	62.8
BENZ(A)PYRENE*	30.1	28.4	13.4	7.24	5.65	52.8	46	75.9
DIBENZ(A,H)ANTHRACENE*	18.6	14.4	10.7	6.65	22.58	25.7	37.9	64.2
BENZO(G,H,I)PERYLENE*	9.29	10	8.12	5.49	0	18.8	32.4	40.9
INDENOPYRENE*	9.43	8	5.85	3.74	15.16	26.9	36.1	60.3
TOTAL PAH	3906	1593	1167	644.8	59.22	26.7	44.8	83.5
TOTAL CPAH	281	284	216	121	0	23.9	44	56.9

NOVEMBER 16, 1994

COMPOUND	INF	CONCENTRATION			% REDUCTION			OVERALL
		R1	R2	R3	R1	R2	R3	% REDUCTION
NAPHTHALENE	38.6	5.88U	5.88U	5.88U	92.4	0	0	92.4
ACENAPHTHYLENE	16.4	2.23U	9.12	2.23U	93.2	0	87.8	93.2
ACENAPHTHENE	112	59.2	6.6	6.6	47.14	89	0	94.1
FLUORENE	276	30.1	8.22	0.874	89.09	73	89.4	99.7
PHENANTHRENE	853	51.5	38.3	3.96	93.96	26	89.7	99.5
ANTHRACENE	181	36.6	4.02	0.625U	79.78	89	92.2	99.8
FLUORANTHENE	1166	629	445	85.3	46.05	29	80.8	92.7
PYRENE	982	431	243	187	56.11	44	23	81
BENZ(A)ANTHRACENE*	123	62.8	34.2	2.5U	48.94	46	96.3	99
CHRYSENE*	67.7	101	49.5	11.2	0	51	77.4	83.5
BENZO(B)FLUORANTHENE*	4.25	35.7	14.6	21	0	59	0	0
BENZO(K)FLUORANTHENE*	18.5	11.6	7.33	10.1	37.3	37	0	45.4
BENZ(A)PYRENE*	30.1	13.3	4.3	8.78	55.81	68	0	70.8
DIBENZ(A,H)ANTHRACENE*	18.6	13.4	8.03	8.91	27.96	40	0	52.1
BENZO(G,H,I)PERYLENE*	9.29	9.76	5.83	5.26	0	40	9.78	43.4
INDENOPYRENE*	9.43	7.48	4.34	4.42	20.68	42	0	53.1
TOTAL PAH	3906	1496	882	356	61.7	41	59.6	90.9
TOTAL CPAH	281	255	128	71	9.253	50	44.5	74.7

U = Number is 1/2 MDL. The new MDL for naphthalene is based on the average relative SD and highest dilution factor used.

(+) = Area counts are over the highest calibration standard.

* = Carcinogenic PAHs.

TABLE 1. (page 4 of 9)

NOVEMBER 22, 1994

COMPOUND	INF	CONCENTRATION			% REDUCTION			OVERALL % REDUCTION
		R1	R2	R3	R1	R2	R3	
NAPHTHALENE	25.1	195	79.2	5.88U	0	59.4	96.3	88.3
ACENAPHTHYLENE	30	25.3	60.8	5.54	15.67	0	90.9	81.5
ACENAPHTHENE	6.6U	145	63.5	6.6U	0	56.2	94.8	0
FLUORENE	378	90.4	27	0.686	76.08	70.1	97.5	99.8
PHENANTHRENE	1377	185	106	9.2	86.56	42.7	91.3	99.3
ANTHRACENE	344	162	11.5	2.12	52.91	92.9	81.6	99.4
FLUORANTHENE	1742+	871	575	169	50	34	70.6	90.3
PYRENE	1526	704	374	167	53.87	46.9	55.3	89.1
BENZ(A)ANTHRACENE*	105	93	67.9	17.5	11.43	27	74.2	83.3
CHRYSENE*	158	95.9	69.1	25.2	39.3	27.9	63.5	84.1
BENZO(B)FLUORANTHENE*	56.3	30.5	21.4	16.3	45.83	29.8	23.8	71
BENZO(K)FLUORANTHENE*	29	21.5	11.9	10.6	25.86	44.7	10.9	63.4
BENZ(A)PYRENE*	31.8	26.2	7.57	10.7	17.61	71.1	0	66.4
DIBENZ(A,H)ANTHRACENE*	25.9	16.3	10.8	7.85	37.07	33.7	27.3	69.7
BENZO(G,H,I)PERYLENE*	15	10.6	7.13	5.42	29.33	32.7	24	63.9
INDENOPYRENE*	13	8.17	5.2	4.13	37.15	36.4	20.6	68.2
TOTAL PAH	5859	2620	1496	457	55.28	42.9	69.5	92.2
TOTAL CPAH	434	302	201	98	30.41	33.4	51.2	77.4

DECEMBER 2, 1994

COMPOUND	INF	CONCENTRATION			% REDUCTION			OVERALL % REDUCTION
		R1	R2	R3	R1	R2	R3	
NAPHTHALENE	25.1	5.88U	5.88U	5.88U	88.29	0	0	88.3
ACENAPHTHYLENE	30	17.2	7.9	12.9	42.67	54	0	57
ACENAPHTHENE	3.3	84.6	45.4	6.6	0	46	92.7	0
FLUORENE	378	243	28.9	30.9	35.71	88	0	91.8
PHENANTHRENE	1377	499	153	109	63.76	69	28.8	92.1
ANTHRACENE	344	201	8.36	17	41.57	96	0	95.1
FLUORANTHENE	1742	1426	683	456	18.14	52	33.2	73.8
PYRENE	1526	1216	448	342	20.31	63	23.7	77.6
BENZ(A)ANTHRACENE*	105	116	44.5	35.5	0.00	62	20.2	66.2
CHRYSENE*	158	148	84.2	56	6.33	43	33.5	64.6
BENZO(B)FLUORANTHENE*	56.3	52	22.3	17.8	7.64	57	20.2	68.4
BENZO(K)FLUORANTHENE*	29	18.4	12.4	10.2	36.55	33	17.7	64.8
BENZ(A)PYRENE*	31.8	24.8	5.48	7.63	22.01	78	0	76
DIBENZ(A,H)ANTHRACENE*	25.9	5.2	9.23	2.39	79.92	0	74.1	90.8
BENZO(G,H,I)PERYLENE*	15	13.3	7	5.86	11.33	47	16.3	60.9
INDENOPYRENE*	13	11.3	4.65	4.44	13.08	59	4.52	65.8
TOTAL PAH	5859	4079	1567	1114	30.38	62	28.9	81
TOTAL CPAH	434	389	190	140	10.37	51	26.3	67.7

U = Number is 1/2 MDL. The new MDL for naphthalene is based on the average relative
     SD and highest dilution factor used.
(+) = Area counts are over the highest calibration standard.
   * = Carcinogenic PAHs.

TABLE 1. (page 5 of 9)

DECEMBER 8, 1994

COMPOUND	INF	CONCENTRATION			% REDUCTION			OVERALL % REDUCTION
		R1	R2	R3	R1	R2	R3	
NAPHTHALENE	25.1	5.88	5.88	5.88	88.29	0	0	76.6
ACENAPHTHYLENE	30	13.1	81.3	20	56.33	0	75.4	33.3
ACENAPHTHENE	3.3	97	13.6	18.4	0	86	0	0
FLUORENE	378	107	31.8	46.4	71.69	70.3	0	87.7
PHENANTHRENE	1377	156	113	139	88.67	27.6	0	89.9
ANTHRACENE	344	7.1	7.74	22.7	97.94	0	0	93.4
FLUORANTHENE	1742	981	615	513	43.69	37.3	16.6	70.6
PYRENE	1526	863	392	405	43.45	54.6	0	73.5
BENZ(A)ANTHRACENE*	105	84	36.1	37.3	20	57	0	64.5
CHRYSENE*	158	118	75.7	69.6	25.32	35.8	8.06	55.9
BENZO(B)FLUORANTHENE*	56.3	33.8	15.5	19	39.96	54.1	0	66.3
BENZO(K)FLUORANTHENE*	29	10.7	6.5	8.82	63.1	39.3	0	69.6
BENZ(A)PYRENE*	31.8	20.3	4.22	8.52	36.16	79.2	0	73.2
DIBENZ(A,H)ANTHRACENE*	25.9	11.5	7.15	9.5	55.6	37.8	0	63.3
BENZO(G,H,I)PERYLENE*	15	11.9	6.17	7.73	20.67	48.2	0	48.5
INDENOPYRENE*	13	9.31	4.04	5.16	28.38	56.6	0	60.3
TOTAL PAH	5859	2527	1413	1333	56.87	44.1	5.66	77.2
TOTAL CPAH	434	300	155	166	30.88	48.3	0	61.8

DECEMBER 15, 1994

COMPOUND	INF	CONCENTRATION			% REDUCTION			OVERALL % REDUCTION
		R1	R2	R3	R1	R2	R3	
NAPHTHALENE	25.1	5.88	5.88	5.88	88.3	0	0	76.6
ACENAPHTHYLENE	30	11.1	43.2	9.65	63	0	77.7	67.8
ACENAPHTHENE	3.3	99.4	22.3	15.2	0	78	31.8	0
FLUORENE	378	69.6	17.7	27.5	81.6	75	0	92.7
PHENANTHRENE	1377	26.5	47.4	88.4	98.1	0	0	93.6
ANTHRACENE	344	57.5	6.16	13.3	83.3	89	0	96.1
FLUORANTHENE	1742	848	699	448	51.3	18	35.9	74.3
PYRENE	1526	435	350	217	71.5	20	38	85.8
BENZ(A)ANTHRACENE*	105	76.8	57.4	36.9	26.9	25	35.7	64.9
CHRYSENE*	158	121	118	69.2	23.4	2.5	41.4	56.2
BENZO(B)FLUORANTHENE*	56.3	31.7	29.6	15.3	43.7	6.6	48.3	72.8
BENZO(K)FLUORANTHENE*	29	13.2	10.3	8.12	54.5	22	21.2	72
BENZ(A)PYRENE*	31.8	17.9	5.47	6.21	43.7	69	0	80.5
DIBENZ(A,H)ANTHRACENE*	25.9	10.2	11.3	8.59	60.6	0	24	66.8
BENZO(G,H,I)PERYLENE*	15	12.3	10.5	8.67	18	15	17.4	42.2
INDENOPYRENE*	13	8.86	6.11	5.88	31.8	31	3.76	54.8
TOTAL PAH	5859	1842	1437	981	68.6	22	31.7	83.3
TOTAL CPAH	434	292	249	159	32.7	15	36.1	63.4

U = Number is 1/2 MDL. The new MDL for naphthalene is based on the average relative
    SD and highest dilution factor used.
(+) = Area counts are over the highest calibration standard.
  * = Carcinogenic PAHs.

TABLE 1. (page 6 of 9)

DECEMBER 30, 1994

COMPOUND	INF	CONCENTRATION			% REDUCTION			OVERALL % REDUCTION
		R1	R2	R3	R1	R2	R3	
NAPHTHALENE	25.1	0	0	0	100	0	0	100
ACENAPHTHYLENE	30	5.02	8.05	16.2	83.27	0	0	46
ACENAPHTHENE	3.3	0	0	0	100	0	0	100
FLUORENE	378	43.3	15.3	16.3	88.54	64.7	0	95.7
PHENANTHRENE	1377	185	70.6	47.7	86.56	61.8	32.4	96.5
ANTHRACENE	344	55.4	6.09	7.43	83.9	89	0	97.8
FLUORANTHENE	1742	640	378	358	63.26	40.9	5.29	79.4
PYRENE	1526	490	195	194	67.89	60.2	0.51	87.3
BENZ(A)ANTHRACENE*	105	50.1	23.5	19.5	52.29	53.1	17	81.4
CHRYSENE*	158	84.9	58.8	56.7	46.27	30.7	3.57	64.1
BENZO(B)FLUORANTHENE*	56.3	26.6	15.7	14.9	52.75	41	5.1	73.5
BENZO(K)FLUORANTHENE*	29	19.4	8.91	10.1	33.1	54.1	0	65.2
BENZ(A)PYRENE*	31.8	18.8	5.17	5.78	40.88	72.5	0	81.8
DIBENZ(A,H)ANTHRACENE*	25.9	17.4	11.7	6.32	32.82	32.8	46	75.6
BENZO(G,H,I)PERYLENE*	15	25.7	14.1	6.61	0	45.1	53.1	55.9
INDENOPYRENE*	13	1.9	1.61	0.71	85.38	15.3	55.9	94.5
TOTAL PAH	5859	1664	813	760	71.6	51.1	6.52	87
TOTAL CPAH	434	245	139	121	43.55	43.3	12.9	72.1

JANUARY 5, 1995

COMPOUND	INF	CONCENTRATION			% REDUCTION			OVERALL % REDUCTION
		R1	R2	R3	R1	R2	R3	
NAPHTHALENE	25.1	0	0	0	100	0	0	100
ACENAPHTHYLENE	30	7.48	7.76	24.6	75.1	0	0	18
ACENAPHTHENE	3.3	116	0	0	0	100	0	100
FLUORENE	378	52	7.44	9.33	86.2	86	0	97.5
PHENANTHRENE	1377	299	96.2	61.1	78.3	68	36.5	95.6
ANTHRACENE	344	78.5	4.23	4.99	77.2	95	0	98.5
FLUORANTHENE	1742	750	339	374	56.9	55	0	78.5
PYRENE	1526	561	195	206	63.2	65	0	86.5
BENZ(A)ANTHRACENE*	105	75.3	21	32.6	28.3	72	0	69
CHRYSENE*	158	84.1	47.9	53.2	46.8	43	0	66.3
BENZO(B)FLUORANTHENE*	56.3	28.5	13	16.2	49.4	54	0	71.2
BENZO(K)FLUORANTHENE*	29	19	7.46	9.91	34.5	61	0	65.8
BENZ(A)PYRENE*	31.8	18.7	2.8	4.46	41.2	85	0	86
DIBENZ(A,H)ANTHRACENE*	25.9	16.3	7.3	9.33	37.1	55	0	64
BENZO(G,H,I)PERYLENE*	15	26.2	11.2	14	0	57	0	6.67
INDENOPYRENE*	13	0.73	0.36	0.41	94.4	51	0	96.8
TOTAL PAH	5859	2133	761	420	63.6	64	44.8	92.8
TOTAL CPAH	434	269	111	140	38	59	-26	67.7

U = Number is 1/2 MDL. The new MDL for naphthalene is based on the average relative
　　SD and highest dilution factor used.
(+) = Area counts are over the highest calibration standard.
* = Carcinogenic PAHs.

TABLE 1. (page 7 of 9)

JANUARY 26, 1995

COMPOUND	INF	CONCENTRATION			% REDUCTION			OVERALL % REDUCTION
		R1	R2	R3	R1	R2	R3	
NAPHTHALENE	25.1	10.4	6.87	0	58.57	33.9	100	100
ACENAPHTHYLENE	30	12.3	22.7	25	59	0	0	16.7
ACENAPHTHENE	3.3	302	91.8	68.3	0	69.6	25.6	0
FLUORENE	378	24	17.7	41.2	93.65	26.3	0	89.1
PHENANTHRENE	1377	13.7	17.4	80.7	99.01	0	0	94.1
ANTHRACENE	344	48.7	28	29	85.84	42.5	0	91.6
FLUORANTHENE	1742	563	343	272	67.68	39.1	20.7	84.4
PYRENE	1526	527	272	191	65.47	48.4	29.8	87.5
BENZ(A)ANTHRACENE*	105	134	93.5	77.2	0	30.2	17.4	26.5
CHRYSENE*	158	77.6	52.8	44.8	50.89	32	15.2	71.6
BENZO(B)FLUORANTHENE*	56.3	27.3	18.2	16.1	51.51	33.3	11.5	71.4
BENZO(K)FLUORANTHENE*	29	19.9	12.2	10.6	31.38	38.7	13.1	63.4
BENZ(A)PYRENE*	31.8	25.7	6.23	9.02	19.18	75.8	0	71.6
DIBENZ(A,H)ANTHRACENE*	25.9	4.66	3.89	3.14	82.01	16.5	19.3	87.9
BENZO(G,H,I)PERYLENE*	15	3.01	3.01	0.9	79.93	0	70.1	94
INDENOPYRENE*	13	8.44	4.7	4.72	35.08	44.3	0	63.7
TOTAL PAH	5859	1802	994	873.7	69.25	44.8	12.1	85.1
TOTAL CPAH	434	300.6	194.5	166.5	30.74	35.3	14.4	61.6

FEBRUARY 2, 1995

COMPOUND	INF	CONCENTRATION			% REDUCTION			OVERALL % REDUCTION
		R1	R2	R3	R1	R2	R3	
NAPHTHALENE	25.1	9.73	6.38	0	61.2	34	100	100
ACENAPHTHYLENE	30	26.7	21.7	22	11	19	0	26.7
ACENAPHTHENE	3.3	223	135	133	0	39	1.48	0
FLUORENE	378	22.8	22	27.2	94	3.5	0	92.8
PHENANTHRENE	1377	15.1	22.7	48.4	98.9	0	0	96.5
ANTHRACENE	344	41.9	30	28.8	87.8	28	4	91.6
FLUORANTHENE	1742	529	330	320	69.6	38	3.03	81.6
PYRENE	1526	525	301	263	65.6	43	12.6	82.8
BENZ(A)ANTHRACENE*	105	112	74.7	79.9	0	33	0	23.9
CHRYSENE*	158	71	45.9	48.4	55.1	35	0	69.4
BENZO(B)FLUORANTHENE*	56.3	27.1	16.5	17.9	51.9	39	0	68.2
BENZO(K)FLUORANTHENE*	29	18.9	10.7	11.5	34.8	43	0	60.3
BENZ(A)PYRENE*	31.8	24.7	10.3	12.1	22.3	58	0	61.9
DIBENZ(A,H)ANTHRACENE*	25.9	5.43	0	0	79	100	0	100
BENZO(G,H,I)PERYLENE*	15	3.49	0	1.02	76.7	100	0	93.2
INDENOPYRENE*	13	8.35	4.16	5.19	35.8	50	0	60.1
TOTAL PAH	5859	1664	1031	1018	71.6	38	1.22	82.6
TOTAL CPAH	434	271	162.3	176	37.6	40	0	59.4

U = Number is 1/2 MDL. The new MDL for naphthalene is based on the average relative SD and highest dilution factor used.

(+) = Area counts are over the highest calibration standard.

* = Carcinogenic PAHs.

TABLE 1. (page 8 of 9)

FEBRUARY 16, 1995

COMPOUND	INF	CONCENTRATION			% REDUCTION			OVERALL % REDUCTION
		R1	R2	R3	R1	R2	R3	
NAPHTHALENE	25.1	5.88U	5.88U	5.88U	88.3	0	0	88.3
ACENAPHTHYLENE	30	14.54	16.46	1.115	51.53	0	93.2	96.3
ACENAPHTHENE	3.3	101.7	80.36	81.22	0	21	0	0
FLUORENE	378	4.24a	1.48a	4.05a	98.9	65.1	0	98.9
PHENANTHRENE	1377	15.28	12.17	33.13	98.89	20.4	0	97.6
ANTHRACENE	344	11.69	7.07	13.74	96.6	39.5	0	96
FLUORANTHENE	1742	509.9	414.98	548.9	70.73	18.6	0	68.5
PYRENE	1526	280.4	217.22	209.6	81.63	22.5	3.51	86.3
BENZ(A)ANTHRACENE*	105	50.04	41.66	38.99	52.34	16.7	6.41	62.9
CHRYSENE*	158	93.11	72.82	78.87	41.07	21.8	0	50.1
BENZO(B)FLUORANTHENE*	56.3	33.48	23.72	25.32	40.53	29.2	0	55
BENZO(K)FLUORANTHENE*	29	17.24	11.08	14.92	40.55	35.7	0	48.6
BENZ(A)PYRENE*	31.8	19.42	10.99	16.36	38.93	43.4	0	48.6
DIBENZ(A,H)ANTHRACENE*	25.9	8.57	9.68	5.26	66.91	0	45.7	79.7
BENZO(G,H,I)PERYLENE*	15	11.2	8.98	8.49	25.33	19.8	5.46	43.4
INDENOPYRENE*	13	8.46	6.33	6.31	34.92	25.2	0.32	51.5
TOTAL PAH	5859	1185	920.88	1093	79.77	22.3	0	81.3
TOTAL CPAH	434	241.5	185.27	194.5	44.35	23.3	0	55.2

MARCH 2, 1995

COMPOUND	INF	CONCENTRATION			% REDUCTION			OVERALL % REDUCTION
		R1	R2	R3	R1	R2	R3	
NAPHTHALENE	25.1	5.88U	5.88U	5.88	88.3	0	0	88.3
ACENAPHTHYLENE	30	2.23U	6.48	2.23	96.3	0	82.8	96.3
ACENAPHTHENE	3.3	94.4	24.76	106	0	74	0	0
FLUORENE	378	5.47	2.06	9.09	98.6	62	0	97.6
PHENANTHRENE	1377	11.2	11.08	29.1	99.2	0.7	0	97.9
ANTHRACENE	344	7.73	6.55	11.2	97.8	15	0	96.7
FLUORANTHENE	1742	328	250	432	81.2	24	0	75.2
PYRENE	1526	228	149.1	212	85.1	35	0	86.1
BENZ(A)ANTHRACENE*	105	20.9	10.01	44.5	80.1	52	0	57.6
CHRYSENE*	158	43.6	46.17	77.9	72.4	0	0	50.7
BENZO(B)FLUORANTHENE*	56.3	25.7	15.2	24.7	54.3	41	0	56.1
BENZO(K)FLUORANTHENE*	29	15.2	8.98	14.5	47.5	41	0	49.9
BENZ(A)PYRENE*	31.8	19.7	4.21	17	38.1	79	0	46.6
DIBENZ(A,H)ANTHRACENE*	25.9	7.14	3.32	6.75	72.4	54	0	73.9
BENZO(G,H,I)PERYLENE*	15	7.44	4.73	7.18	50.4	36	0	52.1
INDENOPYRENE*	13	8.34	4.09	6.19	35.8	51	0	52.4
TOTAL PAH	5859	829	552.6	1008	85.9	33	0	82.8
TOTAL CPAH	434	146	96.72	199	66.4	34	0	54.2

U = Number is 1/2 MDL. The new MDL for naphthalene is based on the average relative
    SD and highest dilution factor used.
(+) = Area counts are over the highest calibration standard.
  * = Carcinogenic PAHs.

TABLE 1. (page 9 of 9)

MARCH 9, 1995

COMPOUND	INF	CONCENTRATION			% REDUCTION			OVERALL % REDUCTION
		R1	R2	R3	R1	R2	R3	
NAPHTHALENE	25.1	7.24	5.88U	5.55U	71.16	59.4	0	88.3
ACENAPHTHYLENE	30	2.23U	2.23U	2.23U	96.28	0	0	96.3
ACENAPHTHENE	3.3	67.88	26.31	34.16	0	61.2	0	0
FLUORENE	378	3.79	1.31	2.66	99	65.4	0	99.3
PHENANTHRENE	1377	8.83	5.2	5.2	99.36	41.1	0	99.6
ANTHRACENE	344	5.73	3.98	6.13	98.33	30.5	0	98.2
FLUORANTHENE	1742	264.5	179.51	243.7	84.81	32.1	0	86
PYRENE	1526	221	138.5	153.9	85.52	37.3	0	89.9
BENZ(A)ANTHRACENE*	105	10.91	7.14	18.21	89.61	34.6	0	82.7
CHRYSENE*	158	42.05	22.64	40	73.39	46.2	0	74.7
BENZO(B)FLUORANTHENE*	56.3	23.05	13.34	17.36	59.06	42.1	0	69.2
BENZO(K)FLUORANTHENE*	29	14.21	8.12	10.45	51	42.9	0	64
BENZ(A)PYRENE*	31.8	17.61	4.35	11.15	44.62	75.3	0	64.9
DIBENZ(A,H)ANTHRACENE*	25.9	5.73	3.82	3.76	77.88	33.3	1.57	85.5
BENZO(G,H,I)PERYLENE*	15	6.23	4.57	4.62	58.47	26.6	0	69.2
INDENOPYRENE*	13	5.92	3.57	4.24	54.46	39.7	0	67.4
TOTAL PAH	5859	707	430.46	562.6	87.93	39.1	0	90.4
TOTAL CPAH	434	125.7	67.55	109.8	71.03	46.3	0	74.7

MARCH 16, 1995

COMPOUND	INF	CONCENTRATION			% REDUCTION			OVERALL % REDUCTION
		R1	R2	R3	R1	R2	R3	
NAPHTHALENE	25.1	4.21U	4.21U	4.21	91.6	0	0	91.6
ACENAPHTHYLENE	30	3.91U	3.91U	3.91	93.5	0	0	93.5
ACENAPHTHENE	3.3	110	64.79	63.5	0	41	1.98	0
FLUORENE	378	7.32	7.07U	7.07	98.1	52	0	99.1
PHENANTHRENE	1377	13.6	9.41	4.66	99	31	50.5	99.7
ANTHRACENE	344	9.33	5.38	3.99	97.3	42	25.8	98.8
FLUORANTHENE	1742	317	209.8	190	81.8	34	9.34	89.1
PYRENE	1526	237	152.8	125	84.5	35	18	91.8
BENZ(A)ANTHRACENE*	105	18.5	9.41	11.3	82.3	49	0	89.2
CHRYSENE*	158	58.6	34.95	36.1	62.9	40	0	77.1
BENZO(B)FLUORANTHENE*	56.3	32.5	18.56	15.3	42.3	43	17.5	72.8
BENZO(K)FLUORANTHENE*	29	19.7	10.15	9.66	32	49	4.83	66.7
BENZ(A)PYRENE*	31.8	19	7.71	8.28	40.2	59	0	74
DIBENZ(A,H)ANTHRACENE*	25.9	5.93	3.57	3.03	77.1	40	15.1	88.3
BENZO(G,H,I)PERYLENE*	15	9.67	6.39	5.26	35.5	34	17.7	64.9
INDENOPYRENE*	13	7.34	4.57	3.79	43.5	38	17.1	70.8
TOTAL PAH	5859	872	552.5	496	85.1	37	10.3	91.5
TOTAL CPAH	434	169	95.31	93	61	44	2.47	78.6

U = Number is 1/2 MDL. The new MDL for naphthalene is based on the average relative SD and highest dilution factor used.

(+) = Area counts are over the highest calibration standard.

* = Carcinogenic PAHs.

approximately 21 and 29%, respectively. No significant volatilization of PAH was evident in R1. Following operational changes initiated in March, including increasing the HRT from 10 to 20 days, overall PAH and CPAH removal rates increased to 91 and 77%.

R1 and R3 pH averaged 7.0, with R2 pH reduced to an average of 2. Dissolved oxygen in R1 and R3 varied depending on mechanical upsets; however, during optimal periods of operation the dissolved oxygen concentrations in these reactors was greater than 2 mg/L. Ammoniacal nitrogen was added to R1 to maintain a target concentration of greater than 40 mg/L in all reactors. The average ammoniacal nitrogen concentration throughout the system was approximately 80 mg/L. Phosphate addition to the system was not required due to the naturally high concentrations in the soil. The average orthophosphate concentration in all reactors was 150 mg/L.

Total heterotrophic counts in R1 and R3 ranged from 10^8 to 10^9 CFU/mL. Despite severe conditions maintained in R2, total microbial counts measured greater than 10^4 CFU/mL. The enumeration of naphthalene degraders was inconclusive.

No adverse impact on the operation of R2 was noted due to the addition of Fenton's reagent.

CONCLUSIONS AND RECOMMENDATIONS

The demonstration illustrated the potential effectiveness of combined bio-slurry treatment and chemical oxidation for the treatment of PAH-impacted soils. The results of the demonstration will be used to generate pilot- and full-scale cost estimates.

ACKNOWLEDGMENT

This project has been partially funded by the U.S. Environmental Protection Agency under contract no. CR821186-01-01.

REFERENCES

Brown, K. L., and J. Sanseverino. 1993. "Factors Affecting PAH Biodegradation in Liquid/Solids Contact Reactors." *Proceedings of the 86th Annual Meeting and Exhibition of the Air and Waste Management Association*, Denver, CO.

Davila, B., F. Kawahara, and J. Ireland. "Combining Biodegradation and Fenton's Reagent to Treat Creosote Contaminated Soil." In press.

Elizardo, K. 1991. "Fighting Pollution with Hydrogen Peroxide." *Pollution Engineering*, pp. 106-109.

Gauger, W. K., V. J. Srivastava, T. D. Hayes, and D. G. Linz. 1991. "Enhanced Biodegradation of Polyaromatic Hydrocarbons in Manufactured Gas Plant Wastes." *Gas, Oil, Coal, and Environmental Biotechnology Symposium*, Vol. 3. New Orleans, LA. pp. 75-92.

Kelley, R. L., W. K. Gauger, and V. J. Srivastava. 1991. "Application of Fenton's Reagent as a Pretreatment Step in Biological Degradation of Aromatic Compounds." *Gas, Oil, Coal, and Environmental Biotechnology Symposium*, Vol. 3. New Orleans, LA. pp. 105-120.

Sanseverino J., B. M. Applegate, J.M.H. King, and G. S. Sayler. 1993. "Plasmid-mediated Mineralization of Naphthalene, Phenanthrene, and Anthracene." *Applied and Environmental Microbiology, 59*:1931-1937.

Novel Slurry Bioreactor with Efficient Operation and Intermittent Mixing Capabilities

Keith E. Stormo and Lee A. Deobald

ABSTRACT

Designing, developing, and testing our novel soil-slurry reactor has given significant insight into the problems associated with low-energy input reactors for use in aerobic and anaerobic remediation of explosives-contaminated soil. Through continued testing and modification of our initial design on laboratory and pilot scales, we have developed an efficient and effective soil-slurry reactor that can operate with sand at 70% wt/vol (700 g/L), or soil at 50% wt/vol (500 g/L). The bioreactor has four basic modes of operation. The water and air pumps are controlled by a repeat-cycle timer that allows either or both pumps to be operated intermittently or continuously, and to propel the blade through the sand or soil slurry. The fluid exiting the nozzle propels the stirrer, brings solids into suspension from the bottom, and rapidly mixes them. If the circulation is for a short time, the solids are mixed, but not so much that they are fully suspended and sucked into the recycle outlet. When foaming is a problem with a specific matrix, the addition of air can be intermittent or by bubbleless oxygenation with a membrane gas-transfer system. Pumping headspace gasses increases vertical mixing in anaerobic operations and reintroduces volatiles into the slurry for further degradation.

INTRODUCTION

Complete, efficient, and cost-effective disposal of explosives waste residues has been an ongoing problem for federal and private manufacturers of ordnance. The need for a new disposal technology has been heightened by the current emphasis on waste minimization in both the military and private sectors, and by the discovery of these ordnance compounds in soils and groundwaters at military depots and arsenals across the United States.

Although the extent of soil contamination by munitions waste is not precisely known, it has been estimated to be over 5×10^6 tons of soil at over 26 sites (Myler

& Sisk 1991). Because of the toxicity of these materials, it is essential to clean up these sites before migration of the contaminants worsens the problem. The only proven method of remediation of the soils is incineration, the cost of which is estimated to be $300 to $600 per yd^3 of soil (Myler & Sisk 1991). Composting the soil has been proposed as an economical alternative to incineration, with cost estimates of $200 per ton of soil (Weston 1993b). However, recent actual costs are higher; and composting has not been established as a viable, large-scale remediation technology. Slurry reactors have been used for remediating of a number of types of contaminated soils and have shown some significant advantages.

Processes for reacting solid-phase materials include heap leaching for low-grade ore, composting (including explosives remediation), solid-state fermentation, and slurry reactors (including explosives remediation). In all of these processes, particle size and liquid or gas interaction are the primary governing variables. The particles in heap leaching are quite large and the time frame for the piles is extremely long (Andrews 1990). Composting for contaminated soil normally increases the total volume considerably (Weston 1993a) and often includes significant time and material handling.

SLURRY REACTORS

Slurry reactors were initially developed in the 1950s for use in chemical-process industries. Major developments in engineering modeling and in design of three-phase slurry reactors were achieved in the 1960s and 70s (Beenackers & van Swaaij 1986; Chaudhari & Ramachandran 1980). Correlations were developed to determine the mechanical and gas-induced agitation necessary to suspend solid particles and to minimize intraparticle diffusion. Overall reaction rates were derived on the basis of the intrinsic reaction kinetics and mass-transfer effects for reactant desorption and diffusional resistances within solid particles and between phases. Complex reaction pathways were modeled, including product inhibition, consecutive reactions, and complex kinetics.

The majority of this development and modeling work was done for chemical processes where mass-transfer effects were the primary limiting steps. In biological degradation of recalcitrant compounds such as trinitrotoluene (TNT), other factors may be rate limiting and therefore reduce the need for high-energy mixing and/or complete suspension of the soil particles. The variables affecting reaction rate and reactor design have been well described for catalytic slurry reactors (Fogler 1992). Typically these reactors operate at 5 to 10% solids with 100 to 200 mm particles (Beenackers & van Swaaij 1986).

Soil-slurry reactors differ in having larger particles, with strongly adsorbed pollutants, and are operated at a higher solids loading. Some specific reactor designs, such as the EIMCO soil-slurry reactor (McCleary 1992), often require fine pulverization of the soil by multiple grinding in a wet ball mill prior to introduction into the reactor (Lauch et al. 1992; Lewis 1993). This allows complete mixing in their reactor, but a significant capital and operational cost is associated with such a completely fluidized system (Andrews 1990; Casey 1986; Jewell et al.

1981). Soil-slurry reactors are operated much like the slurry reactors used for coal cleaning and leaching metal ores (Beenackers & van Swaaij 1986; Shah 1979).

Considerable literature is available describing slurry reactors for municipal and farm sewage digestion, but the total solids are usually below 10% (Zauner 1989). The density of sewage sludges is much closer to water than is the density of soil and, therefore, the mixing method and design of these stirred reactors can be significantly different from that of soil-slurry reactors. Many sewage digester designs are unstirred, and the predominant mixing mechanism is the CO_2 and CH_4 gas generated in the reactor (Jewell et al. 1981). The mixing occurs as these gas bubbles rise through the slurry. Propeller-type mixers are sometimes added for more thorough mixing and to try to maintain the solids in suspension. The current design of most soil-slurry reactors is to finely pulverize the material and try to keep it in suspension with significant power input to shaft stirrers, aerators, recirculation pumps, or a combination of these methods (Bogart 1991; Brox & Hanify 1991; EPA Fact Sheet 1993). Alternative approaches are not to mix at all (Craigmill et al. 1988), or to mix just occasionally.

In aerobic soil-slurry reactors it is difficult to maintain high oxygen concentrations because of the tendency for gas bubbles to coalesce. Also, since the reactors are usually low in profile, there is a very short liquid-gas contact time and a small surface-area-to-volume ratio of the bubbles. Mechanical agitation is usually required to disperse gas bubbles and give smaller gas bubbles; but, as the solids concentration increases, the agitation effect decreases (Andrews 1990).

The operational and design variables that have major influence on soil-slurry reactor performance are the desorption rate from the solid phase, diffusion in the liquid phase, and oxygen supply for aerobic systems. In both aerobic and anaerobic soil-slurry reactors, the rate-limiting steps should be identified and minimized in the design, while capital and operational cost concerns and residence times for complete degradation are maintained. The following factors are often major considerations in design and operation:

1. Desorption rate depends strongly on soil and pollutant chemical structure. Adsorption isotherms and desorption rates can be determined by sequential batch washing (Linz et al. 1991).
2. Diffusion of pollutants to the aqueous phase depends on particle size and porosity. Desorption rates can be determined as a function of particle size. Also, pollutant distribution should be determined as a function of particle size (Compeau et al. 1991). Particles greater than 400 mm cannot be easily suspended, and large particles need to be treated by washing or a size-reduction operation. The diffusion of intermediates and additional reactants in a consortium-mediated reaction is also of primary importance.
3. Oxygen supply is dependent on aeration rate, solids loading, agitation, and gas residence time; it can be determined from oxygen transfer correlations and the physical properties of the slurry. Operational problems such as diffuser clogging, solids settling, and materials corrosion must be avoided. The design of agitators and aeration diffusers are critical to the performance of soil-slurry reactors, especially at high solids loading.

REACTOR DESIGN SUMMARY

The bioreactor as designed (Figure 1) has four basic modes of operation. The water and air pumps are controlled by a repeat cycle timer that allows either or both pumps to be operated intermittently or continuously.

In the first mode of operation, water is recirculated through the blade, which propels the stirrer through the slurry, bringing fresh liquid in contact with the soil for rapid mixing. This can be anaerobic, or aerobic with a bubbleless oxygenation system. The water exiting the nozzle propels the stirrer and also brings solids up into suspension from the bottom and rapidly mixes them. If the circulation is for a short time, the solids are mixed but not so much that they are fully suspended and sucked into the recycle outlet.

In the second mode of operation, the air pump can be added to the operation, allowing very efficient aeration for aerobic reactions and increasing vertical mixing with the rising bubbles. Because the water and air are both under pressure, the amount of oxygen dissolved in the water can be increased considerably above saturation at atmospheric pressure. For high biological oxygen-demand systems, this will allow significantly higher oxygen mass-transfer rates than for a normally bubbled and stirred reactor. If foaming is a problem with a specific matrix, the addition of air can be intermittent.

In the third mode of operation, the air pump will pump headspace gases into the water recycle flow, thereby increasing the mixing rates and allowing

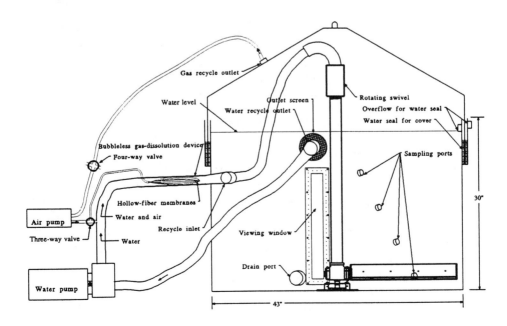

FIGURE 1. Schematic setup of the bioreactor exterior with the associated pumps and ports.

more complete degradation of volatile compounds in the slurry. This will also increase vertical mixing in anaerobic operations and reintroduce volatiles into the slurry for further degradation.

The fourth mode of operation is a combination of the above modes, which can be operated on an intermittent basis to reduce operational costs or to maintain microaerophilic conditions. Also, by switching between the second and third modes with the four-way valve, there will be no excess aeration or lost volatiles. This system can be interfaced easily with a computer for active control of the operating mode.

Nearly all the operational testing of our reactor was done on intermittent basis, which consisted of a cycle of 3.6 s on time for the water and air pumps, and then an off time of 2 min and 9 s. A variety of different blade designs was used during the testing of the slurry reactor. Blade design number 3 (Figure 2) had very good mixing and excellent rotation. The rotation was about 10 degrees per cycle at 20% sand, using a 0.5-in. (1.3-cm) propulsion nozzle. The observations made with blade design 3 showed how important it was to fully fluidize the sand in the immediate area of the blade. Even small stable regions of sand at the blade leading edge could keep the blade from turning. Review of videos and tactile monitoring of the sand at the leading edge caused us to try a different approach for design 4 (Figure 2). This design has only two fluidizing nozzles pointed in a radial direction, one near the hub and the other at the midpoint of the blade, both pointing outward and slightly down. There is one propulsion

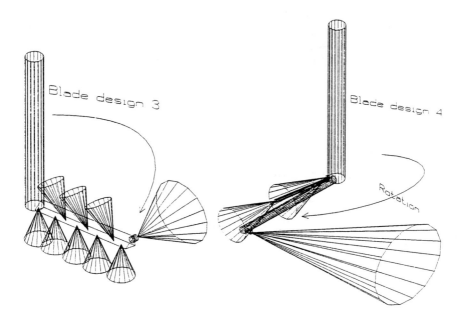

FIGURE 2. Blade designs 3 and 4 in wireframe with simulated flow out of the fluidization nozzles and the rotation nozzle.

nozzle at the outboard end for rotational torque. This last nozzle design has impressive performance characteristics. The level of the sand throughout the bottom of the reactor is more level, without a hill just ahead of the blade as was the case with earlier blade designs. The blade rotated about 25 degrees for each cycle and would go around about once every 30 min. There was very good mixing and good rotation even at 70% wt/vol (which is 52% wt/wt), with a rotation of about 15 degrees per cycle. This blade configuration was operated continuously for about 1 month without any plugging or problems of any kind. A relatively flat upper surface of the sand was produced with this blade design in the reactor (Figure 3). The blade turned so much there never was much of a hill ahead of the blade or a depression behind the blade.

Our novel design is both efficient and cost-effective. Initial tests have shown excellent potential for a combination aerobic/anaerobic reactor that will have very efficient mixing while significantly reducing capital and operational costs. We are continuing to test our novel reactor with sands, soils, and slurries. This reactor has the additional benefit of operating in a wide variety of modes to optimize the biological or chemical breakdown of contaminants in water, sludge, soil, or sand slurries.

ACKNOWLEDGMENTS

This work was funded by SBIR contract #N47408-93-7359 from the Naval Construction Battalion Center at Port Hueneme, California. Patent rights were retained by Innovative BioSystems, Inc., and the patent on the bioreactor design is pending. Additional details on the design and performance of this reactor can be obtained from Innovative BioSystems or from the final report of our SBIR Contract.

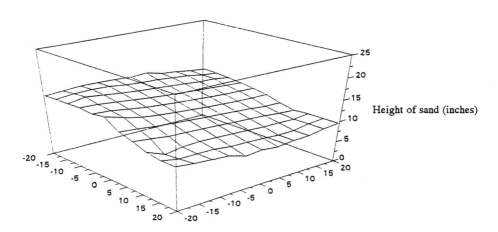

FIGURE 3. Upper sand surface plotted in three dimensions, with blade 4 in the slurry reactor at 70% sand.

REFERENCES

Andrews, G. 1990. "Large-Scale Bioprocessing of Solids." *Biotechnol. Prog.* 6:225-230.

Beenackers, A.A.C.M., and W.P.M. van Swaaij. 1986. "Slurry Reactors, Fundamentals and Applications," pp. 463-538. In H. I. deLasa (Ed.), *Chemical Reactor Design and Technology.* Kluwer Academic Publishers, Boston, MA.

Bogart, J. D. 1991. "Soil and Sludge Treatment Apparatus and Method Including Agitation, Aeration and Recirculation." U.S. Patent 5055204.

Brox, G. H., and D. E. Hanify. 1991. "Bioremediation of Hazardous Wastes in an EIMCO Biolift Slurry Reactor." *Presented at the 5th Annual Bioprocess Engineering Program Symposium*, Atlanta, GA. December 1991.

Casey, T. J. 1986. "Requirements and Methods for Mixing Anaerobic Digesters." In A. M. Bruce, A. Kouzeli-Katsiri, and P. J. Newman (Eds.), *Anaerobic Digestion of Sewage Sludge and Organic Agricultural Wastes.* Elsevier Science Publishing, New York, NY.

Chaudhari, R. V., and P. A. Ramachandran. 1980. "Three Phase Slurry Reactors." *AIChE Journal* 26:177-201.

Compeau, G. C., W. D. Mahaffey, and L. Patras. 1991. "Full-Scale Bioremediation of Contaminated Soil and Water." In G. S. Sayler, R. Fox, and J. W. Blackburn (Eds.), *Environmental Biotechnology for Waste Treatment.* Plenum Press, New York, NY.

Craigmill, A. L. , W. L. Winterlin, and J. N. Seiber. 1988. "Biological Treatment of Waste Disposal Sites," pp. 207-213. In J. S. Bridge and C. R. Dempsey (Eds.), *Pesticide Waste Disposal Technology.* Noyes Data Corp., Park Ridge, NJ.

EPA Fact Sheet. 1993. *SITE Demonstration of the J. R. Simplot Ex Situ Bioremediation Technology for Treatment of Nitroaromatic Contaminants at the Bowers Field Site in Ellensburg, Washington.* U.S. Environmental Protection Agency, Washington, DC. August.

Fogler, H. S. 1992. *Elements of Chemical Reaction Engineering.* Prentice-Hall, Englewood Cliffs, NJ.

Jewell, W. J., S. Dell'Orto, K. J. Fanfoni, T. D. Hayes, A. P. Leuschner, and D. F. Sherman. 1981. "Low Cost Methane Generation on Small Farms." In D. L. Wise (Ed.), *Fuel Gas Production From Biomass.* CRC Press, Boca Raton, FL.

Lauch, R. P., J. G. Herrmann, W. R. Mahaffey, A. B. Jones, M. Dosani, and J. Hessling. 1992. "Removal of Creosote From Soil by Bioslurry Reactors." *Environ. Prog.* 11:265-271.

Lewis, R. F. 1993. "SITE Demonstration of Slurry-Phase Biodegradation of PAH Contaminated Soil." *Air and Waste.* 43:503-508.

Linz, D. G., E. F. Neuhauser, and A. C. Middleton. 1991. "Perspectives on Bioremediation in the Gas Industry." In S. G. Sayler, R. Fox, and J. W. Blackburn (Eds.), *Environmental Biotechnology for Waste Treatment.* Plenum Press, New York, NY.

McCleary, H. 1992. "EIMCO Reactors: Load 'em Up, Move it Out." *The Bioremediation Report,* 1(9).

Myler, C. A., and W. Sisk. 1991. "Bioremediation of Explosives Contaminated Soils," pp. 137-146. In G. S. Sayler (Ed.), *Environmental Biotechnology for Water Treatment.* Plenum Press, New York, NY.

Shah, Y. T. 1979. "*Gas-Liquid-Solid Reactor Design.*" McGraw-Hill, Inc., New York, NY.

Weston, Roy F. 1993a. *Windrow Composting Demonstration for Explosives-Contaminated Soils at the Umatilla Depot Activity, Hermiston, Oregon.* Final Report No. CETHA-TS-CR-9304350 for the U.S. Army Environmental Center.

Weston, Roy F. 1993b. *Windrow Composting Engineering/Economic Evaluation.* Final Report No. CETHA-TS-CR-93050 for the U.S. Army Environmental Center.

Zauner, E. 1989. "Anaerobic Digestion of Agroindustrial Solid Waste." In A. Mizrahi (Ed.), *Biological Waste Treatment.* Alan R. Liss, Inc., New York, NY.

Enhanced Treatment of Refinery Soils with "Open-System" Slurry Reactors

James W. Blackburn, M. Kathryn Lee, and William C. Horn

ABSTRACT

Refinery site cleanups of residual hydrocarbons arising from long-term operations have become a concern. Because contaminated soil has been generated over many years from spills of many types of materials, it is often difficult to identify the actual spilled material. Because many of these materials are weathered, the less degradable fractions can predominate, creating a challenge for bioremedial process solutions. "Open-system" slurry reactors were run with an aged refinery soil after a 6-month period of field bioremediation in which 23% TPH removal resulted. The open system (a system where the liquid medium was replaced daily and the solids were retained in the reactor for 2 weeks) achieved 60 to 80% total petroleum hydrocarbon (TPH) removal based on the initial, prefield bioremediation soil concentration. A process concept twice as effective as other bioremediation schemes has been devised that takes advantage of the formation and removal of small black particulate solids in an "open" or continuous slurry reactor configuration. These small black particles are chemically or biologically produced in the open system and with their small size and low density are easily elutriated from the bioreactor as the liquid medium is changed. A statistically designed experiment has determined optimal values of nutrients, temperature, and mixing.

INTRODUCTION

Building on an early lead that suggested high contamination removals (60 to 80% TPH by EPA Method 418.1) could be achieved in "open" reactor systems, lab studies were designed to investigate the potential for further remediation of soils already field-bioremediated. Soil that had undergone months of field plot bioremediation, reducing its contamination by approximately 23%, was used as the starting material for this work. Of particular interest was the observation in earlier experiments of fine black particles, high in TPH, forming in "open-system" bioreactors. A fractional factorial experiment was completed to investigate the effects of agitation, temperature, and nutrients on additional biological removal of organics beyond the postfield-bioremediation soil endpoints at the time of sampling.

Open and closed systems are terms used in the biological and environmental sciences that refer to systems where a particular phase or physical compartment in a multiphase or multicomponent system flows into and from the system. In this paper, open systems refer to systems where both air and water phases flow through the reactor but soil is contained. The operation included daily settling of the solids with decantation and replacement of the aqueous medium. Any solids or "fines" suspended in the aqueous phase were removed from the system with the decantation of the fluid. These suspended solids were handled in one of three ways: near-complete solids recycle, partial solids recycle, and no solids recycle back to the reactor.

This work provided evidence that, for at least one soil, organic particles are produced in this process, not merely released from soil aggregate particles. There is some prior support for this finding. Unspecified radical oxidation of polycyclic aromatic hydrocarbons (PAHs) leads to a wide variety of metabolic products, including phenolic oxidation products, polymerizates, and conjugates (Bollag and Loll 1983). Essentially 100% of radiolabeled benzo(a)pyrene added to a soil containing *Phanerochaete chrysosporium* (a white-rot fungus) was incorporated into the soil organic matter and lost its chemical identity (Qiu and McFarland 1991). In work on composting systems, radiolabeled anthracene was extracted sequentially with sonicated ethyl acetate; then an alkaline hydrolysis for humic acid extraction was performed. With high label recoveries, the composting system enhanced mineralization, but also generated up to 40% of bound residues (indistinguishably incorporated into the soil organic matter). A free radical pathway was proposed (Mahro et al. 1994). In work done with ^{14}C-phenanthrene in manufactured gas plant (MGP) soils, 15 to 20% of the labeled carbon from the added compound was converted to soil-bound residues in batch flask bioremediation experiments ranging from 72 to 240 h (Blackburn et al. 1991). Although no reports specifically propose that organic-rich particles are formed during bioremediation, published reports detail losses of PAH compounds into soil-bound fractions.

EXPERIMENTAL METHODS

Each reactor system consisted of a 1.0-L indented and clamped kettle reaction flask with an O-ring seal and a reaction kettle cover with four 24/40 necks. A Model TS2010 Lightnin' Mixer equipped with a stainless steel stirring shaft and turbine blade was positioned in the center neck. The other ports were equipped with a glass joint/Teflon™ stopcock gas vent, a rubber septum, and an air inlet tube.

A general operational sequence for all of the open-system reactor experiments began with addition of soil and aqueous medium containing nutrients to the reactor. Mixing (agitation) and aeration were started, beginning the approximate 21-h reaction phase. At the end of the reaction phase, agitation and aeration were stopped and soil and solids were allowed to settle for 2 to 3 h. A measured amount of liquid effluent representing a relatively solid-free aqueous phase containing small-diameter suspended particles was removed. The amount removed varied slightly from experiment to experiment. Fresh replacement aqueous medium was added to the reactors. An additional volume of water was added back to the reactor to replace the effluent removed.

For experiments requiring near complete solids recycle, the aqueous phase removed from the reactors was centrifuged to pellet the solids, and the solids were then returned to the reactor; 15 experiments were run in this way. In experiments with partial solids recycle, the aqueous phase was allowed to settle overnight while the reaction continued with fresh medium and water. The next day, that aqueous phase was decanted from the fine solids and filtered, and the filter cake was transferred to the reactor. Some solids remained on the filter and were not returned to the reactor. Mixing and aeration were then resumed. Six experiments were run in this way. For experiments with no solids recycle, solids were centrifuged and dried. These solids were not returned to the reactor. Four experiments were run in this way.

The 15 fractional factorial experiments were designed to optimize TPH removal by biodegradation for the three operating variables of agitation, temperature, and nutrient concentration. The experimental design enabled the generation of response surfaces of TPH as a function of the independent variables. Combinations of the three variables — agitation (200, 350, and 500 rpm), temperature (22.5, 29.7, 37°C) and nutrient level (10, 30, or 50:1 C:N or P) were used to identify optimal reaction conditions. In a related closed-system or batch experiment, liquids were not removed from the reactor until the experiment's end at 2 weeks. Nutrients were added daily as concentrated solutions to simulate the dosing of the open systems. The closed reactor experienced the same agitation and operating conditions (21 h on and 3 h off) cycle that was part of the open-system experiment.

RESULTS

Improved TPH Removal from Open Systems with No Solids Recycle

Figure 1 presents the TPH resulting from the three types of experimental operation: (1) near-total solids recycle, (2) partial solids recycle, and (3) no solids recycle. Also shown is the initial concentration of TPH from many samples of homogenized field soil and the average concentration after many months of field plot bioremediation. Field bioremediation of this soil resulted in about 23% TPH removal at the time of this sampling. Typical relative mean standard deviations of replicates of each of these homogenized samples were in the 5% range.

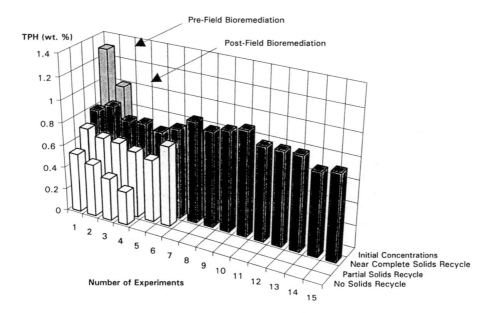

FIGURE 1. TPH results from three modes of open-system treatment.

To investigate the limits of additional biodegradation to the postfield-bioremediated soil, and to identify the main factors responsible, a 15-experiment fractional factorial design was implemented in the near-total-solids recycle mode. Statistically significant ($P = 0.05$) additional biodegradation of TPH in the open-system reactors of about 15 to 20% was revealed. Partial solids recycle resulted in an additional TPH removal with total solids recycle reaching 60 to 80% TPH removal based on the initial, prefield-bioremediation concentrations. Student's t tests reveal that these three data groups were significantly different from one another at the 95% confidence level. As can be seen by the means and variability of these data groups, the solids handling technique was a much more important factor than agitation, temperature, or nutrient concentration. High removals can be achieved if the suspended solids in the open-system effluent are removed and are not recycled to the reactor.

A Concept for an Open-System Slurry Bioreactor Process

Figure 2 shows a proposed treatment concept for removal of TPH from soils. Soil, nutrient buffer, and water are mixed in a reactor at atmospheric pressure for an undetermined time, but, based on this work, a soil residence time of less than 2 weeks. Water residence time will probably be much shorter, perhaps approaching 1 to 2 days. The soil/water ratio in this work was 10%, but ratios up to 30% should be possible.

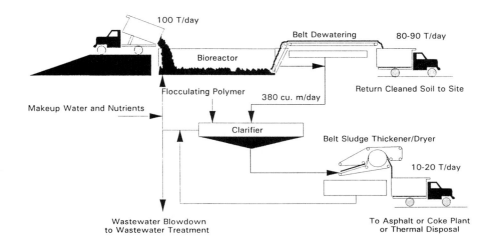

FIGURE 2. A concept for an open-system slurry bioreactor process.

The slurry mixture is sent to a separator where solids are settled and removed. These treated soil solids are low in TPH (~0.3%). They consist of about 80 to 90% of the total mass of initial contaminated soil. The aqueous phase containing the black particles is sent to a clarifier where the particles are settled. A polymer additive could be needed to create efficient separation. The underflow is high in TPH (~2.2% or more) and may have a relatively high heat content. Disposal options could include off-site disposal, but recycle to refinery slop tanks or heavy ends processes may also be possible. Water from this process may be sent to a wastewater treatment facility and potentially recycled back to the process.

Optimal Agitation, Temperature, and Nutrients for Biodegradation

The 15 experiments with near complete solids recycle were analyzed to establish the effects of agitation, temperature, and nutrient levels on biodegradation of an already bioremediated soil. TPH losses could be attributed to biodegradation because other organic carbon losses were negligible. Volatilization in this weathered, bioremediated soil was shown to be negligible in earlier work, and losses of dissolved organic carbon in the daily effluent, ranging from 10 to 30 mg/L, also were negligible.

Figures 3 and 4 present response surface plots of linear statistical models that were derived from the results shown in Figure 1. With one dependent variable (TPH) and two independent variables shown (agitation and C:N ratio, or temperature and agitation), it seems clear that temperatures near 30°C were more effective than lower or higher temperatures, carbon:nitrogen ratios of near 30:1 were better than either 10:1 or 50:1, and reactor mixing speeds of 200 rpm were better than higher speeds in TPH biodegradation. Upper confidence limits for

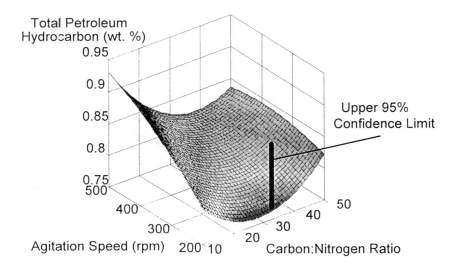

FIGURE 3. **Response surface of TPH vs. agitation and nutrient ratio for the fractional factorial experiment.**

selected points in these highly controlled experiments are shown to stress that variability in these tests was still fairly large relative to the differences seen and the response model should be taken as an expected value with a rather large associated variability range.

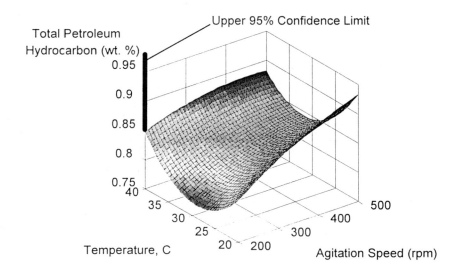

FIGURE 4. **Response surface of TPH vs. agitation and temperature for the fractional factorial experiment.**

Particles Produced During
Open System Operation

Other bioremediation work has identified black particles in contaminated soils. However, an important part of this process is the realization that black organic-rich particles are generated during the open system operation. Proof of this assertion can be seen in the following two photographs. Figure 5 shows a centrifuge pellet from the open-system reactor aqueous effluent in near total solids recycle mode with black particles sedimenting last. This photo demonstrates that the black particles are a separate phase of low density. Figure 6 shows the aqueous effluent from a similar closed or batch run where the aqueous effluent was decanted at 2 weeks. No black particles were present in any batch experiments.

The black particles were not found in the batch runs, operated under similar conditions as the open system runs, and thus did not result from attrition of soil aggregates and must have been formed during the bioreactor operation. In the open systems, black particles were noticed at the 3-day time point but not under 3 days. These apparent kinetics also support the observation of chemical or biological particle production during open-system bioremediation.

CONCLUSIONS

Work with open-system reactors on postbioremediated, aged refinery soil has demonstrated that:

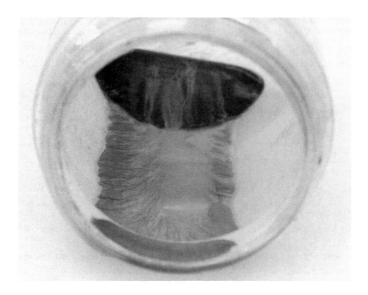

FIGURE 5. Centrifuge pellet from open-system effluent.

FIGURE 6.   Centrifuge pellet from batch reactor effluent.

1. Additional biodegradation can be achieved after field bioremediation, but the extent is limited.
2. A process based on additional biodegradation in an open-system mode with no recycle of effluent solids can achieve high removals, better than other bioremediation techniques.
3. Solids handling (recycling or removal of organic particles) is the most significant operating variable, but temperature, nutrients, and agitation can affect biodegradation. Temperatures near 30°C, nutrients in the 30 carbon:nitrogen range, and agitation of 200 rpm in the 1-L reactor achieve the greatest TPH biodegradation.
4. Effluent solids in the open system are generated during the bioremediation cycle and not merely released from aggregate attrition.

REFERENCES

Blackburn, J. W., P. M. DiGrazia, and J. Sanseverino. 1991. *Treatability and Scale-up Protocols for Polynuclear Aromatic Hydrocarbon Bioremediation of Manufactured Gas Plant Sites.* GRI Report 91/0193, Chicago, IL.

Bollag, J. M., and M. J. Loll. 1983. "Incorporation of xenobiotics into soil humus." *Experientia* 39: 1221-1231.

Mahro, B., G. Schaefer, and M. Kästner. 1994. "Pathways of microbial degradation of polycyclic aromatic hydrocarbons in soil." In R. E. Hinchee, A. Leeson, L. Semprini, and S. K. Ong (Eds.), *Bioremediation of Chlorinated and Polycyclic Aromatic Hydrocarbon Compounds,* pp. 203-217. Lewis, Boca Raton, FL.

Qiu, X., and M. J. McFarland. 1991. *J. Hazard. Wastes Hazard. Mater* 8: 115-118.

Slurry Biotreatment of Organic-Contaminated Solids

John A. Glaser, J.-W. Jim Tzeng, and Paul T. McCauley

ABSTRACT

The issue of environmentally acceptable end points for treatment is beginning to be pursued in earnest. Because analysis of achievable end points is a technology-based assessment, it is important that the performance of the suitable treatment technologies be optimal. The use of slurry bioreactors to treat contaminated solid-phase materials offers considerable potential to reach currently targeted treatment goals (U.S. EPA 1990). An assessment of current slurry bioreactor applications indicates that there is no consensus of thought leading to treatment objectives and process design. This paper presents an overview of current slurry bioreactor practice to develop a basic set of objectives and criteria useful to a more complete definition of slurry practice. In some cases, useful information concerning past applications is exceptionally sparse. Our slurry process evaluation research has assisted the identification of pitfalls and problems associated with the application of this technology.

INTRODUCTION

The treatment of solids (soils, sludges, or sediments) contaminated with organic pollutants is a major challenge to the cleanup of contaminated hazardous waste sites. The use of biological activity to degrade/detoxify contaminated solids shows strong potential as a treatment technology.

Biological treatment of contaminated soil slurries may offer the "optimal" technology for soil bioremediation. All the components required to biotransform the pollutants into desirable end products are present in the slurry treatment process. For aerobic processing, a slurry bioreactor system consists of an aerated suspension of contaminated solids formed from the agitated mixture of solids with water. Examples of slurry treatment have been conducted in reactor systems consisting of mixed tanks or lined lagoons. Slurry systems provide conditions of improved contact between the pollutant and microorganism(s) responsible for the desired biotransformation. Due to the intense input of energy to suspend the solids, treatment is conducted to minimize reactor residence time through optimal mixing technology. Solid-phase biotreatment, such as composting or the use of formal land treatment units, cannot compete with slurry treatment for the extent of treatment in short time frames. The current level of slurry treatment effectiveness demonstrates that the relative impacts of many factors affecting performance are poorly understood.

Slurry bioreactor treatment of solids has not sufficiently progressed to be a durable, reliable, and cost-effective treatment option. The use and design of slurry bioreactors requires further testing of treatment performance before detailed specifications for optimum treatment effectiveness can be achieved. Slurry bioreactors are often considered to be an off-the-shelf technology for conventional, nonhazardous materials (Ross 1990). Little effort has been expended, however, to either understand or optimize the technology for decontamination of hazardous wastes.

A survey of currently available slurry treatment applications shows a diversity of forms responding to a wide variety of process considerations. The pollutant classes considered for treatment span the range from volatiles to nonvolatiles. The preponderance of applications are devoted to semivolatile or nonvolatile treatment. Due to the range of applications there appears to be little attempt to develop a consensus of thought concerning the basic design and application of the technology. Such a state often leads to misapplication of the technology, because uninformed parties are persuaded that the technology can meet the treatment specification of situations that it was not designed to answer.

We have reviewed reported applications to determine the critical factors of operation and engineering that support optimal slurry reactor utilization. We have used three criteria to assemble the components of basic operation: free product and debris exclusion from the feedstock, reduction of feedstock treated in the slurry reactor, and minimization of residence time in reactor.

TREATMENT COMPONENTS

The component parts of reported slurry operation are varied with each application. Many operations have only a single focus of operation which is the slurry reactor itself.

Soil Characterization

The contaminated soil must be evaluated for the soil characteristics that may limit its suitability for slurry treatment. A series of factors contribute to the decision process: (1) particle size distribution and contamination as a function of size distribution, (2) texture/composition (clay, silt, sand), (3) soil nutrients (nitrogen, phosphorous, etc.), (4) pH, (5) cation exchange capacity (CEC), (6) metals, (7) total organic carbon (TOC), (8) free product determination, and (9) microbiological analysis for populations of indigenous degraders.

Feedstock Considerations

Free product associated with the contaminated solids is best removed before size classification. Degradable free product will only extend the residence time of the solids in the reactor and complicate the material handling of the classification scheme. Most of the organic pollutants are hydrophobic and difficult to disperse in water suspension. There have been attempts to slurry treat viscous, hydrophobic sludges that did not perform optimally as a slurry application (Ecology and Environment, Inc. 1995). Because the sludge floated, it adhered to the reactor wall and led to malfunction of the agitation system. Any attempt to slurry-treat such material will require the use of dispersants to increase the surface area of the fluid. If such materials are easily separated, it would be more productive to use other forms of treatment.

Contaminated solids vary widely in particle size distribution. Exclusion of large-dimension materials such as debris and sands is important for the flow of solids through the processing units for material handling and mixing considerations (Griffiths 1995). Some applications have milled large materials into particulate matter of acceptable dimensions (U.S. EPA 1993). However, the contamination of such large-dimension materials is small relative to the surface area; thus, we believe that this processing step is unnecessary. The debris materials are more advantageously treated by means of washing techniques (U.S. EPA 1991). The milling of oversize materials does not necessarily avoid any of the agglomerating tendencies that may be encountered during slurry operation.

The chemistry and physics of pollutant sorption to solids phases is reasonably well understood (Sposito 1984). The energy of pollutant sorption to sand fraction solids having low surfaces areas and silicate attachment chemistries indicates that these materials are most productively treated by washing techniques. Furthermore, the abrasive nature of the sand fraction (+ 200 mesh) will cause excessive wear to moving parts. Even in cases where extensive feedstock preparation was conducted to eliminate the sand fraction, impeller blades were severely worn by small quantities of the sand fraction (Woodhull 1995). The wear may become excessive as the rotational speed of the impeller is increased. Loss of suspension through the excessive wear of the blade could lead to operational failure, because sedimented solids are very difficult to resuspend.

Selection of feedstock can be accomplished after free product and/or debris are removal by initial dry classification through 0.5- or 0.25-in. (1.3- or 0.6-cm) screen. The minus fraction is then processed as slurry through a hydrocyclone to accomplish a rough-cut removal of the sand fraction. Further classification of solids can be conducted by means of wet screening operations. The plus fraction of the classification scheme can be treated through washing technology. This preprocessing scheme offers removal of unsuitable components of the raw feedstream, and any solids aggregation has been dispersed. Inadequate dispersion of the solids can lead to extended release of solids-sorbed contamination, thereby increasing reactor residence time (Glaser et al. 1995a).

Reactor Considerations

Two general treatment configurations are employed to treat contaminated solids in slurry conditions: aerated lagoon and in-vessel (mixing-tank, airlift, and fluidized-bed) reactors. The lagoon designs generally use existing impoundment geometries that are poorly designed for optimal suspension. The lagoons are often rectangular in shape, which permits considerable portions of the solids to deposit in the remote areas of the lagoon. Placement of surface aerators is insufficient to overcome this disadvantage for most applications. These limitations become obstacles to process control for treatment to specified end points.

The in-vessel reactor configurations are generally mixing-tank, airlift, and fluidized-bed designs. These reactors are generally better controlled than the lagoon operations. Hydrodynamics in these reactors are fairly well investigated for optimal control of treatment. Higher control of aeration, suspension mixing, temperature, and emission control are features of these reactors. The geometry of these reactors is a matter of concern because many operations use flat bottom or rectangular tanks that are known for their sedimentation problems. Most operations are conducted in batch mode rather than continuously, although sequential reactor treatment has been employed in a continuous fashion (Castaldi & Ford 1992).

The reactor geometries encountered in the various reviewed reports appear too often to be expedient designs for most situations despite the availability of commercial equipment. The use of rectangular tanks has been troublesome because suspension sedimentation occurs in the corners and, if not aggressively managed, leads to operational problems (Bourquin 1995). Applications using existing lagoons require daily dredging to assist the suspension of sedimented solids (Woodward 1995). Typical monitoring and operating conditions necessary for treatment assessment are dissolved oxygen (>2 mg/L), pH (range of 6 to 9), temperature (recorded daily), treatment duration (variable), nutrients (C:N:P ratio = 100:10:1), and antifoam (as specified).

Suspension Criteria

The degree of particle suspension by impeller agitation is one of the most important factors affecting treatment efficiency. However, the criterion for optimal slurry agitation has not received the required attention. Mixing intensity by the agitation action can be well quantified by the power characteristics of specific impeller systems (Oldshue 1983). The power input from an impeller system dictates not only the homogeneity of the slurry medium but also the degree of particle suspension.

Four flow regimes, i.e., nonsuspension, semisuspension, off-bottom suspension, and complete suspension in increased order of impeller rotational speed, are identified from the viewpoint of particle suspension characteristics. A flow regime map, as shown in Figure 1, can be constructed by using power input as the primary parameter. Identification of flow regime under specific operating condition is of particular importance to the assessment of treatment efficiency

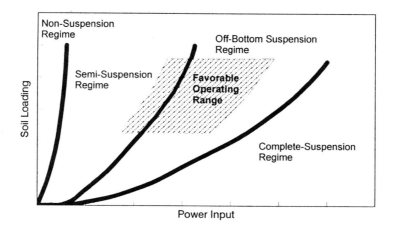

FIGURE 1. Flow regime map of particle suspension in soil-slurry bioreactor.

and energy consumption. The criterion for optimal mixing of slurries has not received the required attention. Five criteria have been advanced for the chemical processing industry: (1) maximum uniformity of suspension; (2) complete off-bottom suspension; (3) complete on-bottom motion of all particles; (4) filleting, but no progressive fillet formation; and (5) height of suspension.

It should be noted that selecting one of the above criteria for specific applications requires evaluation. Success in the operation of the slurry bioreactor relies on appropriate engineering designs of agitation systems and thorough consideration of the solids properties, including the following: (1) cumulative particle size distribution, (2) percent solids, (3) percent suspension, (4) weight percent ultimate suspended solids, and (5) percent ultimate weight-percent settled solids. Operation under the complete suspension regime for achieving maximum uniformity of particle suspension in the slurry may not be necessary because the treatment efficiency may be marginally increased. This is due to the fact that mass transfer resistance between particles and the bulk liquid phase is not the only rate-limiting step in the soil-slurry treatment process.

The location of the impellers (e.g., bottom clearance) greatly affects the particle suspension. A substantial reduction in power input required for both on-bottom and off-bottom particle suspension is obtained as an impeller is placed near the tank bottom. Conventional designs of agitated tanks, with bottom clearance the same dimension as the impeller diameter or tank radius, is not adequate for particle suspension applications.

Aeration

A variety of air (or oxygen) supply devices are available for aerating the slurry medium. With appropriate aerator selection, the dissolved oxygen (DO) content in the slurry system can be maintained at suitable levels for biological activities.

Although low DO is rarely encountered as an operational problem for small-scale operations, the performance of large-scale units could be affected. The momentum induced by aeration can alter the reactor hydrodynamics resulting in internal flow circulation patterns unfavorable to bubble distribution. Power drawn by impellers is reduced as the air flowrate increases in aerated mixing tanks (Nienow 1990). The power reduction associated with impeller performance under aeration conditions may reduce effective mixing in agitated slurry resulting in poorer solids suspension. Clearly, the aeration device is a major component of the slurry treatment technology. Fundamental design criteria on aerators and their subsequent effects on mixing and hydrodynamics in the slurry medium must be investigated further.

Abiotic Loss Mechanisms

Foam formation during slurry treatment has been encountered several times in field applications. Generally, foam production is unpredictable and occurs very quickly. For most field operations, foaming is viewed as an undesirable situation. The operator has only a few ways to respond to these conditions before the foam leaves the reactor to contaminate other areas of the site. Reduction of air or oxygen flow or changes to lessen the rotational speed of the impellers to control foam conditions are the only options at field scale. The use of antifoaming agents helps to suppress the foam, but the foam is reentrained in the slurry. Foam may represent a useful separation component to slurry operations. The concentration of total PAH (t-PAH) constituents in the foam was observed to be 5 times more concentrated than the slurry (Glaser et al. 1995b). The formation of scale in the reactors is another means of contaminant removal. A recent small-scale study showed the scale to contain 40% more of the t-PAH constituents than the suspended slurry (see Figure 2). Because this evaluation was conducted in small-scale equipment there could be a bias due to the higher surface-to-liquid ratio associated with equipment of this scale.

Alternative Forms of Oxygen

In the French Limited lagoon treatment, cryogenic oxygen was used to reduce gasflow through the system because volatiles were included as targeted contaminants for slurry treatment (Bergman et al. 1994; Storms 1994). Reduced gasflows have been suggested as a means to lessen evaporative cooling of the slurry. Hydrogen peroxide has been used as a pretreatment for soils contaminated with town gas waste materials (Liu et al. 1994). Because the contaminants of concern are high-molecular-weight aromatic substrates, this strategy tries to initiate oxidative transformation to the substrates there are expected to be more easily degraded by microorganisms in the slurry system.

Posttreatment

The solid phase must be separated from the water, at the completion of treatment. Dewatering of the suspensions has not received the necessary attention.

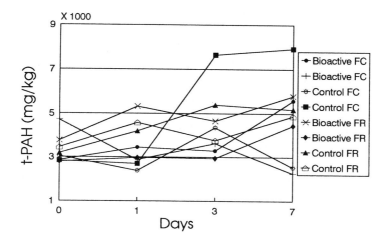

FIGURE 2. Treatability results of bioremediation of soils contaminated with PAHs using bench-scale slurry bioreactors.

For large projects, dewatering/treatment beds with leachate collection and treatment systems have been used. Smaller solids volume treatment may find filter press operations useful.

RECOMMENDATIONS

Key advantages recognized in the use of slurry reactors are more rapid treatment rates, greater degree of process flexibility, better waste containment, and reduced space requirements. The soil solids, sludges, or sediments selected as candidates for slurry treatment should receive minimum pretreatment. Size reduction by milling is a costly enterprise with dubious payoff. The volume of solids must be minimized to economize energy requirements related to the suspension of solids in the reactor. Sand and larger size fractions are excluded because they are more easily treated by soil washing and they cause the suspension equipment to deteriorate. Any free product should be removed to obtain better overall treatment efficiency in the slurry bioreactor. Due to the energy requirements for suspension, minimum residence time in the reactor is desirable. Lower soil loading and longer treatment duration may favor treatment to lower concentrations in the soil. Soil loading and aeration have considerable effects on power consumption and minimum suspension speed. In addition to efforts on designing effective forms of defoam removal mechanisms to minimize operational problems, research efforts are needed to assess the foaming potential, mechanisms, and properties of contaminated soils. Assessment of the biomass and its activity throughout the treatment process remain to be done.

ACKNOWLEDGMENTS

This research was supported in part by an appointment to the Postgraduate Research Program at the RREL/U.S. Environmental Protection Agency administered by the Oak Ridge Inst. for Sci. & Edu. through an interagency agreement between the U.S. DOE and RREL/EPA.

REFERENCES

Bergman Jr., T. J., J. M. Greene, and T. R. Davis. 1994. "An In-Situ Slurry Phase Bioremediation Case with Emphasis on Selection and Design of a Pure Oxygen Dissolution System." Presented at the Fuel Bioremediation Workshop, Naval Facilities Engineering Services Center, Port Hueneme, CA.

Bourquin, A. 1995. Private communication.

Castaldi, F. J., and D. L. Ford. 1992. "Slurry Bioremediation of Petrochemical Waste Sludges." *Water Science and Technology* 25: 207-212.

Ecology and Environment, Inc. 1995. "Aerobic Biodegradability Selection, Treatability Study Report for Southern Shipbuilding." Draft Report to EPA Region 6.

Glaser, J. A., M. A. Dosani, J. S. Platt, and P. T. McCauley. 1995a. Unpublished research.

Glaser, J. A., P. T. McCauley, M. A. Dosani, J. S. Platt, and E. R. Krishnan. 1995b. "Evaluation of Abiotic Fate Mechanisms in Soil Slurry Bioreactor Treatment." U.S. Environmental Protection Agency 21st Annual Risk Reduction Engineering Laboratory Research Symposium, Cincinnati, OH.

Griffiths, R. A. 1995. "Soil Washing Technology and Practice." *Journal of Hazardous Materials* 40, 175-189.

Liu, B. Y., S. P. Pradham, V. J. Srivastava, J. R. Pope, T. D. Hayes, D. G. Linz, C. Proulx, D. E. Jerger, and P. M. Woodhull. 1994. "An Evaluation of Slurry-Phase Bioremediation of MPG Soils." IGT Symposium on Gas, Oil and Environ. Biotech., Colorado Springs, CO.

Nienow, A. W. 1990. "Gas Dispersion Performance in Fermenter Operation," *Chem. Eng. Prog.* 86: 61-71.

Oldshue, J. Y. 1983. *Fluid Mixing Technology*, McGraw-Hill, New York, NY.

Ross, D. 1990. "Slurry-Phase Bioremediation: Case Studies and Cost Comparisons." *Remediation* 1: 61-74.

Sposito, G. 1984. *The Surface Chemistry of Soils*, Oxford University Press, New York, NY.

Storms, G. E. 1994. "Oxygen Dissolution Technologies for Bioremediation Application." Presented at the Fuel Bioremediation Workshop, Naval Facilities Engineering Services Center, Port Hueneme, CA.

U.S. EPA. 1990. *Engineering Bulletin: Slurry Biodegradation*, EPA/540/2-90/016, Cincinnati, OH.

U.S. EPA. 1991. *Technology Evaluation Report: Design and Development of a Pilot-Scale Debris Decontamination System*, EPA/540/5-91/006a.

U.S. EPA. 1993. *Pilot-Scale Demonstration of a Slurry-Phase Biological Reactor for Creosote Contaminated Soil*, EPA/540/A5-91/009.

Woodhull, P. M. 1995. Private communication.

Woodward, R. 1995. Private communication.

Blade-Mixing Reactors in the Biological Treatment of Contaminated Soils

Karsten Hupe, Joachim-Christian Lüth,
Jörn Heerenklage, and Rainer Stegmann

ABSTRACT

The application of mixing reactors was expected to have a positive effect on the biological turnover of contaminants, especially for cohesive soils. During investigations using blade-mixing reactors, it appeared to be of utmost importance to inhibit or reduce pellet formation during the dynamic treatment[1] of soils. In this connection, a comparison of the degradation kinetics in static and dynamic reactors is of great interest.

INTRODUCTION

During recent years, bioreactors have increasingly been applied or developed for the biological treatment of contaminated soils. Compared to the windrow technique, the advantage of bioreactors is that all emissions can be controlled and completely recorded (Kleijntjens 1991; Irvine 1994; Nitschke 1994). Most of the bioreactors developed are slurry reactors (Klein 1994). Their advantage is that they can be operated in a simple technical way and pellet formation during "dry treatment" (water content < maximum water capacity of the soil material) is a priori avoided (Parthen 1992). The application of slurry reactors is considered useful especially when problematic substances such as concentrated residues from soil scrubbing (sediment filter cake) (Elias & Wiesmann 1993; Schuster 1994; Bhandari et al. 1994) have to be treated.

EXPERIMENTAL MATERIALS AND METHODS

For investigations of the biological turnover of contaminants, synthetically contaminated soils (A_h-horizon, clay/peat) were used (Miehlich 1991; Goetz &

---

[1] In this context, the treatment of materials in mixing systems is called dynamic or semidynamic treatment.

Wiechmann 1991). Diesel fuel and lubricating oil were used as contaminants and analyzed in detail (Steinhart et al. 1990; Francke 1990).

The blade-mixing reactors were designed as follows: 3 reactors (5-L volume), run by a motor via a chain drive, operated in parallel. Figure 1 shows the scheme of a combined test setup of fixed-bed and blade-mixing reactors.

The CO_2 content of the exhaust gas was quasicontinuously measured by means of infrared (IR) spectrometry and the volatile organic carbon (VOC) content was measured by means of a flame ionization detector (FID). A personal computer (WorkBench PC 2.0 installed at a PC 386 Mbyte and MS-DOS) was used to control the feed gas and quasicontinuously record the exhaust gas quality data. Additional gas samples were taken from the reactor with a syringe through a septum and directly injected into a gas chromatograph (GC). For the analysis of solid fractions, sampling points were installed.

RESULTS

Optimization of Blade-Mixing Reactors

Preinvestigations showed that the dynamic treatment had a positive effect on contaminant turnover, especially at low temperatures (< 20°C) (Hupe et al. 1993). Considering these results, subsequent tests were carried out at 20°C. In the

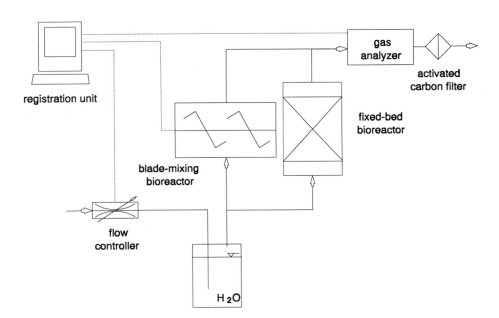

FIGURE 1. Test setup consisting of fixed-bed and blade-mixing reactor with continuous data recording.

preinvestigations on the dynamic treatment of soils in blade-mixing reactors, emphasis was placed on reducing pellet formation. It was found that pellet formation could be reduced when the water content was relatively low [<55% WC_{max} (maximum water capacity)], the portion of organic structural material (e.g., compost) was high, the structural material contained only a low portion of fines ($d_p < 630$ μm; <10%), and the rotational speed of the mixer was low (8 rpm).

Figure 2 shows the behavior of various test materials (Table 1) depending on the compost content after being mixed in a blade-mixing reactor.

Influence of Dynamic Treatment

Test series using blade-mixing reactors showed that the effect of the dynamic treatment of contaminated soils varies depending on the type of soil and contamination. Whereas the turnover of diesel fuel in a soil (A_h-horizon and clay/peat)-compost mixture was enhanced by the dynamic treatment, the opposite effect could be observed when lubricating oil was treated. The curves in Figure 3 represent the CO_2 concentrations in the exhaust gas from the bioreactors during two test series. The effect of static treatment (fixed-bed bioreactor) was compared with that of dynamic treatment (blade-mixing bioreactor). Two different soil types were tested, one of an A_h-horizon contaminated with diesel fuel and one clay/peat sample contaminated with lubricating oil. The soil samples were contaminated with oil (1% of weight), compost (20% of weight, degree of maturity: V) was added, and the water content was adjusted at 50% WC_{max}. The samples were weighed into the reactors (static: fixed-bed reactor; dynamic: blade-mixing reactor). The CO_2 production was continuously measured during the treatment.

CONCLUSION

Based on the investigations carried out in the Research Center SFB 188 of the DFG (German Research Society) "Treatment of Contaminated Soils," the application of mixing reactors in general cannot be recommended for biological treatment of dry soil materials (water content < maximum water capacity of the soil material) because

- It is difficult to avoid pellet formation, which leads to the reduction of the microbial activity and, as a consequence, to the reduction of the contaminant turnover.
- The higher treatment costs resulting from higher investment and operating costs cannot be justified due to the relatively short reduction of the treatment period (only a couple of days can be saved during a treatment period over several months).

Although the dynamic bioreactors did not show the expected advantages in the "dry treatment," it may be useful to apply the mixers to mechanically pretreat soils prior to biological treatment. Ploughshare mixers with installed

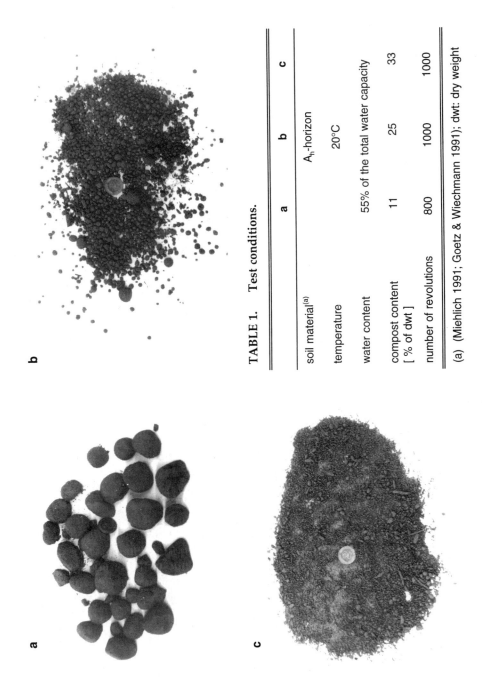

TABLE 1. Test conditions.

	a	b	c
soil material[a]	A_h-horizon		
temperature	20°C		
water content	55% of the total water capacity		
compost content [% of dwt]	11	25	33
number of revolutions	800	1000	1000

(a) (Miehlich 1991; Goetz & Wiechmann 1991); dwt: dry weight

FIGURE 2. Pellet formation dependent upon type of soil and contamination.

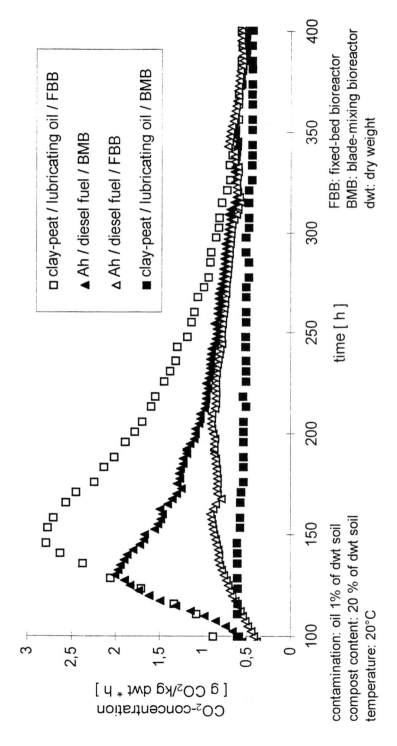

contamination: oil 1% of dwt soil
compost content: 20 % of dwt soil
temperature: 20°C

FBB: fixed-bed bioreactor
BMB: blade-mixing bioreactor
dwt: dry weight

FIGURE 3. Influence of mixing in a blade-mixing reactor dependent on soil type and contamination at 20°C.

cutterheads are well suited to the pretreatment of soils or to the treatment of soils during remedial actions for

- Disagglomeration
- Homogenization
- Disintegration (aeration)
- Admixing of additives.

Care must be taken to adjust the water content at about 50% of the maximum water capacity. In many cases, the water content can be reduced and the disintegration of the soil materials can be reached by adding compost and sawdust. However, it may not be useful to apply the mixers when mycel-forming organisms are of significant influence on degradation. An investigation of process optimization is necessary.

REFERENCES

Bhandari, A., D. C. Dove, and J. T. Novak. 1994. "Soil washing and biotreatment of petroleum-contaminated soils." *Journal of Environmental Engineering* 120(5): 1151-1169.

Elias, F., and U. Wiesmann. 1993. "Biological treatment of wastes from a soil washing plant in a 4-stage reactor cascade on a laboratory scale." In Gesellschaft für Umweltkompatible Prozeßtechnik mbH (Eds.), *II Euro-Forum Altlasten*. Saarbrücken, Germany. pp. 209-222.

Francke, W. 1990. "Elementaranalyse des Diesel- und des Schmieröles. Protokoll der 8. Analytiksitzung des SFB 188 vom 6.12.1990." University of Hamburg, Germany, unpublished.

Goetz, D., and H. Wiechmann. 1991. "Forschungsvorhaben D2 'Einfluß von Ölkontaminationen auf bodenphysikalische und -mechanische Eigenschaften von kontaminierten Standorten'." In Reinigung kontaminierter Böden, Arbeits- und Ergebnisbericht des SFB 188 für den Förderungszeitraum 1989-91. Hamburg, Germany, unpublished. pp. 355-390.

Hupe, K., J. Heerenklage, S. Lotter, and R. Stegmann. 1993. "Anwendung von Testsystemen zur Bilanzierung und Optimierung des biologischen Schadstoffabbaus." In R. Stegmann (Ed.), *Bodenreinigung: Biologische und chemisch-physikalische Verfahrensentwicklung unter Berücksichtigung der bodenkundlichen, analytischen und rechtlichen Bewertung*. Economica, Bonn, Germany. pp. 97-119.

Irvine, R. L. 1994. "Soil Bioreactors — Perspectives in the USA." Oral presentation at 418th DECHEMA-Kolloquium 'Biologische Bodenreinigung in Reaktoren' on March 4, 1994, Frankfurt a.M., Germany, unpublished.

Kleijntjens, R. H. 1991. "Biotechnological Slurry Process for the Decontamination of Excavated Polluted Soils." Ph.D. Thesis, Technical University of Delft, the Netherlands.

Klein, J. 1994. "Entwicklungsperspektiven von Reaktorverfahren zur biologischen Bodenreinigung." Protokoll der 8. Sitzung des Interdisziplinären Arbeitskreises 'Umweltbiotechnologie-Boden' on March 4, 1994, Frankfurt a.M., Germany, unpublished.

Miehlich, G. 1991. "Forschungsvorhaben D1 'Veränderung bodenchemischer Eigenschaften durch Ölverunreinigung und -dekontamination.'" In Reinigung kontaminierter Böden, Arbeits- und Ergebnisbericht des SFB 188 für den Förderungszeitraum 1989-91. Hamburg, Germany, unpublished. pp. 325-354.

Nitschke, V. 1994. "Entwicklung eines Verfahrens zur mikrobiologischen Reinigung feinkörniger, mit polyzyklischen aromatischen Kohlenwasserstoffen belasteter Böden." Ph.D. Thesis, University of Paderborn, Germany.

Parthen, J. 1992. "Untersuchungen zur mikrobiologischen Sanierung von mit PAK verunreinigten feinkörnigen Böden im Drehtrommelreaktor." Ph.D. Thesis, University of Hannover, Germany.

Schuster, E. 1994. "Minimierung von Reststoffen bei der Bodenreinigung durch kombinierte Systeme." In K. Alef, W. Blum, S. Schwarz, A. Riss, H. Fiedler, and O. Hutzinger (Eds.), *Eco-Informa-'94: Bodenkontamination, Bodensanierung, Bodeninformationssysteme.* Umweltbundesamt, Wien, Austria. pp. 123-131.

Steinhart, H., W. Herbel, J. Bundt, and A. Paschke. 1990. "Strukturtypentrennung von Dieselölkraftstoff mittels Festphasenextraktion." Protokoll der 6. Arbeitsgruppensitzung 'Analytik' des SFB 188 vom 4.4.1990," Hamburg, Germany, unpublished.

Reclamation of the Fine-Particle Fraction in Hydrocarbon-Contaminated Soils

Karl Schmid and Hans H. Hahn

ABSTRACT

This paper describes an advanced treatment system for fine-particle sludges as the residual contaminated fraction emerging from the soil washing process. The study was conducted with hydrocarbon (HC)-contaminated sludges from a field soil and a laboratory soil. The maximum HC content of the sludges was 5 g HC/100 g dry weight. In the first stage, the fine-particle sludges were mixed with surfactants and the adsorbed HCs were desorbed. Screening tests determined the surfactants that facilitated complete desorption of the HCs. Ethoxylate (EO) surfactants and terminal-blocked EO surfactants with a hydrophile-lipophile balance (HLB)-number of 11 to 12 produced the optimum desorption rate. After desorption, the liquid phase, contaminated with surfactants and HCs, was separated from the solids. Subsequently, phase separation by increasing temperature was performed. The HCs were reduced to a residual content of 0.5 g/L in the liquid phase. The liquid phase, with its residue of surfactants and HCs, was recycled to the treatment system. For advanced treatment of fine-particle sludges with adsorbed surfactants of up to 4% of their dry weight, further treatment in bioreactors was necessary. This treatment step facilitated the complete reduction of the soil's toxic potential, measured as bacterial toxicity of the soil eluate, within 2 days by bacterial degradation of the surfactants.

INTRODUCTION

Soil washing is a proven technology for economical reclamation of contaminated soils as long as the content of the fine-particle fraction does not exceed 25 to 30% of the processed soil. The development of techniques for the advanced treatment of contaminated fine-particle sludges is required.

The various available technologies for reclaiming fine-particle sludges in bioreactors (Blank-Huber et al. 1992, Luyben & Kleijntjens 1992) have two velocity-limiting factors in common:

- Biological reclamation requires an extended treatment period in which a considerable residual contamination remains, because soil washing produces an increased concentration of nondegradable contaminants such as iso- and cycloalkanes.
- The adsorption of the HCs onto the organic soil material (van Afferden et al. 1992) or to mineral surfaces (Haus 1993) may reduce bioavailability.

Therefore, the study described in this paper used surfactants for the advanced desorption of HCs from the fine-particle fraction. The pretreated fine-particle sludges had to be postprocessed in bioreactors to reduce the amount of adsorbed surfactants.

EXPERIMENTAL METHODS AND MATERIALS

Soil Types

Laboratory Soil. A silt-clay-loam of an organic C concentration of 0.5% and soakable clay minerals of 17% was chosen for laboratory work. The dry soil was contaminated with a synthetic engine oil (43 mg/g soil). The oil contained only isoalkanes and cycloalkanes, no *n*-alkanes or polycyclic aromatic hydrocarbons (PAHs).

Field Soil. Soil washing was applied to a gravel soil. The fine-particle fraction, which could be characterized as silt-loam, was used in the study. The content of soakable clay minerals and organic C was 11%, respectively. Contamination amounted to 16 mg/g being composed of two fractions — a mean distillation fraction and a grease fraction. Due to the long exposure to contamination, no individual substances except phytane and pristane could be identified. It is quite plausible that the contaminants are isoalkanes and cycloalkanes; PAHs of a concentration of 0.1 mg/g fine-particle sludge have been found (total EPA-PAH).

Surfactants

The screening tests used technical-grade surfactants. The average chain length of the hydrophobe group (number of C-atoms) and the average EO number of the EO-surfactants are provided in the Results section. All surfactants used in the study had linear C chains. The surfactants listed in Tables 1 through 3 were tested. The following test conditions were used for the screening tests.

- 0.1 L of a 2.5% surfactant solution
- 10 g of contaminated pilot soil
- 24 h shaking
- Total HC concentrations in the solid and liquid phase were determined according to Schmid (Schmid and Hahn 1995).

TABLE 1. General screening of surfactants.

	Hydrophilic Group	C-chain Length	HLB
Perenin AS	Sulfosuccinate	14	n.b.
APG 600	1, 6 Glycosidresidues	13	> 10
G 1087	6 EO residues	18:1	9.5
Atlox	Ca-Sulfonate	?	10.7
OS 6	6 EO residues	18:1	9.5
SAS 60	Na-Sulfonate	12	n.b.
TR 8468	Ionics and nonionics	?	?
Triton n 101	8 EO residues	Phenyl, C9	13.4
T3	6 EO with terminal CH_3	14	11.7

Emulsion Destabilization

The liquid phase was separated and the temperature was increased for phase separation. The phase was mixed for 1 h and left alone until separation occurred. Subsequently HC and chemical oxygen demand (COD) concentrations of the liquid subphase were determined and from the amount of the total COD the surfactant concentration was calculated, according to the following equation:

$$\text{Surfactant (mg/L)} = COD_{\text{surfactant}} * K$$

Where: $COD_{\text{surfactant}} = COD_{\text{total}} - COD_{\text{HC}}$
COD_{HC} = calculation from the stoichiometrical oxygen demand
K = constant, determined by calibration

TABLE 2. HLB screening.

	No. of EO Residues	C-chain Length	Molecular Weight	HLB
Synperonic a 2	2	14	284	6.2
Synperonic a 3	3	14	329	8.0
T3-x	6	14	461	11.4
Renex	7	14	505	12.2
Synperonic a 9	9	14	593	13.3
Synperonic a 11	11	14	681	14.2
Synperonic a 20	20	14	1,077	16.3

TABLE 3. Screening depending on molecular weight.

	No. of EO Residues	C-chain Length	Molecular Weight	HLB
Dehydol 04	4	8	289	12.2
Dehydol 980	6	12	433	12.2
T3-x	6	14	461	11.5
Renex	7	14	505	12.2
Ta 8	8	17	591	11.9
Synperonic a 9	9	14	593	13.4
Dehydol 100	9	15	607	13.1
Brij 76	10	18	693	12.7

Biological Reclamation in Bioreactors

To treat the fine-particle fraction after application of surfactants, pH-static operated slurry reactors were used. For this purpose a 10% (weight percent) soil suspension was produced, enriched with nutrients (C:N:P ratio of 100:10:3), continuously mixed and aerated.

The development of biomass was determined by total Kjeldahl nitrogen (TKN) and the degradation of surfactants was determined by COD.

Bioluminescence Tests

Soil eluates were produced by shaking a 1% soil suspension in a 2% NaCl solution for 24 hours. Measurements were performed by determining the effective concentration, causing a 20% decrease in light output (EC 20) within 15 min. Subsequently, the dilution factor (DF), which produced a decrease in light output of less than 20%, will be given.

RESULTS

Screening Tests

Table 1 lists the examined surfactants. As can be seen in Figure 1, the substituted ethoxylate T3 produced the best desorption results for the pilot soil. A test series for determining the optimum HLB range for desorption of HCs by means of the surfactants of a homologous series with constant C chain, but variable ethoxylate number (Table 2), defined an optimum in the HLB range between 11 and 12. The desorption effectiveness of the examined pilot soil was >75% in an HLB range between 11 and 12 of the surfactants. Any deviation

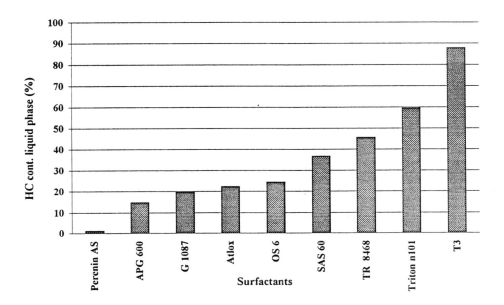

FIGURE 1. Screening test.

toward higher or lower HLBs resulted in a drastic reduction of desorption effectiveness (HLB 8: 5%; HLB 14: 27%).

The surfactants listed in Table 3 were used to investigate the relationship between desorption effectiveness and molecular size with constant HLB (Figure 2). The observed values are mean values resulting from three test series. The wide distribution of the observed values resulted from the use of technical-grade products as well as the fact that the surfactants used in this study show HLB values ranging from 11.5 to 13.4. The effect produced by the molecular size were overlapped by the HLB-caused effects. Nevertheless, a considerable increase in the amount of desorbed HCs was observed when surfactants of a smaller molecular size were used. The tests clearly indicated that ethoxylates with a HLB of 11 and a molecular weight of 350 to 450 produce the best desorption rate. For this reason, terminal-blocked ethoxylate T3 was used in all following tests. T3 has very favorable desorption qualities and produces considerably less lather than other comparable nonterminal-blocked ethoxylates.

Soil Washing Series

Several washing processes were performed to reduce the amount of surfactants. For this purpose the fine-particle suspension was treated as explained under Screening Tests except that only 0.5% surfactant was added. After an incubation of 24 h, the fine-particle sludge was centrifuged and again added to a new 0.5% surfactant solution. This process was repeated three times for

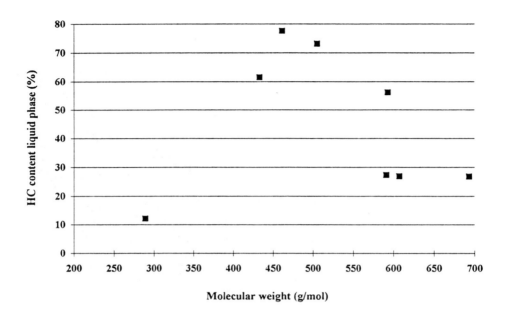

FIGURE 2. Relationship between effectiveness of desorption and the molecular size of the surfactants.

the pilot soil. Finally, there remained a residual contamination of 3.9 mg/g in the soil. After the washing process was repeated two times for the field soil, this soil type did not show any HC residue, while the initial contamination amounted to 17 mg/g soil.

Emulsion Destabilization

Because the desorption behavior of ethoxylates depends on temperature, we investigated whether a rise in temperature would produce phase separation. Figure 3 depicts the dependence on temperature of the HC and surfactant concentrations in the liquid phase. At a temperature beyond 53°C, the aqueous phase showed a surfactant content of only <2,000 mg/L and an HC concentration of <500 mg/L. The separation of a surfactant and an HC-rich phase was observed.

Biological Posttreatment

Batch tests examined the dependence on time of the degradation of surfactants in the liquid phase, soil-adsorbed surfactants, and HCs as depicted in Figure 4. The concentration of surfactants in the solid phase was reduced from 4,000 mg/L to 600 mg/L within 3 days (given 100 g solid phase/L), the content of surfactants in the liquid phase was reduced to 25 mg/L. The adsorbed HCs were degraded with a low rate, the content was reduced from 14 mg HC/g soil to 9.5 mg HC/g soil within 20 days. The DF to reach the EC 20 of the soil eluates

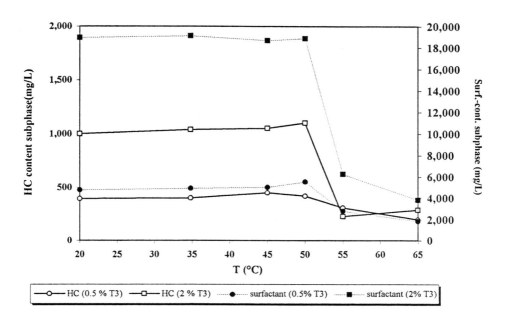

FIGURE 3. Emulsion destabilization.

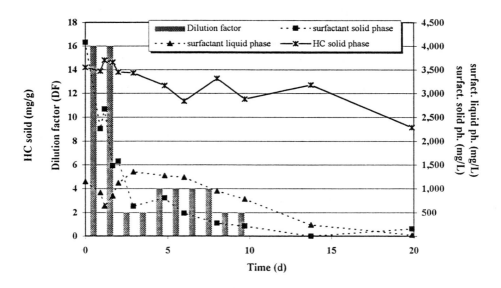

FIGURE 4. Degradation of HCs and surfactants, and development of the eluate's toxicity on time.

was reduced from 16 to 2 within 2 days. The eluate's toxicity was decreased along with the decreasing surfactant content. The HC concentration did not affect toxicity.

DISCUSSION

The screening tests show that nonionic surfactants are a suitable substance to desorb HCs from the soil. A comparison of all examined surfactants shows that ethoxylates in an HLB range of 11 to 12 and a molecular weight ranging from 350 to 500 produce the highest desorption effectiveness. Repeated washing with lower surfactant concentrations produces the same results as a single use of high concentrations, thus reducing the amount of required surfactants. One side effect of the phase separation produced for ethoxylates by a rise in temperature was that surfactants could be partially recycled. The optimum is that 40% of used surfactants can be recycled back.

After a 0.5% surfactant solution was used three times surfactants are distributed in the contaminated pilot soil is as follows:

> Solid phase: 25%
> Floating surfactant and HC-rich phase: 35%
> Liquid phase: 40%

The examined T3 surfactant produced a surfactant-rich phase, which could be segregated by emulsion destabilization at a temperature $>53°C$.

As is known from the literature (Rosen 1989), these surfactant-rich and HC-rich mesophases show the lowest boundary tension in the system. Low boundary tensions are a prerequisite for desorption of contaminants from the surface. These conditions reduce the stability of the phase and increase readsorption. One prerequisite for stable desorption of contaminants from the surface is the generation of steric or electrical barriers on the surface of the emulsion drops. Therefore, the use of a surfactant mixture is required, where the individual components have been adapted to the different individual demands:

- Substances for the reduction of boundary tensions
- Substances for the emulsification of HCs
- Substances for the generation of steric or electrical barriers.

If the composition of suitable surfactant mixtures allows use of ethoxylates in a three-phase system and if readsorption is to be avoided, a clear increase in desorption effectiveness can be expected.

An advanced reclamation of the fine-particle sludges of the pilot soil, i.e., the biological degradation of adsorbed surfactants, is possible with a short-term biological treatment in a slurry reactor. The concentration of surfactants in the solid phase is reduced by 85% to 500 mg/L given 100 g soil/L within 3 days. Toxicity of the eluate, is reduced from DF 16 to DF 2 within 2 days.

CONCLUSION

The study clearly indicates that the use of surfactants allows the reclamation of HC-contaminated fine-particle fractions. Further studies are required to develop adapted surfactant mixtures. The main objective must be to develop surfactant mixtures, that result in the greatest possible reduction of boundary tensions while allowing a stable emulsification of the contaminants. The desorption rates observed until now can be increased by the use of these mixtures. The study presented in this paper shows that combined surfactant treatment and biological posttreatment of fine-particle sludges can be established as an operable system.

ACKNOWLEDGMENTS

This research was funded by the Projektträger Wasser-Abfall-Boden. The authors gratefully acknowledge the invaluable help of Miss Schönherr who did a large part of the analyses.

REFERENCES

Blank-Huber, M., E. Huber, S. Huber, J. Hutter, and R. Heiss. 1992. "Development of a mobile soil cleaning plant for mineral oil contaminated soil material, combining mechanical washing with sewage sanitation and biological degradation." In Dechema (Ed.), *International Symposium Karlsruhe Soil Decontamination Using Biological Processes*. Frankfurt am Main. pp. 713-718.
Haus, R. 1993. "Mikrogefügeänderungen toniger Böden nach Kohlenwasserstoffkontamination und Tensideinsatz." Dissertation Inst. f. angewandte Geologie, Universität Karlsruhe, Works Vol. 25.
Luyben, A. M., and R. H. Kleijntjens. 1992. "Bioreactor design for soil decontamination." In Dechema (Ed.), *International Symposium Karlsruhe Soil Decontamination Using Biological Processes*. Frankfurt am Main. pp. 195-204.
Rosen, M. J. 1989. *Surfactants and Interfacial Phenomena*. John Wiley & Sons, New York, NY.
Schmid, K., and H. H. Hahn. 1995. "Entwicklung eines Verfahrens zur biologischen Behandlung von Feinkornextrakten aus der Bodenwäsche bei Kohlenwasserstoff-Altlasten." *Altlastenspektrum* 4(2).
van Afferden, M., M. Beyer, and J. Klein. 1992. "Significance of Bioavailability for the Microbial Remediation of PAH-Contaminated Soils." In Dechema (Ed.), *International Symposium Soil Decontamination Using Biological Processes*. Frankfurt am Main. pp. 605-610.

Prediction of Bioremediation Rates at Multiple Fixed-Site Soil Centers

Gary R. Hater, Roger B. Green, Tricia Solsrud, David A. Bower, Jr., and John A. Barbush

ABSTRACT

The measurement of petroleum hydrocarbon biodegradation rates for field-scale soil remediations typically is approached by evaluating reductions in soil hydrocarbon concentrations during treatment until a threshold value is achieved. The ability to predict when soil sampling and laboratory analysis is warranted is an important aspect of cost-effective remediation activities. Monitoring of soil gas CO_2 and O_2 has been widely used to estimate hydrocarbon degradation by aerobic bacteria. In this study, O_2, CO_2, volatile organic compounds (VOCs), methane, and temperature were monitored over the course of operating three vacuum-heap soil remediation facilities. The vacuum heaps, located in landfills, commercially treat petroleum-contaminated soil. From initial concentrations ranging from 994 mg/kg to 4,500 mg/kg, petroleum hydrocarbon levels were reduced to below regulatory levels in less than 60 days. An observed decrease in O_2 utilization and CO_2 production over time suggests that these parameters are useful in predicting the course of remediation. However, biodegradation rates calculated from soil gas measurements significantly underestimated the mass of petroleum removal calculated from soil analysis. The measured rate of biodegradation appeared to be proportional to the initial hydrocarbon concentration.

INTRODUCTION

Waste Management, Inc. operates seven fixed-site soil bioremediation facilities in four states. Six of the facilities are commercial operations that accept petroleum-contaminated soil for bioremediation. To date, these facilities have successfully treated more than 350,000 tons of petroleum-contaminated soil to concentrations below regulatory requirements. Remediated soil is used for landfill construction activities or, in some cases, returned to customers for reuse.

The augmented vacuum-heap process used at these facilities consists of aerating a pile containing up to 60,000 tons of soil inoculated with a petroleum-acclimated bacterial culture and a nitrogen and phosphorus source. After construction is complete, the soil piles are covered with a synthetic tarp and air is recirculated through the pile with a series of pipes connected to a blower. This air-recycle design minimizes volatile emissions and returns heat and moisture to the pile while providing sufficient oxygen for aerobic biodegradation (Hater et al. 1994). A cross-sectional view of a typical pile is shown in Figure 1.

In an effort to minimize soil sampling cost by optimizing the sampling events, CO_2, O_2, and VOCs were monitored during the course of treatment. Gas samples were analyzed using portable field instruments. It was anticipated that changes in O_2 use and CO_2 production would predict when soil sampling would be appropriate. Correlations between CO_2, O_2, and gas-phase hydrocarbons were evaluated also.

The piping system used to distribute air throughout the pile is essentially a closed-loop system. However, pressure differentials in the pile result in some flux of air into and out of the soil. This has made interpreting O_2 utilization and CO_2 production in soil piles difficult (Wurdemann et al. 1994). In an attempt to estimate the magnitude of this flux, a helium tracer gas test was performed on an operating pile.

METHODS

Soil Sampling and Analysis

A weighted mean initial concentration of total petroleum hydrocarbons (TPH) for each pile was determined from analytical results submitted by customers during the profiling process. Closure sampling was performed by dividing the

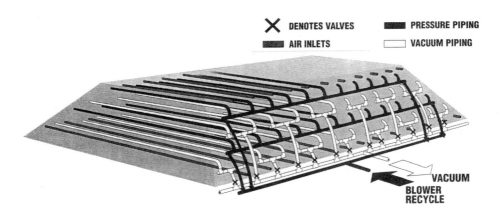

FIGURE 1. Cross-sectional view of a vacuum heap showing air distribution piping.

pile into sample grids with dimensions of 5 m × 5 m × 2 m. To reduce bias, grids were numbered and sample locations were selected using a random number table. Soil samples were collected from the appropriate grids with a split-spoon sampler.

Soil Gas Sampling and Analysis

Samples from the Ohio site were analyzed for TPH by EPA Methods 418.1 and 8015 modified. Samples from the two Wisconsin sites were analyzed using Wisconsin diesel-range organics (DRO) and gasoline-range organics (GRO) methods. Percent moisture analysis also was performed on selected samples collected at different depths during operation.

Soil gas samples were taken from the air distribution piping header connected to the blower. Samples were collected in Tedlar™ bags and analyzed for O_2 (GasTech GT-201), CO_2 (Riken 411-A), VOCs, and nonmethane hydrocarbons (Foxboro TVA-1000 FID). Samples containing analyte concentrations exceeding the linear range of the instrument were diluted with nitrogen, O_2 and CO_2 or ultrapure air and VOCs, and reanalyzed. All analyses were completed within 24 h of sampling. Airflow and temperature were measured using an in-line electronic flowmeter and thermocouple.

Tracer Gas Test

To estimate air flux at the surface of the pile, a helium gas tracer test was performed on an operating vacuum heap configured to recycle air through the pile. The test was conducted by injecting a pulse of helium gas into the air return line at a rate of 0.85 m^3/min for 10 min. The helium content of the air drawn from the pile was monitored with a Marks model 9820 helium detector calibrated with 1.0%, 0.5%, and 0.1% helium in air standards. Samples were collected at the air-recirculation header approximately every 1.5 min for 2 h.

RESULTS AND DISCUSSION

The treatment time required to reach regulatory limits for petroleum hydrocarbons at the three sites ranged from 27 to 56 days. Initial and final petroleum hydrocarbon levels for each site are listed in Table 1. The TPH value listed for the sites in Wisconsin are the sums of measured DRO and GRO values. Contrary to expectations, soil moisture content, which ranged from 12 to 15% across the sites, did not show vertical stratification or significant changes over the course of the treatment.

CO_2 production and O_2 utilization for each site showed a decreasing trend, with final rates generally 30% or less of the initial rates (Figure 2). CO_2 production and O_2 use were highest within the first few days after the start of aeration and then tapered off. The large initial rates may be partially attributed to the accumulation of CO_2 and depletion of O_2 in the pile before aeration. A significant

TABLE 1. Parameter correlations and biodegradation rates.

Pearson Correlation Parameter Pairs	Milwaukee, WI		Whitelaw, WI		Northwood, OH	
	r	(p)	r	(p)	r	(p)
CO_2 — O_2	-0.47	>0.05	-0.79	<0.001	-0.24	>0.05
CO_2 — total VOC	0.46	>0.05	0.66	<0.001	0.80	<0.001
CO_2 — CH_4 VOC	0.07	>0.05	0.59	<0.005	0.86	<0.001
CO_2 — non-CH_4 VOC	0.32	>0.05	0.64	<0.005	0.74	<0.001
O_2 — total VOC	0.13	>0.05	-0.38	>0.05	0.26	>0.05
O_2 — CH_4 VOC	0.07	>0.05	-0.31	>0.05	0.14	>0.05
O_2 — non-CH_4 VOC	0.19	>0.05	-0.61	<0.005	0.32	>0.05
Initial Petroleum Hydrocarbons (mg/kg)	1910.0		994.2		4580.6	
Final Petroleum Hydrocarbons (mg/kg)	197.0		108.4		65.8	
Biodegradation Rate from Soil (mg/kg/d)	38.1		32.8		112.9	
Biodegradation Rate from O_2 (mg/kg/d)	8.5		1.9		5.9	
Biodegradation Rate from CO_2 (mg/kg/d)	12.5		3.9		7.4	
Soil Moisture	12±2, 13±0.1, 14±4		13.5±0.5, 12±4		13±4	
Contaminants	diesel, gasoline, oil		diesel, gasoline		diesel, gasoline, oil	

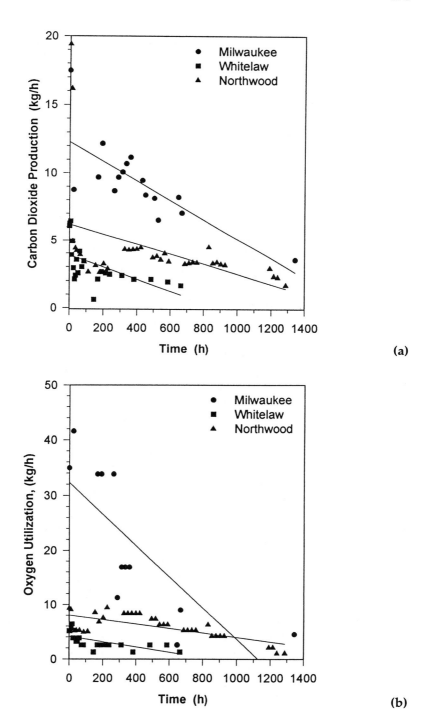

FIGURE 2. (a) CO_2 production and (b) O_2 utilization rates.

correlation between CO_2 and O_2 concentrations was found only for the Whitelaw site, ($r = -0.79$, $p = <0.001$). Significant correlations between CO_2 and the VOC parameters were found at the Northwood site and the Whitelaw site (Table 1). No significant correlations were found between recirculation air temperature and any of the other parameters. This may be because changes in recirculation air temperature were small and diurnal variations were not investigated.

To relate O_2 utilization and CO_2 production soil hydrocarbon reductions, a stoichiometric relationship for fuel biodegradation is required. Goldsmith and Balderson (1989) used the following equation to describe the biological oxidation of diesel fuel:

$$C_{13}H_{28} + 8.6\ O_2 + 2.4\ NH_4 \rightarrow 2.4\ C_5H_7O_2N + CO_2 + 10.4\ H_2O \qquad (1)$$

$C_{13}H_{28}$ is a weighted average of hydrocarbons in diesel fuel and $C_5H_7O_2N$ represents a typical cell. Assuming that the stoichiometric cell yield is 71% and carbon not incorporated into the biomass goes to CO_2, equation 1 becomes:

$$C_{13}H_{28} + 11.23\ O_2 + 1.846\ NH_4 \rightarrow 1.846\ C_5H_7O_2N +$$
$$3.77\ CO_2 + 11.231\ H_2O \qquad (2)$$

Biodegradation rates calculated using this equation are listed in Table 1. Rates based on petroleum hydrocarbon reductions ranged from 30 to 112 mg/kg/day. Larger initial soil concentrations appear to have resulted in higher removal rates. In contrast, rates based on cumulative O_2 use and CO_2 production ranged from 1.85 to 10.0 mg/kg/day. For each site, estimates based on CO_2 production yielded a higher rate than O_2 utilization.

The discrepancy between rates derived from soil samples and rates derived from air samples may be attributed to the following difficulties. First, the variability of petroleum hydrocarbon analytical methods is well known. This variability may account for some of the difference between soil- and gas-derived rates, because initial TPH concentrations were calculated using values from several laboratories. Second, although Equation 2 provides for the mineralization and incorporation into biomass of hydrocarbons, it does not account for partial degradation or incorporation into soil organic matter. Third, air flux into and out of the soil pile would reduce the apparent rate of O_2 use and CO_2 production. However, results of the helium tracer gas test suggest that this last consideration may not be significant.

Data from the tracer test are depicted in Figure 3. The peak in helium concentration represents the advective mixing of air containing the injected helium with air being displaced from the pore spaces of the pile as air is recycled. After approximately 50 min a constant rate of decay in the helium concentration is observed, suggesting that helium gas has been distributed throughout the pore spaces of the pile. This period is taken to be the empirical retention time for one pore volume. The theoretical retention time (i.e., the effective pile pore volume/airflow rate) was calculated to be 99 min, assuming a soil porosity of

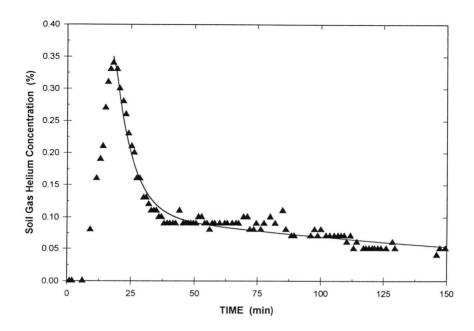

FIGURE 3. Helium tracer gas test results.

0.35, water-filled porosity of 0.15, a flowrate of 14.16 m³/min, and a pile volume of 7,020 m³. The difference in the theoretical and the empirical values may be attributed to the vacuum and pressure piping configuration of the vacuum heap allowing for a more rapid equilibration of the helium concentration. Assuming the reduction in helium concentration over time is a function of airflow into and out of the pile (i.e., helium is not irreversibly adsorbed onto soil and the rate of decrease is not a diffusion-limited process), the flux of air at the pile surface can be estimated. The exchange of air required to effect the concentration decrease after equilibration is calculated as:

$$J_a = kQ \qquad (3)$$

where J_a is the air flux at pile surface, k is the slope calculated from the linear portion of the curve in Figure 2 and is equal to 3.43×10^{-4}, and Q is the volumetric flowrate of the recycled air. For example, at a flowrate of 14.16 m³/min, $J_a = 0.0048$ m³/min. This rate of air exchange would not significantly increase the O_2 use and CO_2 production rates calculated.

SUMMARY

The removal rate of TPH in three vacuum heaps appeared to be proportional to the initial TPH concentration. O_2 and CO_2 monitoring significantly underestimated the mass of hydrocarbon removal indicated by soil analysis for three soil piles. This suggests O_2 use and CO_2 production may not be suitable surrogates for soil testing. However, O_2 utilization and CO_2 production rates decreased by more than 70% from the start of aeration to the time of closure sampling. Therefore, O_2 and CO_2 monitoring can be used as a simple and inexpensive way to track treatment progress in aerated soil piles. This type of monitoring can reduce the frequency of soil sampling required to demonstrate completion of the remediation.

REFERENCES

Goldsmith, C. D., Jr., and R. K. Balderson. 1989. "Biokinetic Constants of a Mixed Microbial Culture with Model Diesel Fuel." *Hazardous Waste & Hazardous Materials*, pp. 145-152. Mary Ann Liebert, Inc., Publishers.

Hater, G. R., J. S. Stark, R. E. Feltz, and A. Y. Li. 1994. "Advances in Vacuum Heap Bioremediation at a Fixed-Site Facility: Emissions and Bioremediation Rate." In R. E. Hinchee, B. C. Alleman, R. E. Hoeppel, and R. N. Miller (Eds.), *Hydrocarbon Bioremediation*, pp. 368-373. Lewis Publishers, Boca Raton, FL.

Wurdemann, H., M. Wittmaier, U. Rinkel, and H. H. Hanert. 1994. "A Simple Method for Determining Deficiency of Oxygen During Soil Remediation." In R. E. Hinchee, B. C. Alleman, R. E. Hoeppel, and R. N. Miller (Eds.), *Hydrocarbon Bioremediation*, pp. 454-458. Lewis Publishers, Boca Raton, FL.

Bioremediation of Ethylbenzene- and Styrene-Contaminated Soil Using Biopiles

Netta Benazon, David W. Belanger,
Dirk B. Scheurlen, and Mark J. Lesky

ABSTRACT

Two 26-m(L) × 12-m(W) × 2.5-m(H) biopiles (one active and one passive), containing 500 m^3 of soil contaminated with ethylbenzene, styrene, and total petroleum hydrocarbons (TPH), were constructed and operated over the winter and spring months. In the active pile, a 99% reduction in ethylbenzene and styrene concentrations and a 97% reduction in total petroleum hydrocarbons concentrations was achieved within 85 days of operation. Based on mass balance calculations, 98% of the mass removed was attributed to biodegradation; the remaining mass loss was by volatilization. Similar reductions in contaminant concentrations were achieved in the passive pile following 168 days of treatment at depths ranging from 0 to 0.8 m below the biopile surface (60% of total soil volume). Very little or no treatment occurred at greater depths because bioremediation was limited by surface diffusion of oxygen into the biopile. Based on the results, both an active or a passive system could be used to remediate soils from a polystyrene plant. The passive pile would need a longer treatment time and more space because the height would have to be limited to 0.8 m. However, the cost of remediation would be lower because off-gas treatment and associated monitoring and maintenance would not be required.

INTRODUCTION

A number of studies have focused on evaluating the performance of active biopiles (Hater et al. 1994, Samson et al. 1994). However, relatively few studies have examined the effectiveness of passive aeration. A passive system has a cost advantage because off-gas treatment and associated monitoring and maintenance are not required. The purpose of this study was to compare the performance of active versus passive aeration.

METHODOLOGY

Two 26-m(L) × 12-m(W) × 2.5-m(H) biopiles, each consisting of 500 m³ (750 tonnes) of soil, were constructed on an asphalt pad equipped with a leachate collection system. One of the biopiles (active pile) was equipped with an active aeration system consisting of four pipes at the base of the pile, which were manifolded together and connected to the vacuum side of a blower (Figure 1). A recycle line, installed after the blower, diverted 75% of the off-gas from the blower back into the pile. The remaining 25% was conveyed to an off-gas treatment system, consisting of two activated carbon tanks in series, before being discharged to the atmosphere. The other biopile (passive pile) contained a passive aeration system consisting of four pipes installed at the base of the biopile and three at an intermediate depth of 1.5 m below the surface of the pile (Figure 1). This system was designed to allow oxygen to enter the biopile by diffusion and free convection. The biopiles were covered with a 0.3 m layer of woodchips and a 20 mil polyethylene cover. A cross section representing both biopiles is presented in Figure 2.

The biopiles were operated from December 1993 to July 1994 (200 days). Soil temperature, organic vapor, and oxygen concentrations in the soil gas were measured during the course of treatment according to the following schedule: daily for the first 2 weeks; weekly for the next 2 months; and biweekly for the remaining treatment period. Temperature measurements were made using thermocouples installed at 0.8 and 1.7 m below the surface of the biopiles at the locations shown in Figures 1 and 2. Total organic vapor (TOV) and oxygen

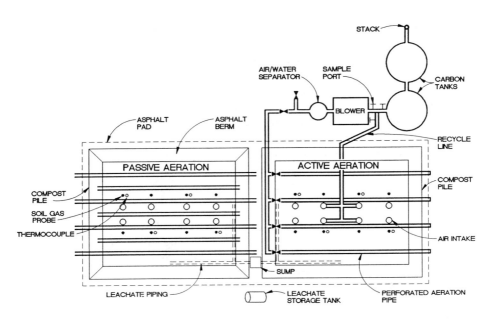

FIGURE 1. Plan view of biopiles.

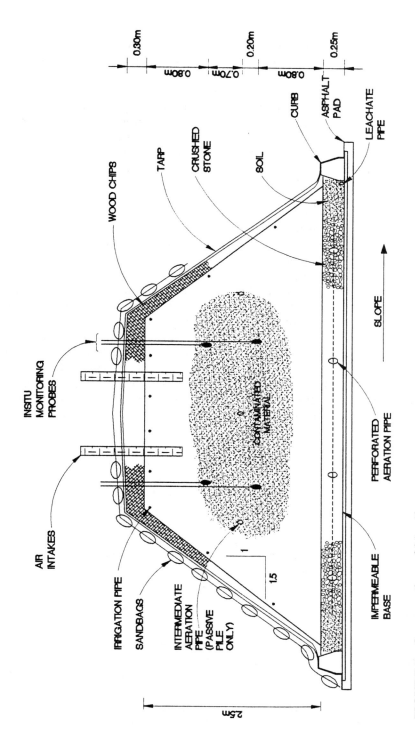

FIGURE 2. Cross section of biopiles.

concentrations in piles were measured by collecting a sample of soil gas from soil gas probes installed at 0.8 and 1.7 m below the surface of the biopiles (Figures 1 and 2) and analyzing using a GasTech Hydrocarbon Super Surveyor, Model 1314, Newark, California (GasTech).

TOV and oxygen concentrations, as well as air velocity measurements, were determined at the off-gas line from the biopile, the recycle line, and the influent to the air treatment system (see Figure 1 for sampling port locations) at a frequency of approximately every two weeks during the course of treatment. TOV levels were measured using an organic vapor analyzer (OVA) meter equipped with a flame ionization detector. The flowrate measurements were taken using a hot wire anemometer. These discrete measurements were conducted to provide gross estimates of the mass of TOV extracted, recycled, and treated. The aeration rate was adjusted to maintain oxygen levels at >10% in the pile. A flowrate of approximately 200 m³/h (120 CFM) was used during the first 40 days of operation, which was then reduced to 100 m³/h (60 CFM) because of a decrease in oxygen consumption. The flow was increased back to 200 m³/h when it was discovered that the inner core (soil at depths between 2 and 2.5 m below the biopile surface) was not remediating because oxygen was not reaching this area.

Contaminant concentrations in soil were measured on days 0, 84, and 168 by collecting 6 to 8 randomly located soil samples from each pile. A hand auger was used to collect soil samples at depths <1 m; samples at deeper depths were collected using a portable Pionjar vibratory soil drill equipped with a split-spoon sampler. The samples were analyzed for ethylbenzene, styrene, and the purgeable portion of TPH by gas chromatography/mass spectrophotometry (GC/MS) purge and trap using a modified EPA 624 analysis. Samples for TPH extractable analysis were prepared by sonification followed by a solvent extraction and GC/MS analysis.

The biodegradation rate was determined in the active pile by conducting oxygen uptake tests on days 5, 40, 65, and 119. The tests were performed by shutting off the blower and measuring the decrease in oxygen concentration with time in the soil gas using the soil gas probes and the GasTech. The slope of the oxygen versus time graph is proportional to the biodegradation rate. An average oxygen uptake rate was determined by calculating the mean of the slopes obtained from the individual soil gas probes. Oxygen uptake tests could not be conducted in the passive pile because oxygen levels were too low (0 to <5%) and there was no effective means of cutting off the oxygen supply.

RESULTS

The temperature monitoring results (Figure 3) show that the highest temperature reached in the active pile (23°C) was greater than that in the passive pile (15°C), indicating a higher biodegradation rate in the active system. Ambient temperature, which dropped to −14°C during the first month of operation (January), appeared not to have a negative impact on the soil temperature. The temperature increase observed after 100 days of operation (April) is believed

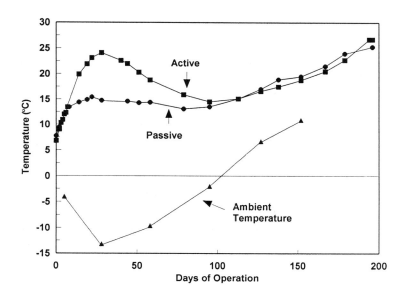

FIGURE 3. Average temperature versus time.

to be related to the increase in ambient temperatures rather than to biodegradation because oxygen uptake tests indicated there was very little biological activity taking place at this time.

The TOV concentrations in the active pile increased initially from 670 to 1,320 ppmv and then dropped sharply to less than 500 ppmv following the first 30 days of treatment (Figure 4). The initial increase in concentration can be attributed to the increase in soil temperature, while the decrease was caused by the drop in soil contaminant concentrations as bioremediation progressed. In contrast, the TOV levels in the passive pile did not begin decreasing until the 100th day of operation and the decrease was much less dramatic, indicating a much slower remediation rate.

The average oxygen concentration in the active pile was approximately 14% during the first 30 days of operation (Figure 5). Concentrations then began to increase gradually to approximately 18% as treatment progressed. In the passive pile, oxygen concentrations were less than 5% throughout most of the treatment period, although concentrations increased slightly following 80 days of operation, particularly in the shallow soil depths where remediation likely had occurred.

The biopile off-gas sampling results are presented in Table 1 along with the estimated calculated mass of TOV extracted, recycled, and treated. An estimated 817 kg of TOV was extracted from the biopile, of which approximately 609 kg of the total extracted was recycled back to the biopile and an estimated 208 kg of TOV was treated by the air treatment system.

In the active pile, initial concentrations were as follows: ethylbenzene ranged from 376 to 2,650 mg/kg, styrene varied from <5.5 to 338 mg/kg and TPH ranged

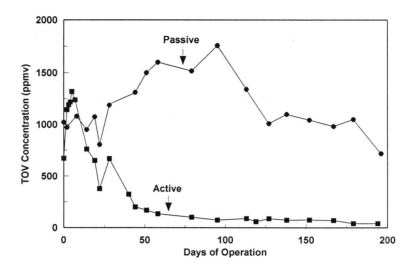

FIGURE 4. Average TOV concentration versus time.

from 3,190 to 27,100 mg/kg (Table 2). Following 84 days of treatment, concentrations of ethylbenzene and styrene at depths between 0 and approximately 1.5 to 2 m below the biopile surface dropped to <1 mg/kg, while concentrations of TPH decreased to levels ranging from 79 to 1,000 mg/kg. These results

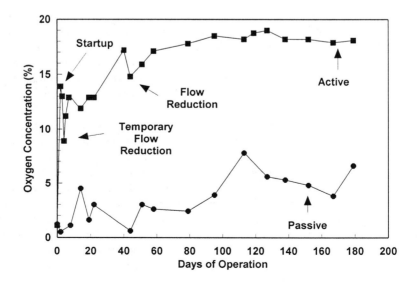

FIGURE 5. Average oxygen concentration versus time.

TABLE 1. Biopile off-gas monitoring results.

Day of Operation	Duration Days	Oxygen Conc. (ppmv)	TOV Conc. (ppmv)	TOV Conc. (mg/m³)	Extracted Flow (m³/h)	Extracted Mass (kg)	Recycled Flow (m³/h)	Recycled Mass (kg)	Treated Flow (m³/h)	Treated Mass (kg)
1	3	18.8	1,000	4.3E+03	155	48	0	0	149	46
4	1	18.0	900	3.9E+03	85	8	0	0	82	8
5	1	18.5	900	3.9E+03	140	12	0	0	135	13
6	21	18.0	800	3.5E+03	204	359	172	303	32	56
29	15	17.5	390	1.7E+03	190	116	144	88	46	28
44	21	18.3	290	1.3E+03	108	70	82	54	26	17
65	30	19.8	135	5.8E+02	117	49	90	38	27	11
96	11	20.3	70	3.0E+02	120	10	93	7	27	2
107	6	19.8	70	3.0E+02	120	5	94	4	26	1
113	14	19.8	70	3.0E+02	230	23	188	19	42	4
127	11	20.0	80	3.5E+02	230	21	188	17	42	4
138	14	19.5	95	4.1E+02	233	31	188	26	45	6
152	15	20.0	75	3.2E+02	236	27	191	22	45	5
167	13	19.8	60	2.6E+02	224	18	182	15	42	3
179	17	20.5	20	8.6E+01	370	13	299	10	71	3
196	6	20.5	30	1.3E+02	370	7	299	6	71	1
TOTAL	199					817		609		208

TABLE 2. Soil analysis results.

	Day 0				Day 84				Day 168			
	Ethylbenzene	Styrene	TPH[a]	Depth[b]	Ethylbenzene	Styrene	TPH[a]	Depth[b]	Ethylbenzene	Styrene	TPH[a]	Depth[b]
Active Pile	2,050	338	24,900	1	0.002	<0.0006	526	0.2	<0.0008	<0.0011	651	1.2
	2,650	65.8	27,100	1.2	<0.0004	<0.0006	79	0.7	<0.0008	<0.0011	335	0.4
	2,250	220	20,200	0.6	0.041	0.016	462	1.2	<0.0008	<0.0011	988	0
	1,220	<5.5	14,100	1.4	2,710	256	17,700	2.3	<0.0008	<0.0011	922	0.3
	1,600	233	10,500	1.5	3,950	412	31,700	2.3	<0.0008	<0.0011	1,310	1.5
	376	44.9	3,190	0.6	0.2	<0.055	1,000	0.7	1,810	362	30,300	2.1
									4,060	222	53,000	2.4
									2,340	982	46,000	2.2
Passive Pile	1,800	24.9	22,500	0.3	0.3	<0.011	368	0	<0.0039	<0.0055	1,010	0.2
	592	79.6	3,500	1.2	2.4	<0.550	767	0.3	1,680	250	24,800	1.1
	909	30.7	9,090	0.7	0.009	<0.00055	233	0.2	1,770	150	29,000	1.2
	2,190	365	30,000	1.8	4,460	530	53,400	1.4	<0.0253	<0.0011	1,580	0.8
	858	21.1	12,100	0.8	1,020	87.9	10,600	1.3	2,310	78.6	35,500	2.2
	1,660	<11	20,500	2.2	0.651	0.061	807	0.6	4,920	511	63,000	1.4
									<0.0039	<0.0055	1,500	1.6
									<0.0008	<0.0011	353	0.5

(a) TPH — total petroleum hydrocarbons.
(b) Depth below biopile surface.

represent average decreases of 99% for ethylbenzene and styrene and 97% for TPH. At depths ranging between 2 and 2.5 m below the surface of the biopile, however, concentrations remained similar to startup levels. Sampling results obtained on day 168 showed little change.

In the passive pile, initial concentrations were as follows: ethylbenzene ranged from 592 to 2,190 mg/kg, styrene varied from <11 to 365 mg/kg, and TPH ranged from 9,090 to 30,000 mg/kg (Table 2). After 84 days of treatment, ethylbenzene and styrene levels decreased to <5 mg/kg and TPH levels dropped to <1,000 mg/kg at depths between 0 and approximately 0.6 to 0.8 m below the biopile surface. Concentrations remained similar to startup levels at greater depths. Following 168 days of treatment, lower concentrations of ethylbenzene and styrene (<<1 mg/kg) were achieved at the 0 to 0.8 m depth range; however, no changes in concentration occurred at depths greater than approximately 0.8 m.

Oxygen uptake test results (Figure 6), conducted at depths of 0 to 2 m below the biopile surface, showed that the oxygen consumption rate was highest on day 5 of operation. Oxygen consumption declined substantially after 40, 65, and 119 days of operation, indicating a decrease in the biodegradation rate as the food source (contaminants) was depleted.

DISCUSSION

A mass balance calculation was performed on the active pile at depths between 0 and 2 m below the biopile surface. Biodegradation rates were estimated using the slope obtained from the oxygen uptake curves at days 5, 40, 65, and 119 and assuming a dry density of 1.5 kg/m³, a void volume of 33%, and 3.5 kg oxygen consumed per kg of TPH (Table 3). An approximate duration

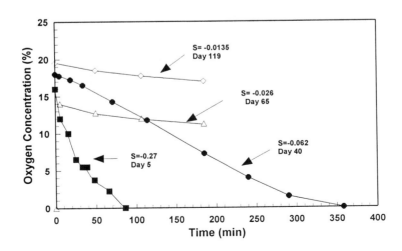

FIGURE 6. Oxygen uptake tests (0 to 2 m depth).

for each biodegradation rate was assigned based on trends in soil temperature, as well as oxygen and TOV concentrations in the soil gas (Table 3). The product of the biodegradation rate and the duration provided an estimate of the mass of contamination removed by biodegradation over a particular treatment period. The mass of contamination volatilized and treated by the air treatment system was calculated to be 347 mg/kg (208 kg TOV volatilized/6×10^5 kg soil at depths between 0 and 2 m). The total mass removed during the treatment period was estimated to be 15,630 mg TPH/kg soil (Table 3), 98% of which was attributable to biodegradation. The average initial TPH concentration was 16,665 mg/kg. Therefore, if 15,600 mg/kg was removed, the final TPH concentration would be 1,035 mg/kg. TPH concentrations at the end of the treatment period were similar to this value (Table 2). Because the biodegradation rate was measured indirectly using the oxygen consumption rates, and the operating period under each rate could only be estimated, the above mass balance calculation is approximate.

As mentioned previously, very little or no biodegradation occurred in the inner core, which represented 20% of the volume of soil in the pile. It is believed that as oxygen was drawn into the pile, it was consumed by the soil bacteria at shallower depths so the air reaching the inner core of the pile was depleted of oxygen. A second reason could be that air may have been preferentially channelling around and not through the inner core. This would explain why the concentration of oxygen in the off-gas from the biopile was generally higher than the concentrations measured in the soil gas probes.

A mass balance calculation similar to that conducted for the active pile could not be performed on the passive pile because oxygen uptake tests could not be conducted. However, based on the soil analysis (Table 2) and on the soil gas

TABLE 3. Mass balance calculation (active pile).[a]

Process	Day of Operation	Slope (% O_2/min)	Approximate Biodegradation Rate (mg/kg/day)	Approximate Operating Period[b] (Days)	Approximate Mass TPH Removed (mg/kg)
Biodegradation	5	−0.27	325	33	10,725
	40	−0.062	75	20	1,500
	65	−0.0256	31	47	1,458
	119	−0.0135	16	100	1,600
Volatilization/ Activated Carbon Treatment				200	347
Total					15,630

(a) Starting concentration = 16,665 mg/kg.
(b) Based on results of oxygen uptake tests and on trends in soil temperature, as well as O_2 and VOC concentrations in the soil gas.

TABLE 4. TOV concentrations in passive pile soil gas (day 200).

TOV Concentration	
0.8 m[a]	**1.7 m**[a]
190	1,500
150	285
800	1,700
260	310
1,400	3,900
160	170
110	4,400
130	235

(a) Depth below biopile surface.

monitoring, which generally showed low vapor levels in the shallow compared to the deep soil gas probes (Table 4), it appears that the soil was remediated between 0 and approximately 0.8 m below the surface of the biopile. This surface layer corresponds to a soil volume of 300 m^3 or approximately 60% of the soil in the pile. It is not possible to distinguish between mass loss due to volatilization and due to biodegradation. However, the mass loss due to volatilization is believed to be minimal because the pile was covered and the results of GasTech TOV monitoring around the biopile and near the aeration pipes gave readings of 0 ppm.

CONCLUSIONS

In the active pile, most of the mass removal was achieved by biodegradation in the first 85 days of operation. The achievable cleanup level was found to be <1 mg/kg for ethylbenzene and styrene and <1,500 mg/kg TPH. Some minor design modifications, such as placement of a set of aeration pipes within the inner core, could be implemented in future active piles to enhance bioremediation in this area. This could potentially lead to a treatment time of under 100 days.

Similar cleanup levels were achieved in a passive pile at depths between 0 and 0.8 m below the biopile surface; however, the remediation time was longer than the active pile. Very little if any treatment occurred at depths greater than 0.8 m, a direct result of oxygen being metabolized before it could diffuse into the inner core. Therefore, future passive biopiles should be kept to a maximum height of 0.8 m.

Passive piles have the advantage of being less costly to operate than active piles because off-gas treatment and associated monitoring and maintenance would not be required; however, they require more time and space to implement.

REFERENCES

Hater, G. R., J. S. Stark, R. E. Feltz, and A. Y. Li. 1994. "Advances in Vacuum Heap Bioremediation at a Fixed-Site Facility: Emissions and Bioremediation Rate." In R. E. Hinchee, B. C. Alleman, R. E. Hoeppel, and R. N. Miller (Eds.), *Hydrocarbon Bioremediation*. Lewis Publishers, Boca Raton, FL. pp. 368-373.

Samson, R., C. W. Greer, T. Hawke, R. Desrochers, C. H. Nelson, and M. St-Cyr. 1994. "Monitoring of an Aboveground Bioreactor at a Petroleum Refinery Site Using Radiorespirometry and Gene Probes: Effects of Winter Conditions and Clayey Soil." In R. E. Hinchee, B. C. Alleman, R. E. Hoeppel, and R. N. Miller (Eds.), *Hydrocarbon Bioremediation*. Lewis Publishers, Boca Raton, FL. pp. 329-333.

Biotreatment of Aqueous Extract From Chlorobenzene-Contaminated Soil

*Serge R. Guiot, Jean-Claude Frigon, Ruxandra Albu-Cimpoia,
Stéphane Deschamps, Xin Q. Zhou, Jalal Hawari,
Sylvie Sanschagrin, and Réjean Samson*

ABSTRACT

Chlorobenzene-contaminated groundwater originating from soil steam flooding (soil washing) was considered for treatment in a biological reactor. The present study addresses optimal conditions for biotreatment of 1,2-dichlorobenzene, 1,3-dichlorobenzene, 1,4-dichlorobenzene, and 1,2,4-trichlorobenzene at average initial concentrations (mg/L) of 25, 2.1, 3.9, and 0.4, respectively. Microcosm tests demonstrated that soil indigenous populations were able to mineralize all contaminants, with a 57 to 72% CO_2 recovery. Assays were performed in a mechanically stirred bioreactor that was inoculated with contaminated soil and municipal activated sludge. Contaminants such as tri- and dichlorobenzene are easily degraded under aerobic conditions. Concentrations of all contaminants were less than 10 µg/L in the released effluent after 30 to 45 days of reaction with soil at 5% (w/w) and dissolved O_2 at 80% of saturation. Within the bioreactor operating range, the contaminant did not volatilize significantly (between 0.2 and 2.8% of the total amount eliminated). Biodegradation was preceded by a transient adsorption of the compounds. The specific rate of dichlorobenzene degradation was improved four-fold when the soil inoculum was supplemented with municipal activated sludge.

INTRODUCTION

In many cases, soil remediation might be cost effectively realized by associating a physical or chemical pretreatment with the bioprocessing of the resulting product. The present study addresses the biological degradation of groundwater-borne chlorobenzenes originating from steam flooding of a contaminated soil.

The principal contaminants found were 1,2-dichlorobenzene (1,2-DCB), 1,3-dichlorobenzene (1,3-DCB), 1,4-dichlorobenzene (1,4-DCB), and 1,2,4-trichloroben-zene (1,2,4-TCB), whose concentrations (mg/L) ranged from 6 to 38, 0.3 to 4, 0.7 to 5.6, and 0.2 to 1.9, respectively.

A wide variety of aerobic bacteria (e.g., *Acinetobacter, Arthrobacter, Acetobacter, Alcaligenes, Bacillus, Flavobacterium,* and *Pseudomonas* spp.) have been shown to degrade mono- and dichlorobenzene as a sole source of carbon (Häggblom 1990). Usually mixed populations have been used as inocula because different micro-organisms present complementary hydrocarbon-degrading capabilities. Adapted communities, such as indigenous microbial populations of contaminated sites, were found to exhibit higher biodegradation rates than communities with no history of contamination (Leahy & Colwell 1990).

The objectives of the study are (1) to test the biodegradability of the contami-nants by the indigenous population of the contaminated soil; (2) to test perfor-mance in a mechanically stirred bioreactor as a function of the soil concentration, nutrients, and the level of dissolved O_2 (DO); and (3) to test alternative sources of inoculants, such as urban activated sludge.

MATERIALS AND METHODS

The soil was sampled from a pile excavated from a site contaminated mainly by chlorobenzenes. The sample was homogenized in a splitter and screened through a 2-mm mesh sieve (moisture, 9 to 10%; organic content, 2.6%, w/w).

All microcosm studies were carried out in 100-mL serum bottles, equipped with a glass tube containing 0.5 mL of 1.0 M KOH as a CO_2 trap and a Mininert™ valve. The serum bottles were inoculated with 50 mL of soil suspension (5%, w/v) prepared either in mineral salt medium (MSM) at pH 7.2 (Greer et al. 1990) or in actual groundwater. Once pulsed with ^{14}C uniformly labeled (UL)-1,2-DCB and ^{14}C-UL-1,4-DCB (Sigma Chemical Co., St. Louis, Missouri) and cold 1,2-DCB and 1,4-DCB, to have 50 and 25 mg/L respectively, the bottles were incubated at 25°C with agitation at 150 rpm. Abiotic controls were prepared by adding 0.2% (w/v) sodium azide. Pure oxygen was replenished in the headspace upon every KOH sampling. Radioactivity in the KOH was measured regularly using a scintillation counter (Packard Tri-Carb model 4530, Downers, Illinois).

The bioreactor used for the experiments was a mechanically stirred (225 rpm) 4.5-L vessel made of steel and glass, with automated regulation of pH and DO by air sparging (Celligen, New Brunswick Scientific, Edison, New Jersey). The bioreactor was operated at 25 ± 2°C. The air flowrate was measured with a wet test gas meter (GCA Precision Scientific, Chicago, Illinois).

After inoculation of soil into the reactor and while agitating, aliquots were taken to verify the amount of soil in the mixed liquor (ML). Over the course of this study, five batch experiments were conducted:

1. 5% (w/w) soil suspension, measured organic content of 1.25 g volatile solid per kg of soil-liquid mix (VS/kg$_{ML}$), over 80% oxygen saturation, 1,2-DCB,

1,3-DCB, 1,4-DCB, and 1,2,4-TCB concentrations (mg/L): 38, 2.8, 5.6, and 0.22, respectively.

2. 10% (w/w) soil suspension, 2.0 g VS/kg$_{ML}$ measured, 80% oxygen saturation, 1,2-DCB, 1,3-DCB, 1,4-DCB, and 1,2,4-TCB concentrations (mg/L): 25.7, 2.1, 4.8, and 1.7, respectively.

3. 5% (w/w) soil suspension, 1.1 g VS/kg$_{ML}$ measured, nutrients (mg/L): yeast extract (4), $(NH_4)_2HCO_3$ (61), $(NH_4)_2SO_4$ (7), KH_2PO_4 (22), over 80% oxygen saturation, 1,2-DCB, 1,3-DCB, 1,4-DCB, and 1,2,4-TCB concentrations (mg/L): 25.5, 2.1, 5.2, and 1.9, respectively.

4. 5% (w/w) soil suspension, 1.0 g VS/kg$_{ML}$ measured, 10% oxygen saturation, 1,2-DCB, 1,3-DCB, 1,4-DCB, and 1,2,4-TCB concentrations (mg/L): 5.7, 0.28, 0.7, and 0.22, respectively.

5. 10% (w/w) soil suspension (1.8 g VS/kg$_{ML}$ measured) added with municipal activated sludge (1.8 g VS/kg$_{ML}$), 80% oxygen saturation, 1,2-DCB, 1,3-DCB, 1,4-DCB, and 1,2,4-TCB concentrations (mg/L): 17.8, 4.6, 5.1, and 1.5, respectively.

For both the filtered liquid and the ML samples, chlorobenzenes were analyzed on a gas chromatograph (GC) (Sigma 2000, Perkin-Elmer, Norwalk, Connecticut) with a flame ionization detector (FID), coupled to a Tekmar 7050 headspace sampler (Cincinnati, Ohio). Samples (5 mL) with concentrations lower than 100 μg/L were purged and trapped by a Tekmar LSC 2000 & 2016 sampler (Cincinnati, Ohio). The trapped material was then desorbed and quantified with a Varian (Mississauga, Ontario, Canada) GC/MS (mass spectrometer) equipped with a Saturn II ion trap detector. Volatilized contaminants trapped in the charcoal cartridge (Supelco Inc., Bellafonte, Pennsylvania) placed on the air exit line were extracted by gently stirring the charcoal (0.145 g) with 1 mL of CS_2 (Aldrich Chemical Company Inc, Milwaukee, Wisconsin) for 2 min. Extracts were then analyzed on the Sigma 2000 GC/FID. Recovery was calculated using a known concentration of chlorobenzene, which was allowed to stay in contact with virgin charcoal for 12 h.

RESULTS AND DISCUSSION

The microcosm tests provide a quick answer about whether soil microorganisms are capable of mineralizing a specific compound. The results showed a high mineralization activity of the soil suspension for the two compounds studied. After approximately 30 days, 57% and 62% of 1,2-DCB and 1,4-DCB were mineralized to CO_2. The initial mineralization rates of soil suspension in MSM were 3.3% of CO_2/d evolved for 1,2-DCB, and 3.6% CO_2/d for 1,4-DCB, during the first week. However, the suspension in groundwater showed half of these values, i.e., 1.6 and 1.4% CO_2/d for 1,2-DCB and 1,4-DCB, respectively. These results suggest that the soil might be used as an inoculant for a bioreactor, and that the mineralization rate might be improved by addition of nutrients like nitrogen, phosphorus, and trace metals (provided by MSM).

In the bioreactor assays, residual concentrations of the contaminants were measured at several time intervals in the supernatant liquid (effluent after decantation), in the soil-liquid mix (ML), and in the charcoal trap. In the control assay (soil suspension: 5% w/w), DO was maintained at 84 (±16)% of saturation. Typically, contaminant removal from the ML followed a hyperbolic curve, with a steep initial slope. In most cases, all three DCBs were removed at slightly faster rates. More than 92% of the studied contaminants were removed within the first 10 days while the removal rate characteristically slowed down with substrate depletion (Figure 1A). To release an effluent with less than 10 µg/L of contaminant (Figure 1B), 30 to 47 days were required.

In most cases, contaminant removal was faster in the supernatant liquid than in the total ML. A mass balance performed on the liquid alone and the

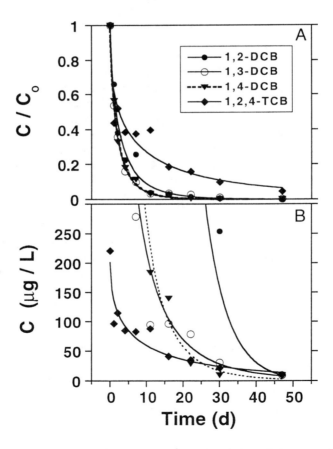

FIGURE 1. Change in the relative concentration (A) and the actual concentration in the lowest range (B) of 1,2-DCB, 1,3-DCB, 1,4-DCB, and 1,2,4-TCB, as a function of time. Initial concentrations (µg/L): 38, 400, 2,800, 5,600, and 200, respectively. Soil suspension: 5% (w/w).

ML showed that, for the first day, transient physical adsorption onto the soil accounted for up to 60% of the compounds that were removed from the bulk liquid. The adsorbed contaminants were then degraded by the microorganisms, as the difference between the contaminant concentrations in the liquid and the mixed liquor was reduced progressively. This indicated that the biological degradation occurred at a greater rate than the rate of physical adsorption once the contaminants became scarcer in the liquid. The residual fraction of contaminants that was adsorbed on soil or sludge particles at the end of the tests ranged from less than 0.1 to 4% of the total amount eliminated. These data are comparable to the observations of Mueller et al. (1991) for treatment of chlorinated compounds in an aerated reactor where the residual amount adsorbed on biomass (0.12 g VS/L) was 0.5% of the initial amount after 30 days.

Chlorobenzenes are volatile compounds and, therefore, are expected to be stripped off proportionally to the air flowrate. In fact, the contaminant lost by volatilization was the highest in the reactor, being supplied with, on average, 5.7 L_{air}/L_{rx}·d, although this loss was minimal (between 0.2 and 3% of the initial mass). However, it should be realized that this aeration rate was extremely low. To reduce volatilization, aeration was performed at a lower rate (1.7 L_{air}/L_{rx}·d; DO at 10% of saturation). Volatilization was 2- and 28-fold less for 1,4-DCB and 1,2,4-TCB, respectively. However, volatilization of 1,2-DCB and 1,3-DCB was not reduced as compared with figures obtained at the higher aeration rate.

To compare the overall degradation processes through the different bioreactor tests, characterization of each removal profile by a single kinetic parameter was carried out. A first-order empirical equation with respect to substrate concentration ($-dC/dt = K_1 \cdot C$) was preferred to a zero order one. This allowed maximization of the relative weight of the lowest range data onto the parameter estimation. This is important since the prediction of degradation is more critical at the lowest concentration range. The K_1 values were estimated using the least squares best fitting of the contaminant total concentration (in soil-liquid mix) as a logarithmic function of time. As expected, kinetics also depended on the biomass concentration because, in one experimental run, doubling the soil suspension (10% w/w soil added, 2 g VS/kg_{ML} measured) resulted in a two-fold increase in contaminant removal. K_1 has thus been substituted by the product of a specific constant k_1 and the biomass content, X. The biomass content was assumed not to vary significantly over the experimental interval. The suspension organic content was found to remain relatively constant, as its change remained within the same order of magnitude as the accuracy of the measurement method. This is consistent with the fact that the organic content of the inoculant was two orders of magnitude greater than the total organic carbon assimilable from the liquid (≈ 15 mg/L). The specific constant k_1 is calculated by dividing K_1 by the measured value of X. The above equation, integrated between the time and concentration limits of 0 and t, and of C_o and C, respectively, is thus: $C = C_o \cdot \exp(-k_1 \cdot X \cdot t)$.

Kinetics for all bioreactor experiments and all contaminants are compared in Figure 2. Although addition of nutrients graphically appeared to improve the removal rate of most of the contaminants, this was not reflected by the kinetics

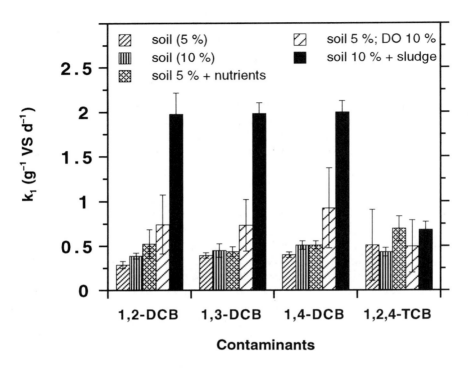

FIGURE 2. Comparison of the first-order kinetics constant for all experiments and contaminants.

constants. Nutrients might have influenced the initial slope of removal and thus had less impact on the first order constant value. This could indicate that the nutrient presence is critical only when the substrate is not limiting.

The most interesting results were obtained when the soil inoculant was supplemented with municipal activated sludge (soil:sludge 1:1, organic content basis). This led to specific kinetics constant values four-fold higher than those of the previous assays, except for the 1,2,4-TCB (Figure 2). In the latter assay, the time to lower contaminant concentration below 5 µg/L was reduced to less than 3 days. This might be related to various factors: the sludge fluid content in many nutrients and trace elements, the high number of microorganisms, the large variety of microbial species, or the adaptation of the microbes to a broad range of chemicals. The latter point is supported by the fact that many chemicals are found in the wastewater (fossil-fuel-processing wastes, urban runoff water, spillages, effluents from industries pretreatment facilities, etc.) and eventually end up in large urban treatment plants (Cardinal & Stenstrom 1991).

Reactor efficiencies in dichlorobezene biodegradation have been simulated as a function of the operating time for two types and two contents of biomass, using an average of the above-estimated kinetics constants (Figure 3). The first curve simulates the removal efficiency of a reactor inoculated with acclimated

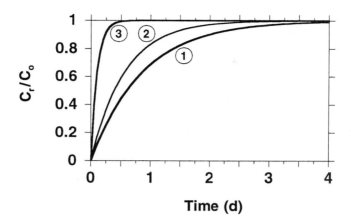

FIGURE 3. Predictive curve of the relative removal of the dichlorobenzenes for different inocula. Equation used: relative concentration removed $(C_r/C_o) = 1 - \exp(-k_1 \cdot X \cdot t)$. Curve 1: acclimated soil (10%), with X = 2.6 g VS/L and $k_1 = 0.45$ g^{-1} VS d^{-1}. Curve 2: soil (15%), with X = 3.9 g VS/L and $k_1 = 0.45$ g^{-1} VS d^{-1}, or mixed soil-activated sludge, with X = 0.9 g VS/L and $k_1 = 2$ g^{-1} VS d^{-1}. Curve 3: mixed soil-activated sludge, with X = 5 g VS/L and $k_1 = 2$ g^{-1} VS d^{-1}.

soil (10% w/v, k_1: 0.45 g^{-1} VS d^{-1}). The second curve corresponds to a 15% suspension of the same soil. This curve would be identical to that of a 2% suspension of the same soil added with 0.5 g VS/L of urban activated sludge (k_1: 2 g^{-1} VS d^{-1}). The third curve simulates the removal efficiency by a mixed suspension of acclimated soil and activated sludge totaling 5 g VS/L. This clearly depicts the high efficiency that can be attained in a mixed soil-activated sludge reactor, operated with a relatively low hydraulic residence time of 0.5 d.

CONCLUSIONS

The main conclusions of this study are (1) contaminants such as tri- and dichlorobenzenes are easily degraded under aerobic conditions; (2) within the bioreactor operating range, the contaminants did not volatilize; (3) biodegradation was preceded with a transient adsorption of the compounds; and (4) urban activated sludge can be used to improve the contaminant degradation rate.

REFERENCES

Cardinal, L., and M. K. Stenstrom. 1991. "Enhanced Biodegradation of Polyaromatic Hydrocarbons in the Activated Sludge Process." *Res. J. Wat. Pollut. Control Fed.* 63(7): 950-957.

Greer, C. W., J. A. Hawari, and R. Samson. 1990. "Influence of Environmental Factors on 2,4-Dichlorophenoxyacetic Acid Degradation by *Pseudomonas cepecia* Isolated From Peat." *Arch. Microb. 154*: 317-322.

Häggblom, M. 1990. "Mechanisms of Bacterial Degradation and Transformation of Chlorinated Monoaromatic Compounds." *J. Bas. Microbiol. 30*(2): 115-141.

Leahy, J. G., and R. R. Colwell. 1990. "Microbial Degradation of Hydrocarbons in the Environment." *Microbiol. Rev. 54*(3): 305-315.

Mueller, J. G., S. E. Lantz, B. O. Blattmann, and P. J. Chapman. 1991. "Bench-Scale Evaluation of Alternative Biological Treatment Processes for the Remediation of Pentachlorophenol- and Creosote-Contaminated Materials: Slurry-Phase Bioremediation." *Envir. Sci. Technol. 25*: 1055-1061.

Case Study: Bioremediation in the Aleutian Islands

Kara J. Steward and H. Donald Laford

ABSTRACT

This case study describes the design, construction, and operation of a bioremediation pile on Adak Island, which is located in the Aleutian Island chain. Approximately 1,900 m³ of petroleum-contaminated soil were placed in the bioremediation pile. The natural bioremediation process was enhanced by an oxygen and nutrient addition system to stimulate microbial activity. Despite the harsh weather on the island, after the first 6 months of operation, laboratory analyses of soil samples indicated a significant (80%) reduction in diesel concentrations.

INTRODUCTION

Adak Island is located off the coast of Alaska, near the western end of the Aleutian Island chain, approximately 1,930 km west of Anchorage, Alaska. Adak Island is approximately 51 km long and 34 km wide (725 square km). The U.S. Navy occupies almost 26,000 hectares of the northern half of the island. Adak Island's maritime climate is characterized by persistently overcast skies, high-velocity winds, and frequent, often violent, cyclonic storms originating in the northern Pacific Ocean and Bering Sea. The mean annual precipitation on the island is about 162 cm; mean monthly temperatures range from 0°C to 10°C; and wind velocity averages 28 km/h with gusts exceeding 185 km/h.

Many underground storage tanks have been removed from Navy property on the island. The soil below and surrounding the former tanks was contaminated with petroleum; it was excavated and stockpiled in accordance with state regulations. The cost of transport of these soils off the islands was prohibitive; therefore, alternatives were investigated. A treatability study performed on soil samples collected from the stockpiled soil indicated that the type and quantity of microbes in the soil were adequate for effective bioremediation. This study indicated the need for nutrient amendment. Other soil remediation alternatives that were evaluated included solidification, soil washing, low-temperature thermal desorbing, and chemical stabilization. The Navy decided to undertake on-site bioremediation of the petroleum-contaminated soils. The purpose of the bioremediation was to reduce the petroleum concentration of the soil to a level that is acceptable for use as backfill material or a landfill cover. The bioremediation process was undertaken in order to proactively reduce the volume of contaminated materials that would need to be transported off island. The operation of the bioremediation pile did not include testing the system (e.g., systematically operating with no air, no heat, or no nutrients).

DESIGN

An air, moisture, and nutrient addition system was designed to enhance the natural bioremediation process by introducing oxygen, water, and nutrients into the soil pile to stimulate microbial activity (Figure 1). The site selected for construction of the bioremediation pile was removed from residential and operation areas, thereby reducing the number of personnel exposed to the contaminated soil and minimizing the likelihood of unauthorized visitors. Furthermore, the location ensured that prevailing winds would carry emissions away from the island population center. Electricity and water were available at the site.

The bioremediation pile was sited on a concrete slab, providing an impermeable, level surface for construction. The bottom liner consisted of 0.76-mm-thick high-density polyethylene (HDPE) that would be able to withstand abrasion during construction and would be impermeable to petroleum hydrocarbons and

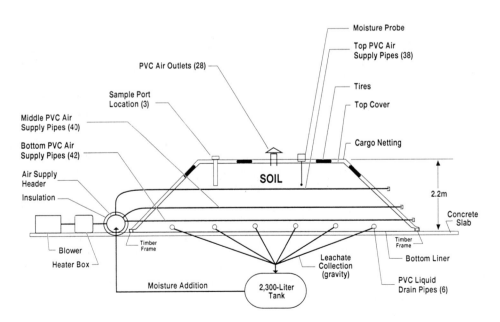

FIGURE 1. Bioremediation pile design.

leachate. The thickness of this liner exceeded the minimum thickness requirements for extruded fabric liners in the state guidance manual. The bottom liner
(65 m by 14 m) was seam welded to form a continuous waterproof membrane.
The bottom liner was secured by wrapping it over a 6-inch × 6-inch timber that
was secured to the concrete slab. The top cover, which was designed to shed
rainwater, was also specified as 0.76-mm-thick HDPE and seam welded. Because
of the high-velocity winds on the island, the top cover was secured with tires
and rope cargo netting, which were tied down to the concrete slab.

Leachate in the sump area of the pile was collected by perforated polyvinyl
chloride (PVC) pipes wrapped in geotech cloth to filter out soil fines. The leachate system drained by gravity to a 2,300-L tank adjacent to the pile. Three layers
of perforated filter-wrapped PVC pipe (1.3 cm diameter) were laid horizontally
across the pile to provide aeration, moisture, and nutrients. The pipes were
spaced approximately 1.52 m on center to ensure a uniform distribution of air.
Exposed piping was protected from the cold by fiberglass wrap insulation and
electric heat tapes that automatically turn on when the ambient temperature
drops below 2°C.

Before the soil was placed in the pile, it was screened (5- and 7.6-cm screens)
to remove rock and debris. Approximately 1,150 kg of ammonium nitrate, 47 kg
of monosodium phosphate, and 130 kg of disodium phosphate were distributed
on the surface of the soil for each 0.61 m of lift. The soil was aerated via a 20-hp
commercial roots-type blower forcing 156 L/sec of air at 0.7 kg/cm² into the
piping in the pile. The blower was included to provide sufficient aeration of
the stockpiled soil. Valve and gauge connections were provided to adjust and
balance the airflow for an even distribution in the pile. A total of 28 air vents,
each 15 cm in diameter, were installed in the pile and discharged to the atmosphere. An 11.4-kW heater was installed in the air supply system to raise the
temperature of the supply air. The soil temperature was monitored by six stationary thermocouples built into the soil pile.

Because of the potential for the airflow to dry out the soil in the pile, water
spray nozzles were installed in the air distribution pipe to add water to the pile.
Soil moisture was controlled via a moisture probe in the soil that would start
and stop the pump feeding water into the spray nozzles. Soil moisture inside
the pile was maintained at a level between 10 and 18%. This humidifier system
also provided a way to add nutrients to the soil.

CONSTRUCTION

The bioremediation pile was built in November and December 1993. Construction was sequenced to avoid damage to the liner and cover. First the timber
and angle iron hold-down frame was installed. After the frame was cleaned
to remove any sharp objects, the bottom liner was installed and fastened to
the 15- by 15-cm timbers. A limited quantity of contaminated soil was placed
on the bottom liner to prevent the liner from lifting as a result of the wind.

The bottom level piping for air supply and leachate collection was installed. Contaminated soil and nutrients were placed loosely on the pile using a front-end loader and backhoe bucket. The soil was not compacted and equipment was not operated on the soil pile.

Petroleum-contaminated soil was screened and placed in the pile along with the intermediate- and upper-level air supply piping. Approximately 1,900 m³ of soil were used to create the bioremediation pile, which measures 65 m by 14 m to a height of 2.2 m. After all of the soil was placed in the pile, the cover, vents, sample ports, cargo netting, and tires were installed. The mechanical and electrical connections for the blower and heater were then completed. The air supply was blown through the heater box into a 10-cm PVC main header that branched off to the piping in the pile. The moisture control system was tested and calibrated, and the air supply to each distribution pipe was balanced.

MONITORING

On a monthly basis, the condition of the pile and ancillary equipment was monitored and soil samples were collected. Because of the remote location, more frequent monitoring was not possible. Monthly monitoring included visual inspection of the cargo netting, the HDPE cover, the piping, and the leachate collection tank. The blower air filter and lubricating oil were changed monthly. The system design was intended to allow of addition of moisture and nutrient as required based on monthly analyses.

Soil samples were collected each month and submitted to the laboratory for direct total microbial count and analyses for ammonia-nitrogen, nitrate-nitrogen, and phosphate. Also each month, field tests were conducted for soil moisture, pH, and temperature. On a quarterly basis, soil samples were tested for gasoline-range organics, aromatic volatile organics (benzene, ethylbenzene, toluene, and xylenes), diesel-range organics, and total petroleum hydrocarbons. Soil pH (7 to 8.25), temperature (0 to 15°C), and moisture (10 to 15%) levels were optimal for microbial activity.

Figure 2 presents a comparison of the microbial count and diesel concentrations over the duration of monitoring. Concentrations of petroleum decreased after the first 6 months of operation. This comparison shows the greatest decrease in diesel concentration occurred during the highest microbial count. Additionally, reduction of diesel concentrations does not continue after the microbial count decrease in May. The microbial count significantly decreased after the fifth month of operation, possibly because of the decrease in available ammonia-nitrogen. Figure 3 shows a comparison of nutrient levels and microbial count over the duration of monitoring. The available ammonia-nitrogen levels appear to affect the microbial count. During the months of April, May, and June, additional nutrients were supplied to the bioremediation pile using the existing system. Based on the nutrient levels after these dates, the nutrient addition system was not effective.

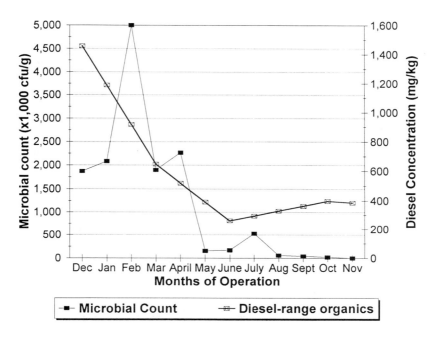

FIGURE 2. Comparison of microbial count and diesel concentration.

LESSONS LEARNED

The impacts of weather conditions on the bioremediation pile included frequent power outages during the winter, wind damage, rain damage, and dangerous working conditions. The initial electric supply to the blower was a standard manual magnetic starter design. Frequent power outages caused the motor breaker to trip and the blower to stop. Because the equipment was inspected only periodically, the blower remained off for days before the motor was manually restarted. The corrective action was to add an automatic restart circuit. Strong winds blew the rain hoods off the vent pipes allowing rainwater into the pile. The corrective action was to securely fasten the rain hoods in place.

The moisture controller key pad had a weatherproof rating but was not waterproof. Therefore, heavy rains damaged the unit. The corrective action was to place the unit inside a waterproof box. The surface of the HDPE cover was slippery when wet. However, the rope cargo netting provided a footing for ascending and descending the pile for sample collection and inspections. Mechanical PVC pipe and glue joint fittings were used for air distribution. Wind, pipe movement, and severe weather conditions caused leaks in the air supply piping resulting in reduced efficiency. Corrective action was taken during pipeline joint repair to ensure secure PVC welds. The water spray nozzles became plugged with rust from the water tank. The corrective action was to install an in-line filter.

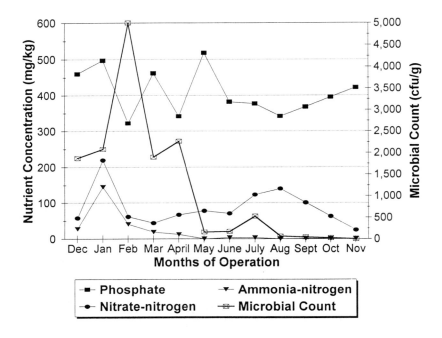

FIGURE 3.   Comparison of microbial count and nutrient concentration.

Only the concentrations of diesel-range organics were above the state soil cleanup level for all sample events. Reduction of the diesel concentration to 300 mg/kg corresponds to the success rate experienced in the original treatability study. The bioremediation pile system design was reevaluated at the end of this study and the existing system was recommended to be used with a modified nutrient addition system.

Economics of Biofiltration for Remediation Projects

Jerry M. Yudelson and Paul D. Tinari

ABSTRACT

Biofilters with granular activated carbon (GAC) filter backup units offer substantial savings compared to conventional GAC filters and catalytic/ thermal oxidation (Catox) units in controlling emissions of volatile organic compounds (VOCs) from petroleum remediation projects. Provided that the biofilter supplier is willing to satisfy the client's and consultant's risk-management concerns, biofilters offer a new method for reducing the cost of remediation projects, with savings of up to $10,000 (24%) per facility in 24-month projects and up to $16,000 (32%) per facility in 36-month projects for simple gas station remediation projects. Savings will be greater for longer projects and projects with higher average contaminant loadings.

INTRODUCTION

Biofiltration is a new form of a proven odor-control technology, used in Europe since the 1960s and recently adapted for use in controlling emissions of (VOCs) in remediation projects. Biofilters contain colonies of microorganisms that effectively metabolize petroleum hydrocarbon contaminants from remediation projects, including those of greatest concern, the BTEX family (benzene, toluene, ethylbenzene, and xylenes). Biofilters are designed to compete economically with GAC filters and Catox destruction technologies, in remediation projects using soil vapor extraction (SVE) or air-stripping technology to remove VOCs from contaminated sites.

This paper deals exclusively with the use of biofilters in remediation projects. Many states (e.g., California and Washington) are moving to control VOC emissions on remediation projects, and this requirement is expected to be in force nationwide within the next 2 to 3 years.

FEATURES AND BENEFITS OF ALTERNATIVE
METHODS FOR CONTROLLING VOC EMISSIONS
FROM REMEDIATION PROJECTS

A number of methods are available for controlling VOC emissions from soil vapor extraction and air stripping of contaminated groundwater. The most widespread approaches use GAC filtration or catalytic/thermal oxidation (Catox). Table 1 contrasts the important features of each approach, along with biofiltration. From the perspective of regulators and owners concerned with future liability, the best method provides for complete destruction of VOCs at the remediation site with the least possible energy cost. Each of the three methods has drawbacks in this regard. In addition, from a client's (and consultant's) perspective, the best method allows the client to meet regulatory obligations at the least possible cost, consistent with safety and reliability of operation. Other client selection criteria might include flexibility of the method, initial cost, total project cost, areal extent ("footprint") of the technology, and adaptability to other client projects.

The purpose of this paper is to present the relative economic costs and benefits of the newer biofiltration technology, in contrast with the other two methods. However, the Catox method is more expensive than the GAC method, in terms

TABLE 1. Features of major VOC control methods.

Method/Feature	Biofilters	GAC Filters	Catalytic/Thermal Oxidation
1. On-site destruction?	Yes	No	Yes
2. Works at ambient temperatures?	Yes	Yes	No
3. Provides complete destruction?	No[a]	No[b]	Yes
4. Easily adaptable to varying contaminant levels?	Yes[c]	Yes	Yes
5. Requires significant external energy use?	No	Yes[d]	Yes

(a) Biofilters generally are not 100% efficient at removing VOCs and must be backed up with GAC filters.
(b) GAC filters adsorb contaminants, but do not destroy them. When the adsorptive capacity is used up, the spent GAC along with the contaminants must be transported elsewhere for contaminant destruction.
(c) Biofilters are generally sized to handle combinations of contaminant levels and flowrates. Higher contaminant loadings can be tolerated, but overall removal rates ($g/m^3/h$) will decrease.
(d) GAC filters require external energy to destroy adsorbed contaminants.

of both initial cost and total project cost. Catox technologies typically are used today when complete on-site destruction is required. However, the relatively high cost of this choice leads most clients to select the GAC option. Therefore, this paper focuses on the relative cost effectiveness of biofiltration compared to GAC filtration alone.

ECONOMIC COMPARISON OF BIOFILTERS WITH GAC FILTERS

In providing an economic contrast of biofilters with GAC filters, we consider only the option of biofilters that are "backed up" with a small GAC filter. This is because biofilters do not yet provide 100% contaminant destruction and, therefore, are not recognized as the best available control technology (BACT) by air quality regulators.

Biofilters have the following characteristics:

- Reduction of 80% to 90% in carbon use from GAC filters alone
- Low operating cost compared with GAC filters
- No significant external energy use
- Higher capital cost than GAC filters.

Biofilters have a higher capital cost and a lower operating cost than GAC filters; thus, there will clearly be a point in time at which biofilters are a more economic option. Because most remediation projects using soil vapor extraction or a "pump and treat" technology tend to last at least 2 to 3 years, the economic comparison will be based on projects of that duration.

Tables 2 and 3 give the basic economic and technical variables used in the economic comparison of the two VOC-control technologies.

Results of Economic Analysis

Direct Purchase. For a 24-month project, the biofilter at 90% efficiency (with GAC backup) provides a savings in total project costs of 24% over a conventional GAC filter, assuming the client purchases the unit. For a 36-month project, the savings are 32%.

Lease Option. If the client prefers not to purchase and own the biofilter, a lease option is available at a rate of 5% of the capital cost per month for a 24-month lease, and 3.5% of the capital cost per month for a 36-month lease. Because of the cost of financing, the biofilter is relatively more expensive in this situation; however, monthly service costs are included in the lease price. In the case of a lease for a 24-month project, the biofilter provides a savings in total project costs of 19% over a conventional GAC filter. For a 36-month project, the savings are 26%. Table 4 provides an overview of the economic analysis.

TABLE 2. Key economic variables.

Key Variables	Biofilters	GAC Filters
1. Capital cost	$27,500[a]	$11,000[b]
2. Operating & maintenance (O & M) cost	$125/mo	$250/mo
3. Cost of carbon	$0.85/lb ($1.87/kg)	$0.85/lb[c] ($1.87/kg)
4. Cost of carbon replacement	$2.50/lb ($5.50/kg) (1,000-lb unit)	$2.50/lb[d] ($5.50/kg) (1,000-lb unit)

(a) Cost delivered and installed from World Envirotech anywhere in Oregon and Washington. Biofilter provides approximately 4 m³ of filter volume.
(b) Quoted cost on a gas station remediation project in Seattle, Washington, based on two 1,000-lb (454-kg) carbon canisters fully loaded.
(c) Current market costs for large carbon users, based on telephone survey in August 1994 from major West Coast suppliers.
(d) Depends on the size of the carbon unit, because there are high fixed costs associated with professional time and transportation for each regeneration cycle. A 1,000-lb (454-kg) carbon unit is typical in these projects. Smaller carbon units have higher unit regeneration costs.

Sensitivity Analysis

In any economic comparison, it is useful to look at changes in outcome as a function of changes in key input variables. In the case of the biofilter/GAC comparison, the key variables are the biofilter efficiency and the cost of carbon regeneration. A detailed examination of these changes is beyond the scope of this paper, but the general results can be summarized as follows:

1. Reducing biofilter efficiency from 90% to 75% reduces carbon savings by 8% (nominal) and $3,452 in the 24-month project, and by 10% (nominal) and $4,704 in the 36-month project, owing to greater carbon use.
2. Reducing the size of the carbon canister from 1,000 lb to 750 lb (with a reduced capital cost of $3,500 per unit) increases the biofilter benefit by 1% (nominal) in the 24-month project and by 2% in the 36-month project. In this case, the reduced capital cost of the carbon units is offset by the higher cost of the need for more frequent carbon regeneration.

SUMMARY

Biofilters (with GAC filter backup units) offer substantial savings in comparison with conventional GAC filters and Catox units in controlling emissions of VOCs from petroleum remediation projects.

TABLE 3. Key technical variables.

Key Variables	Biofilters	GAC Filters
1. Contaminant loading rate	4 lb/day (1.8 kg/day)	4 lb/day (1.8 kg/day)
2. Reduction in loading rate	10% each 6 months	10% each 6 months
3. Contaminant adsorption rate	85%	0.3 lb/lb (0.3 kg/kg)
4. Typical flowrate	50 to 150 SCFM (25 to 75 L/s)	50 to 150 SCFM (25 to 75 L/s)

Provided that the biofilter supplier is willing to satisfy the client's and consultant's risk management concerns, biofilters offer a new method for reducing the cost of remediation projects, with savings of up to $10,000 per facility in 24-month projects, and up to $16,000 per facility in 36-month projects for simple gas station remediation projects. Savings will be greater in longer-term projects and in projects with higher average contaminant loadings (up to about 1,000 ppmv contaminant loading, at which point Catox units begin to provide better economics than do biofilters).

TABLE 4. Overview of economic comparison.

Cost Item	Biofilter with GAC Backup	GAC Filter Alone
1. Capital cost	$27,500	$11,000
2. Total project cost (purchase option)	$30,301 (24 mo) $35,135 (36 mo)	$40,013 (24 mo) $51,354 (36 mo)
3. Biofilter savings	—	24% (24 mo) 32% (36 mo)
4. Total project cost (lease option)	$32,301 (24 mo) $37,785 (36 mo)	$40,013 (24 mo) $51,354 (36 mo)
5. Biofilter savings	—	19% (24 mo) 26% (36 mo)

Respirometric Oxygen Demand Determinations of Laboratory- and Field-Scale Biofilters

Denis Rho, Pierre Mercier, Jean-François Jetté,
Réjean Samson, Jiyu Lei, and Benoit Cyr

ABSTRACT

A biofiltration experiment operated at three inlet concentrations (425, 830, and 1450 mg m^{-3}), showed that the specific oxygen consumption rate was highly correlated (R = 0.938, n = 23) with the toluene elimination capacity. A radiorespirometric test was found to be more sensitive and appropriate for the field-scale biofilter treating gasoline vapors.

INTRODUCTION

The biological activity of a filter bed operated under specific environmental and operating conditions is best described by the elimination capacity (EC) parameter. The presence of inhibitory components, nutrition depletion zones, dry filter bed material, and dead zones has a detrimental effect on the spatial distribution, within the filter bed, of the microbial population responsible for the removal of pollutants from gaseous effluents. Biologically active sections can be localized by establishing a detailed concentration profile along the biofilter (Ergas et al. 1994, Deshusses and Hamer 1993) or by evaluating the spatial distribution of the microbial populations (Ergas et al. 1994). The first method necessitates installing many reliable and convenient sampling ports, whereas the second method requires multiple time-consuming steps (extraction, plating, incubation, etc.). In the present work, the specific oxygen consumption rate (qO_2) of the filter bed, measured with an absolute pressure respirometer, is presented as a simple alternative to these two methods. Waste biodegradability and biokinetic constants determined by respirometric techniques have been used for designing activated sludge and soil remediation site bioprocesses. However, this practice is not common for the design and operation of packed-bed biofilters. A preliminary experiment indicated a possible relationship between the filter bed's qO_2 and the EC (Rho et al. 1994). A more detailed investigation of the relation between these two parameters is presented here.

MATERIAL AND METHODS

Bioreactors and Analytical Methods

Biofilter Laboratory Setup. The biofiltration setup consisted of a humidification unit, a toluene saturator, and a biofilter. The stainless steel 30-L biofilter was made of two cylindrical parts, each 0.5 m high and 0.2 m internal diameter. At the bottom of each stage, a stainless steel grid (mesh 4) was installed, providing two distinct filter beds of about 0.40 m. The filter bed was a commercial composted peat moss and chicken manure blend. Prehumidified air was fed into the biofilter. Six ports, uniformly distributed along the biofilter, allowed for gas sampling as well as temperature and pressure measurement. In packed bed biofilters the water content of the filter material decreased more rapidly where the gas entered the bed (especially at high organic load). Consequently, all experimental runs were operated with gas flowing downwards because it was easier to add water from the top of the reactor. Biofilter bed heights were determined according to the gas flow direction ($h = 0$, at the gas inlet). All biofiltration experiments were run at room temperature (20 to 24°C). A comprehensive description of the laboratory setup is presented by Rho et al. (1994).

Field Biofiltration Setup. The biofiltration unit was used to purify off-gases from a bioventing operation (phase 1, days 0 to 80) and a sparging process (phase 2, days 100 to 132) used, respectively, for the decontamination of soil and groundwater contaminated with gasoline hydrocarbons (Lei et al. 1995, 1994). The setup consisted of an air/water separator, a biotrickling filter and, a biofilter. The latter is the unit of interest for this work. The contaminated air was forced to pass through the biotrickling filter before entering, from the bottom, the biofilter filled with 1 m of a compost-based material. The superficial velocity was kept constant at 50 m h^{-1}.

Gas Analysis. The gas samples were collected during the laboratory and field experiments into 250-mL glass sampling tubes from several sampling ports distributed along the biofilters. Concentrations of toluene, benzene, and xylenes were determined using a GC equipped with a FID. All gaseous contaminant concentrations reported herein are expressed in mg m^{-3} at 20°C and 1 atmosphere.

Respirometry. Respirometry of the filter bed was conducted using an absolute pressure respirometer (APR). The lab biofilter was emptied and the filter material was divided into four parts corresponding to four bed sections (0 to 10 cm, 10 to 30 cm, 30 to 40 cm, and 40 to 50 cm). Each section was then separately homogenized before sampling and testing. The remaining filter material was returned to its original location in the reactor. For the field investigation, the filter bed was divided into two equal sections (0 to 50 cm and 50 to 100 cm) from which representative samples were taken by a core sampler. These two procedures maintained the filter bed's profile. The filter material (40 g wet weight) taken from a defined section of the filter bed was placed, immediately after being sampled, in a closed vessel of the APR system and incubated in a thermostated water bath (20°C). The cumulative O_2 supplied and the incubation time were used to construct the respiration curves (mg O_2 consumed vs. time). The specific oxygen consumption rate (qO_2, mg O_2 kg^{-1} dw h^{-1}) was calculated by dividing the slope of these curves by the weight of dry compost used for the test.

Microcosm. In 125-mL serum bottles closed with Mininert valves, biofilter samples (5 g wet weight) were incubated in the presence of a known amount (approximately 100,000 disintegration per minute) of uniformly labeled ^{14}C-toluene or ^{14}C-benzene. In combination with the radioactive substrate, 1 µL of benzene or toluene was also added to the serum bottles. The radiolabeled carbon dioxide was trapped in a potassium hydroxide solution (30%, w/w) that was removed for analysis every 30 min for 3 h using a scintillation counter. The specific mineralization rates were calculated as counts per minutes (CPM) g^{-1} dw h^{-1}.

RESULTS AND DISCUSSION

Laboratory-Scale Biofiltration

A lab-scale toluene biofiltration experiment was run at three inlet concentrations (425 mg m^{-3} ± 34 mg m^{-3}, n = 4; 830 mg m^{-3} ± 23 mg m^{-3}, n = 5; and 1450 mg m^{-3} ± 73 mg m^{-3}, n = 5), for 17 days, 17 days, and 14 days, respectively. The superficial velocity was maintained constant at 25 m h^{-1}. The first parameter to be determined was the EC. Throughout this experiment, several toluene concentration profiles were determined on 14 separate occasions to assess the performance of the reactor. These profiles (Figure 1 shows four profiles provided as examples) were used to determine the "local" EC of four different sections of the filter bed (i.e., 0- to 10-cm, 10- to 30-cm, 30- to 40-cm, and 40- to 50-cm layers). The performance of this filter bed in terms of superficial velocity, toluene concentration, and organic load has been discussed in a previous paper (Rho et al. 1994). The second parameter to be determined was the specific oxygen consumption rate (qO_2). The qO_2 was determined for the 0- to 10-cm layer (6 times) during biofiltration of toluene performed at 425 mg m^{-3} and 830 mg m^{-3}, and for four sections (0- to 10-cm, 10- to 30-cm, 30- to 40-cm, and 40- to 50-cm layers) at 1,450 mg m^{-3} (4 times).

Respiration curves of four different filter bed samples (Figure 1, 1,403 mg m^{-3}) and one control sample (fresh compost) are shown in Figure 2. The O_2 consumption curves were all linear (zero-order reaction rate) over a 20-h incubation period. The slope of these curves was used to calculate the qO_2 parameter. The qO_2 of each layer and the corresponding EC are presented in the insert of Figure 2. The maximum qO_2 (143 mg O_2 kg^{-1} dw h^{-1}) and EC (120 g kg^{-1} dw h^{-1}) were measured for the 10- to 30-cm layer. All other bed sections showed intermediate qO_2 and EC results. They ranged between the maximum qO_2 and EC values and the basal response of the control sample. Moreover, the qO_2 measurements of each bed section were different from each other, contrary to the case where

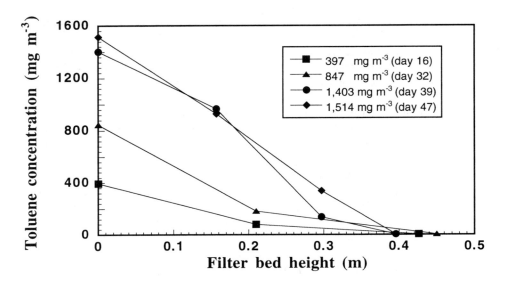

FIGURE 1. Toluene concentration profiles as a function of the filter bed height. Surface load = 25 m h^{-1}.

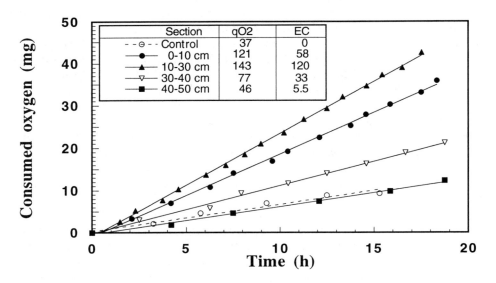

FIGURE 2. Oxygen consumed by filter bed samples taken at different levels in the biofilter. Operating conditions: v= 25 m h^{-1}, toluene inlet concentration = 1,403 mg m^{-3}·qO$_2$ (mg kg^{-1} dw h^{-1}), EC (g m^{-3} h^{-1}).

the toluene concentration profile was linear (1,514 mg m^{-3}, Figure 1). In that case, all qO$_2$ measurements were similar (0 to 10 cm: 148 mg O$_2$ kg^{-1} dw h^{-1}, 10 to 30 cm: 158 mg O$_2$ kg^{-1} dw h^{-1}, 30 to 40 cm: 163 mg O$_2$ kg^{-1} dw h^{-1}).

All EC data, measured for various sections during the toluene biofiltration experiment, were plotted as a function of their respective qO$_2$ values (Figure 3). This plot clearly shows that the specific oxygen consumption rate is highly correlated ($R = 0.938$, $n = 23$) with the elimination capacity. A linear regression fitting resulted in the following equation: EC = 0.77 × qO$_2$ − 27.2. This equation shows that the EC is positively proportional to the qO$_2$. In other words, the qO$_2$ reflects the potential amount of toluene that can be degraded by the microbial population of the respective layer. The relation between the EC and qO$_2$ was valid for ECs ranging from 0 to nearly 120 g m^{-3} h^{-1} and was independent of the experiment protocol.

The x-intercept (35.3 mg O$_2$ kg^{-1} dw h^{-1}) and the slope (0.77 g toluene g^{-1} oxygen) of the above equation. Both have biological meaning. The former corresponds to the basal respiration rate of a filter bed subjected to zero organic load (37 mg kg^{-1} h^{-1}) and the latter potentially represents the specific substrate oxygen yield of the microflora of that filter bed. Based on a stoichiometric equation for bacterial growth on toluene as the only C source, a theoretical yield of 0.4 g toluene g^{-1} oxygen was obtained (calculation not shown). The 2-fold difference between this value and the experimental data could be the consequence of a metabolic shift; the microflora within the respirometer reactor was incubated in the absence of toluene vapors where as within the biofilter toluene vapors were present.

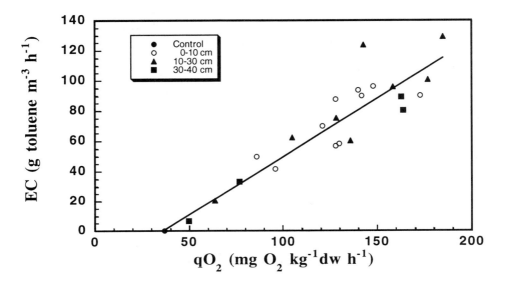

FIGURE 3. Correlation between the elimination capacity and the specific oxygen consumption rate.

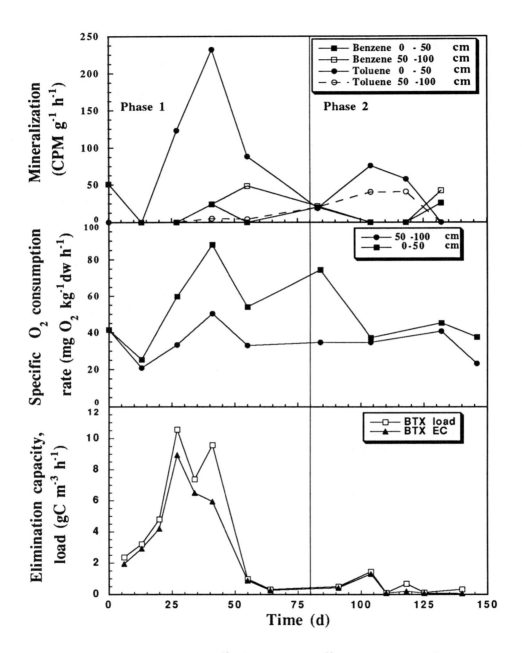

FIGURE 4. Mineralization of ^{14}C-benzene and ^{14}C-toluene, specific oxygen consumption rate, BTX elimination capacity, and load of a field-scale biofilter treating off-gases from soil venting.

Field Biofiltration

A field investigation was conducted by Biogénie Inc. to test the removal of volatile hydrocarbon compounds, principally benzene, toluene, and xylenes (BTX), with a novel biofiltration unit. The results of the BTX-EC and the BTX load are presented as a function of the operation time (Figure 4). A respirometric test and a specific mineralization test were performed on filter bed samples taken in the packed-bed biofilter. The evolution of the qO_2 and the benzene and toluene mineralization rates were in good agreement with the BTX-EC profile.

During phase 1, the BTX load reached a maximum between day 25 and day 40, while the qO_2 and the benzene and toluene mineralization rates peaked at day 40. These rates were highest for the bottom section (0 to 50 cm) since the gas was flowing through that section first. From day 100 to 150 (phase 2), the BTX load was approximately 1 gC m^{-3} h^{-1}, that is 9 times less than the average organic load between day 25 and day 40. During phase 2, the results of the mineralization tests showed a higher toluene-related activity than the benzene-related activity, except at day 127. The mineralization rate of the toluene (0 to 50 cm) was 5 times less than the rate at day 40. Throughout the second phase, the respiration rate was similar to the nonacclimated compost (day 0, 40 mg O_2 kg^{-1} h^{-1}). At low organic load, the radiorespirometric test was therefore a more sensitive and specific method than the respirometric test. The former determines the degradation rate of a target pollutant, whereas the latter determines the oxygen consumption of the whole microflora.

The regression equation obtained from the laboratory experiment cannot be used to predict the elimination capacity from the qO_2 measurements made during the field investigation. The main reasons include the differences between the filter material used, the pollutants to be treated, and the microbial population of each filter bed. All three parameters are known to have a significant impact on the qO_2 parameter. The difficulty of interpreting the qO_2 measurements come from the fact that during the field experiment the volatile organic compounds fed into the biofilter were gasoline vapors.

The laboratory test showed that there is a direct relationship between the EC and the qO_2 measured with the respirometer. Therefore, the results of the respirometry test could be used as a specific indicator of the biofilter's performance or of the physiology of the microbial population present in a specific section of the reactor. The relationship between the qO_2 and EC parameters is specific for the lab-scale biofilter adapted to toluene vapors. For this reason the relationship found is not valid for the field biofilter treating gasoline vapors.

REFERENCES

Deshusses, M. A., and G. Hamer. 1993. "The Removal of Volatile Ketone Mixtures from Air in Biofilters." *Bioprocess Eng.* 9: 141-146.

Ergas, S. J., E. D. Schroeder, and D.P.Y. Chang. 1994. "Control of Air Emissions of Dichloromethane, Trichloroethene and Toluene by Biofiltration." *Proceedings of the 86th Annual Meeting of the Air and Waste Management Association*, Denver, CO.

Lei, J., D. Lord, R. Arneberg, D. Rho, C. Greer, and B. Cyr. 1995. "Biological Treatment of Waste Gas Containing Volatile Hydrocarbons." In R. E. Hinchee, G. D. Sayles, and R. S. Skeen (Eds.), *Biological Unit Processes for Hazardous Waste Treatment*. Battelle Press, Columbus, OH. pp. 275-282.

Lei, J., D. Lord, R. Arneberg, D. Rho, P. Mercier, and B. Cyr. 1994. "Design, Field Operation and Performance of a Biological System to Remove Volatile Hydrocarbons from Waste Gases." *Proceedings of the 4th Annual Symposium on Groundwater & Soil Remediation*, Calgary, Alberta, Canada. pp. 691-696.

Rho, D., P. Mercier, J. F. Jetté, R. Samson, J. Lei, and B. Cyr. 1994. "Performances of a 30-L Biofilter for the Treatment of Contaminated Air with Toluene and Gasoline Vapors: Filter Media Selection, Microbial Activity and Substrate Interactions." *Proceedings of the 4th Annual Symposium on Groundwater & Soil Remediation*, Calgary, Alberta, Canada. pp. 671-682.

Development of Novel Biofilters for Treatment of Volatile Organic Compounds

Dolloff F. Bishop and Rakesh Govind

ABSTRACT

Biofiltration involves contacting a contaminated gas stream with immobilized microorganisms in a contactor to biodegrade the contaminants. It is emerging as an attractive technology for removing low concentrations (i.e., less than 800 ppmv) of volatile organic chemicals (VOCs) from air. Compared with other technologies, biofiltration fully mineralizes the contaminants, is inexpensive and reliable, and requires no posttreatment. In the study described in this paper, four types of media consisting of porous ceramic monoliths with several straight passages were studied to determine the effects of adsorptive and nonadsorptive media on biofilter startup time, dynamic response to step changes in inlet substrate concentration, biofilm adherence, and overall VOC-removal efficiency. Volatile compounds studied were benzene, toluene, ethylbenzene, m-xylene, and o-xylene. Adsorbing media such as activated carbon, when compared with nonadsorbing media such as ceramic, exhibit faster biofilter startup, are more stable to dynamic changes in inlet concentration, and attain higher VOC-removal efficiencies due to better adherence of biofilm on media surfaces.

INTRODUCTION

Technologies for controlling VOCs have become increasingly important because of environmental regulations and their economic impact. VOC air emissions are most commonly treated by adsorption, absorption, or incineration. Biofiltration, as an alternative to these technologies, uses microorganisms to degrade or transform most VOCs (Ottengraf 1986). Biofiltration has been employed for purification of waste gases since the early 1960s. Recently, studies have been conducted on the use of pelletized and ceramic structured media in biofiltration (Utgikar et al. 1991; Govind and Bishop 1993). Conventional applications of biofilters use fine or irregular support media, such as soil, peat, or compost. Compared with conventional biofilters, biofilters with pellets or structured media potentially have significant advantages, including better gas distribution, improved pH control by using buffers in the nutrient solution trickling through the bed, and the ability to remove excess biomass from the media.

In this study, experiments were conducted to understand the role of the support media in air biofiltration. Previous studies (Utgikar 1993; Govind and Bishop 1993) have shown that biofilm support materials may affect the efficacy of biofiltration because (1) the geometry of the medium controls the biofilm's surface area per unit volume; (2) biofilm thickness depends on adhesion between the cells and the surface of the medium (Utgikar 1993); (3) adsorptive media can allow the adsorbed substrate to back-diffuse into the biofilm; and (4) porous media can improve nutrient retention, which keeps the biofilm moist and maintains a supply of essential minerals.

Four types of support media were studied: (1) nonadsorbing cordierite, which is a ceramic material; (2) cordierite coated with a resin; (3) cordierite coated with a 0.1-mm layer of carbon; and (4) cordierite coated with a 0.1-mm layer of activated carbon, with efficient adsorption of substrates. Cordierite was chosen since it was commercially available from Corning Glass Inc., New York, and could be obtained either uncoated or coated with activated carbon and resin materials. All four support media were extruded in straight-passage geometry with 100 cells per square inch.

MATERIALS AND METHODS

Experimental Apparatus

The experimental apparatus consisted of four separate biofilters, with a different medium in each biofilter. Each biofilter was a jacketed glass column with an internal diameter of 2.5 cm and a height of 30 cm. Contaminated air was prepared by injecting contaminant mixtures using a Sage syringe pump (Model 355, Fisher Scientific) into airflow controlled by MKS thermal mass-flow controllers (Controller 1259, Control Module 247). The contaminated airflow was split into four streams using the mass-flow controllers and fed to the four biofilters. The arrangement ensured uniformity of composition and flowrate across the four biofilters. Nutrient solution was sprayed into the top of the biofilters. Ports at the outlet of each of the biofilters and one inlet port enabled the sampling and analysis of influent and effluent gas streams.

Nutrient Solution Composition

The nutrient solution contained the following ingredients: KH_2PO_4 (85 mg/L), K_2HPO_4 (217.5 mg/L), $Na_2HPO_4 \cdot 2H_2O$ (334 mg/L), NH_4Cl (25 mg/L), $MgSO_4 \cdot 7H_2O$ (22.5 mg/L), $CaCl_2$ (27.5 mg/L), $FeCl_3 \cdot 6H_2O$ (0.25 mg/L), $MnSO_4 \cdot H_2O$ (0.0399 mg/L), H_3BO_3 (0.0572 mg/L), $ZnSO_4 \cdot 7H_2O$ (0.0428 mg/L), $(NH_4)_6Mo_7O_{24}$ (0.0347 mg/L), $FeCl_3 \cdot EDTA$ (0.1 mg/L), and yeast extract (0.15 mg/L).

Support Media

Four extruded support media supplied by Corning, Inc., New York, were used in the study: uncoated cordierite, carbon-coated cordierite, resin-coated cordierite, and activated-carbon-coated cordierite. These support media were geometrically identical. Each was a cylinder, 2.5 cm diameter × 2.5 cm high, with 100 channels per square inch (per 6.45 cm^2) of cross-sectional area. The channel side length and wall thickness were 2.19 mm and 0.35 mm, respectively. Each biofilter was packed with 8 cylinders separated by a 0.5-inch (1.3-cm) space, producing a nominal packed volume of 100 cm^3.

Contaminants

The contaminants (substrates) targeted for the study were benzene, toluene, ethylbenzene, *ortho*-xylene, and *meta*-xylene. The desired concentration of the substrates in the gas phase was obtained by premixing the required quantities of the substrates to form a stock solution, and injecting a portion of the solution into the airstream at a controlled rate. Portions of the stock solution were stored in 5-mL glass vials equal to the syringe volume.

Analytical Procedure

The gas samples were analyzed in a Hewlett-Packard 5710A gas chromatograph equipped with 20 ft (6 m) of 1/8-in. (0.3-cm)-diameter column, packed with PT 10% Alltech AT-1000 on Chromosorb W-AW 80/100 and a flame ionization detector. N_2 was used as the carrier gas. CO_2 was measured on a Fisher 1200 Gas Partitioner equipped with a thermal conductivity detector.

RESULTS AND DISCUSSION

Adsorption Capacity of the Support Media

The four columns were operated first as control columns (i.e., no biomass was seeded on the supports) to determine the adsorption capacity of the substrates. For the breakthrough study, the flowrate of air through the column was 50 cm/min (gas-phase residence time of 2 min). Substrate composition was benzene, 100 ppmv; toluene, 500 ppmv; ethylbenzene, 100 ppmv; *ortho*-xylene, 25 ppmv; and *meta*-xylene, 25 ppmv. The nutrient flowrate was 400 mL/day. Inlet and outlet gas samples were analyzed for each control column until complete breakthrough of all the substrates was observed at the outlet of the column, as indicated by equality of inlet and outlet concentrations.

The results of the breakthrough study revealed that the uncoated cordierite and resin-coated cordierite had negligible capacity for the adsorption of substrates as evidenced by the short time (< 2 h) required for the compounds to completely break through the column. Carbon-coated and activated carbon-coated cordierite

had significantly more adsorption capacity than the other media. The breakthrough times and the adsorption capacities of the carbon and activated carbon
media are shown in Table 1. As expected, the activated-carbon-coated cordierite
had much higher capacity than the carbon-coated cordierite for adsorption of
substrates.

Startup of the Biofilters

The four biofilters were seeded with biomass and acclimated to the substrates
in an acclimation bioreactor. The biofilters were seeded by recirculating the
acclimated biomass suspension through the four biofilters until significant quantities of biomass were immobilized on the media, as visually confirmed by
decreased turbidity of the recirculating biomass suspension. Operation of the
biofilters was started with the flowrate of air through the column at 50 cm/min
(gas-phase residence time of 2 min). The substrate composition was benzene,
100 ppmv; toluene, 500 ppmv; ethylbenzene, 100 ppmv; *ortho*-xylene, 25 ppmv;
and *meta*-xylene, 25 ppmv. Nutrient flowrate was 400 mL/day. The removal
efficiency of the biofilters was monitored as a function of time.

The results of the startup study are summarized in Table 2. The uncoated
cordierite required more startup time than the other media. This was mainly
caused by inefficient biofilm attachment on the cordierite surface, as visually
observed using flat plates of cordierite (Utgikar 1993). Toluene attained high
removal efficiency in 6 days, whereas *m*-xylene required 17 days to achieve high
removal efficiency; *m*-xylene required even more time than *o*-xylene to achieve
high removal efficiency. The resin-coated biofilter performance was very similar
to the uncoated cordierite performance, because both media had negligible
adsorption of the compounds. Carbon-coated cordierite performance was better

TABLE 1. Results of the breakthrough study.

Compound	Carbon-coated cordierite		Activated-carbon-coated cordierite	
	Breakthrough time (days)	Adsorption capacity (mg/g of support)	Breakthrough time (days)	Adsorption capacity (mg/g of support)
Benzene	4.0	0.83	18.4	3.82
Toluene	10.1	11.18	26.9	30.05
Ethylbenzene	15.2	3.86	18.0	4.63
m-Xylene	4.0	0.22	17.2	0.95
o-Xylene	4.5	0.25	21.6	1.20

TABLE 2. Biofilter startup time for the four support media.

Biofilter support media	Benzene	Toluene	Ethylbenzene	*m*-Xylene	*o*-Xylene
	Time required to reach 50% removal (days)				
Uncoated Cordierite	7	4	6	12	8
Carbon-Coated Cordierite	4	3	4	8	9
Resin-Coated Cordierite	8	5	6	13	9
Activated-Carbon-Coated Cordierite	2	1	1	4	3
	Time required to reach >90% removal (days)				
Uncoated Cordierite	14	6	8	17	10
Carbon-Coated Cordierite	8	5	6	11	11
Resin-Coated Cordierite	15	7	9	16	12
Activated-Carbon-Coated Cordierite	4	2	2	6	5

than the uncoated and resin-coated media. The fastest startup and highest contaminant adsorption was achieved by the activated-carbon-coated biofilter. The main reasons for superior performance of the activated-carbon-coated biofilter were (1) back-diffusion of the adsorbed contaminants into the biofilm, as shown conclusively by studying biodegradation of trichloroethylene (TCE) with toluene preadsorbed in the activated carbon coating (Govind and Bishop 1993), and (2) better biofilm attachment on the activated-carbon surface, as visually observed using flat plates of cordierite coated with activated carbon (Utgikar 1993).

Dynamic Response of the Biofilters

The biofilters were operated with gas and nutrient flowrates as above, but the contaminant mixture was changed to obtain an airstream containing 100 ppmv of benzene, 500 ppmv of toluene, and 25 ppmv each of *m*- and *o*-xylenes. Biofilters were operated at these conditions until the removal efficiencies of all four biofilters for all components had attained steady-state values. At this point, a disturbance was introduced in the system by doubling the injection rate of the stock solution, which resulted in increasing the inlet concentrations to 200 ppmv benzene, 1,000 ppmv toluene, 50 ppmv *m*-xylene, and 50 ppmv *o*-xylene at an inlet-air flowrate of 50 mL/min (empty-bed gas-phase residence time of 2 min). The response of the biofilters was monitored as a function of time by analyzing the outlet gas streams.

The dynamic response of the four biofilters to sudden increases in inlet concentrations is shown in Table 3. The biofilms on uncoated and resin-coated

TABLE 3. Dynamic response of the four biofilters for a step change in inlet concentration[a].

Biofilter support media	Benzene	Toluene	Ethylbenzene	*m*-Xylene	*o*-Xylene
	Time required to reach 50% removal (days)				
Uncoated Cordierite	5	2	2	6	7
Carbon-Coated Cordierite	3	2	2	4	5
Resin-Coated Cordierite	5	3	3	7	8
Activated-Carbon-Coated Cordierite	2	<1	<1	2	2
	Time required to reach >90% removal (days)				
Uncoated Cordierite	7	3	4	7	9
Carbon-Coated Cordierite	5	3	3	6	7
Resin-Coated Cordierite	8	4	4	8	9
Activated-Carbon-Coated Cordierite	3	<1	<1	3	3

(a) Empty-bed residence time = 2 min.

cordierite required the longest time to respond to the dynamic step change in inlet composition. Both biofilters required 9 days after the step change to reestablish high removal efficiency (>90%) for all compounds. The fastest dynamic response was produced by the activated-carbon-coated biofilter, probably because of increased adsorption of contaminants on the medium's surface, thereby increasing the effective residence time of the contaminants for biodegradation in the biofilter. The dynamic response of the carbon-coated biofilter was slower than the activated-carbon-coated cordierite biofilter, but faster than the uncoated and resin-coated biofilters.

Effect of Operating Parameters

The experimentally determined removal efficiencies of the biofilters for the substrates are shown in Table 4. The removal efficiencies are highest for all biofilters at low inlet-substrate concentrations and high residence times, and lowest at the other extreme of high inlet concentration and low residence time. The best performance was achieved by the activated-carbon medium biofilter. This is most likely related to better biofilm attachment and thicker biofilms due to back-diffusion of adsorbed contaminants into the biofilm. The uncoated and resin-coated cordierite exhibited similar performances, and the carbon-coated medium's performance was intermediate between the uncoated and activated-carbon-coated media.

TABLE 4. Removal efficiencies of the four biofilters as a function of operating conditions.

Biofilter support media	Airflow rate (mL/min)	% Removal efficiency at low inlet concentrations				
		Benzene (100 ppmv)	Toluene (500 ppmv)	Ethylbenzene (100 ppmv)	m-Xylene (25 ppmv)	o-Xylene (25 ppmv)
Uncoated Cordierite	50	60	98	97	81	75
	100	40	90	92	60	42
Carbon-Coated Cordierite	50	85	>99	>99	90	83
	100	72	95	96	77	68
Resin-Coated Cordierite	50	67	97	97	79	72
	100	55	91	92	57	49
Activated-Carbon-Coated Cordierite	50	>99	>99	>99	98	98
	100	>99	>99	>99	95	96

		% Removal efficiency at high inlet concentrations				
		Benzene (200 ppmv)	Toluene (1,000 ppmv)	Ethylbenzene (200 ppmv)	m-Xylene (50 ppmv)	o-Xylene (50 ppmv)
Uncoated Cordierite	50	42	64	67	54	49
	100	29	62	59	32	22
Carbon-Coated Cordierite	50	66	83	80	77	68
	100	50	67	70	52	48
Resin-Coated Cordierite	50	44	60	65	50	47
	100	22	64	54	37	28
Activated-Carbon-Coated Cordierite	50	92	>99	>99	87	85
	100	91	>99	>99	82	84

REFERENCES

Govind, R., and D. F. Bishop. 1993. "Environmental Bioremediation Using Biofilters." Paper presented at Frontiers in Bioprocessing III, Boulder, CO.

Ottengraf, S.P.P. 1986. "Exhaust Gas Purification." In H. J. Rehm and G. Reed (Eds.), *Biotechnology* 8: 301-332. VCH, Weinheim, Germany.

Utgikar, V., R. Govind, Y. G. Shan, R. C. Brenner, and S. I. Safferman. 1991. "Biodegradation of Volatile Organic Compounds in a Biofilter." In D. W. Tedder and F. G. Pohland (Eds.), *Emerging Technologies for Hazardous Waste Management II*, pp. 342-357. ACS Symposium Series 468, ACS, Washington, DC.

Utgikar, V. 1993. "Fundamental Studies on the Biodegradation of Volatile Organic Chemicals in a Biofilter." Ph.D. Thesis, University of Cincinnati, Cincinnati, OH.

Control of Air Emissions From POTWs Using Biofiltration

Todd S. Webster, Joseph S. Devinny,
Edward M. Torres, and Shabbir S. Basrai

ABSTRACT

The University of Southern California (USC), in collaboration with the County Sanitation Districts of Orange County (CSDOC), the South Coast Air Quality Management District (SCAQMD), Southern California Edison (SCE), the Water Environment Research Foundation (WERF), and Huntingdon Environmental Engineering, Inc. (HEEI), is conducting a research project to evaluate the application of biofiltration to remove volatile organic compounds (VOCs), odor-causing air pollutants, and toxics from a publicly owned treatment works (POTW) waste airstream. As part of this project, bench-scale and pilot-scale experiments are being conducted to test the effectiveness of biofiltration and determine the optimum parameters for applying biofiltration to POTWs. Results from the bench-scale experiments demonstrate that biofiltration is effective in reducing the concentration of hydrogen sulfide (H_2S) and total VOCs present in waste airstreams by over 99% and up to 90%, respectively. Average reduction of specific aromatic and carbonyl compounds ranged from 55% to 91%. Removal efficiencies for chlorinated hydrocarbons were variable, ranging from 6% to 88%. Overall, biofiltration appears to be a promising technology for full-scale implementation at POTWs for VOC and odor emission compliance.

INTRODUCTION

The effectiveness of a biofilter is influenced by the properties and character-istics of the support medium, which include porosity, degree of compaction, water retention capabilities, and the ability to host microbial populations. Critical biofilter operational and performance parameters include moisture content, medium pH, medium porosity, and nutrient content.

The end products of biofiltration are water vapor, carbon dioxide, and inor-ganic constituents. Heterotrophic bacteria, which use organic compounds for energy, are primarily responsible for the removal of VOCs. H_2S is degraded by organisms involved in the sulfur cycle (Leson & Winer 1991).

METHODS AND EXPERIMENTAL APPARATUS

The experiment involved two phases of operation. In Phase 1, nine bench-scale biofilters were operated using four different media under differing conditions. A 7-month experimental run was conducted on the bench-scale columns to determine the best media and optimum conditions for use during pilot-scale runs. All reactors, except for one, were operated at a neutral pH. A low-pH reactor was established to demonstrate the long-term effects of a biofilter operation if no pH control were administered. This reactor was used to determine whether VOC and hydrogen sulfide could be removed under low pH conditions. Phase 2, currently in progress, involves a 9-month experimental run in which both bench and pilot reactors are being tested (Table 1).

Bench-scale experiments were initiated using nine acrylic plastic columns, each with an internal diameter of 7.5 cm and a length of 150 cm. The pilot-scale reactors had an internal diameter of 107 cm, were 142 cm in height, and were constructed from fiberglass. In addition, 14 sampling ports, 0.3 cm in diameter, were drilled 7 cm apart along the length of the biofilter reactors. These were used to evaluate reductions in gas concentrations through the filter bed depth. All media for the reactors were seeded with an inoculum of activated sludge,

TABLE 1. Matrix of experiment runs.

Bench-B Pilot-P	Run #1	Bench-B Pilot-P	Run #1
B1	Medium: GAC, pH=7	B1	Medium: GAC, pH=uncontrolled
B2	Medium: GAC, pH=7	B2	Medium: GAC, pH= 7
B3	Medium: Yard Waste Compost, pH=7	B3	Medium: Yard Waste Compost, pH = controlled
B4	Medium: Yard Waste Compost, pH=7	B4	Medium: Yard Waste Compost, pH=7
B5	Medium: Zeolite, pH=7	B7	Medium: Sewage Sludge Compost, pH = uncontrolled
B6	Medium: Zeolite, pH=7	B8	Medium: Sewage Sludge Compost, pH=7
B7	Medium: Sewage Sludge Compost, pH=7	B11	Medium: GAC, inoculated with methylene chloride-degrading organisms
B8	Medium: Sewage Sludge Compost, pH=7	B12	Medium: Yard Waste Compost, inoculated with methylene chloride-degrading organisms
B9	Medium: GAC, pH=1	P1	Medium: GAC
		P2	Medium: Yard Waste Compost

dehydrated organisms, necessary nutrients, and dechlorinated water. Additionally, the low-pH GAC column was inoculated with a solution containing *Thio-
bacillus thiooxidans.*

A positive displacement blower, designed to deliver 7 m³/min of air at 17 kPa
pressure, was used to deliver gas from the headworks off-gas main duct into
the biofilters. The excess flow of inlet gas to the biofilters, or the exhaust from
the biofilters during any process upsets, was vented to the atmosphere after
passing through a knockout drum and carbon canister (Figure 1). A Manostat®
flowmeter and valve were used for monitoring and controlling the airflow
through each reactor. The pressure drop across the column was measured using
a U-Tube manometer. Contaminated air was supplied from a side airstream
from the CSDOC's headworks.

Air Analyses

Air samples were drawn into passivated SUMMA® canisters and through
sampling cartridges for laboratory chemical analyses. Laboratory chemical
analyses included analysis for aldehydes and ketones using EPA Method TO-11,
analysis for volatile organics using EPA Method TO-14, and analysis for total
gaseous nonmethane organic (TGNMO) compounds using South Coast Air Quality Management District (SCAQMD) Modified Method 25.2. Field tests included
on-site monitoring of speciated VOCs using a portable gas chromatograph (GC)
with an argon detector, and H₂S using a Jerome® meter.

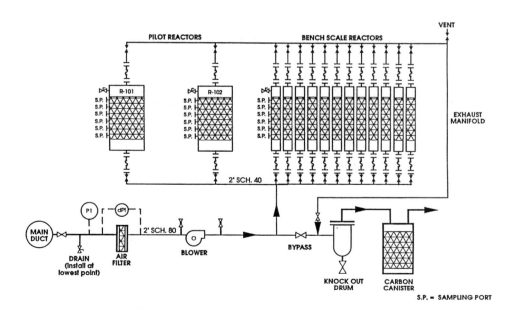

FIGURE 1. Process flow diagram.

Media Analyses

Media parameters measured included pressure drop across the biofilter columns, medium pH, and moisture content. Phospholipid fatty-acid (PLFA) analysis was performed by Microbial Insights, Inc. (Knoxville, Tennessee) on the upper portion of the medium bed to obtain information on viable biomass and the microbial community structure. Biomass was estimated assuming the cell PLFA content known for *Escherichia coli*.

Community structure was estimated as the percent of PLFA in classes associated with certain groups of microorganisms (Table 2).

RESULTS

Treatment Effectiveness

Nine target air pollutants, representing the major chemical groups of toxics and VOCs that were detected in the off-gas streams of POTWs during testing for compliance with California's Air Toxic "Hot Spots" Information and Assessment Act (AB2588) and SCAQMD Rule 1179, were selected for monitoring. Removal efficiencies were averaged over the course of Phase 1 for the two columns of each medium type (Table 3). These results incorporate both adsorptive and biodegradation removal processes. However, the adsorptive capacity of the GAC was calculated to be exhausted after 65 days of continuous operation. All other media were assumed to have lower adsorptive capacities. No breakthrough could be seen which would define exactly when adsorption ended and biodegradation began. For this reason, a total removal is calculated assuming an early removal process (< 65 days) dominated by a combination of adsorption and biodegradation, followed by biodegradation only (> 65 days). Results from Phase 2, which is currently in progress, will be published at a later date.

Both organic and inorganic media were effective in achieving high removal rates for H_2S and total VOCs. Occasional analysis of samples taken from ports at various depths in the column showed near complete removal of hydrogen sulfide within the first 10 cm of the packed beds. Substantial removal of the aromatics and polar hydrocarbons was also seen for all media. The removal of chlorinated

TABLE 2. Community structure.

Class of Fatty Acid	Associated Microorganism
Terminally branched saturates	Gram-positive and sulfate-reducing bacteria
Monoenoics	Gram-negative and some microeucaryotes
Polyenoics	Eucaryotic microorganisms
Branched monoenoics	Sulfate-reducing bacteria
Midchain branched saturates	Gram-positive and sulfate-reducing bacteria

TABLE 3. Average compound removal efficiency for different media.

Compound	Avg. Conc. (ppbv)	Phase 1 Bench-Scale Removal %				
		GAC	YWC[a]	Zeolite	SSC[b]	GAC (pH=1)
Methylene chloride	18	20.87	43.85	26.80	22.01	25.95
Chloroform	18	35.00	26.94	11.35	5.85	29.62
Tetrachloroethene	18	88.84	21.89	11.31	11.15	68.47
Formaldehyde	6	77.59	78.18	76.06	74.88	75.39
Acetone	5	63.39	61.24	60.83	64.41	57.35
Benzene	8	64.85	66.65	54.26	65.14	60.22
Toluene	17	90.59	85.98	83.25	88.00	84.79
TGNMO (ppmv)	2.19	90.49	72.82	83.57	78.13	74.01
Hydrogen sulfide (ppmv)	0.5-5.0	99.92	99.88	99.94	99.88	99.87

(a) YWC is yard waste compost. (b) SSC is sewage sludge compost.

hydrocarbons was below 45%, except in the case of tetrachloroethylene (PCE), which was efficiently removed only in the granular activated carbon (GAC) columns. It is likely that PCE removal occurs as reductive dechlorination in anaerobic zones within the GAC particles, even while the bulk of the biofilter is aerobic (Enzien et al. 1994). The YWC showed some removal of PCE, but because of its nonuniform, less porous structure, may not have promoted the presence of microzones of anaerobic activity as much as the GAC did. The zeolite and sewage sludge compost columns were less effective for every chlorinated hydrocarbon, showing only small removals.

Compounds such as benzene and acetone are readily metabolized by microorganisms. Others, like methylene chloride, are resistant to degradation because few microorganisms possess the enzymes to degrade the carbon-chlorine bonds. Also, compounds with high air-to-water partition coefficients and low adsorption coefficients will pass through the biofilter rapidly, will be present in the solids and water phases in low concentrations, and will not be degraded effectively (Hodge and Devinny 1995).

The GAC column operating at a low pH and the GAC columns operating at neutral pH show similar removal efficiencies for most compounds. This similarity may indicate that acidophilic microorganisms are capable of adequate H_2S and VOC control in biofilters. Continued observance through Phase 2 is needed to further assess this phenomenon.

Media Analyses

Pressure drops for all media were less than 5.5 cm of water. Moisture contents, depending on the media type, ranged between 40 and 68%. Moisture contents were kept constant by direct water addition to the top of the medium.

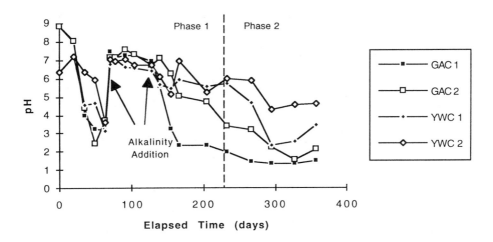

FIGURE 2. pH vs. time.

The pH of the GAC and the pH of the yard waste compost (YWC) media were compared. Results varied with time as H_2S was oxidized to form sulfuric acid (Figure 2). Attempts were made to restore declining pH by adding a sodium bicarbonate wash solution. This was successful until about day 130, after which gradual pH decline was seen in all the reactors despite the bicarbonate additions. The GAC reactors showed a sharper decline in pH than did the YWC reactors, indicating better pH buffering by the organic material.

The microbiological population estimates using the PLFA method showed increasing trends for biomass over time, followed by an approach toward steady state, for both the neutral pH GAC reactors and the YWC reactors (Figure 3).

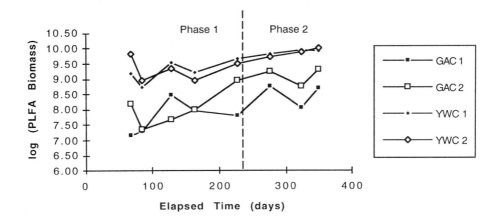

FIGURE 3. Log (PLFA biomass) vs. time.

This is consistent with acclimation of the microbial ecosystem to the new environment (Ergas et al. 1994). Community structure data for the GAC showed that gram-negative populations containing monoenoic fatty acids were the most prevalent of all organisms present in the reactors. These organisms are considered generally faster growing, utilize many carbon sources, and readily adapt to changing environments. Gram-positive organisms containing terminally branched saturates and mid-chain branched saturate acids also increased over time (Figure 4). These organisms are slower growing, degrade more resilient compounds, and can also adapt to changing environmental conditions. The compost reactors show a steadier community structure over time, indicative of the fact that these organisms are indigenous to the media (Figure 5). Their population numbers are also five times that of the GAC reactors. Because both media adequately treat the wastestream, this difference in population numbers may suggest that there are more organisms on the compost than what is necessary to treat the air. Additionally, data suggest that a correlation may exist between declining pH and microbial population numbers. This may be important in defining a low pH range in which microbial populations may exist and still adequately treat the air. Additional data are being compiled during Phase 2 of the experiment so that possible correlations between microbial populations and specific compound removal may be conducted.

CONCLUSIONS

Preliminary evidence indicates biofiltration will be adequate for removing total VOCs to levels required by upcoming mandates. Through the results from Phase 1 of operation, granular activated carbon and yard waste compost have

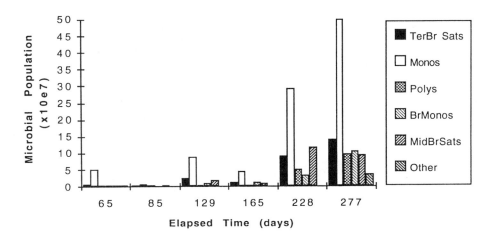

FIGURE 4. Community structure data vs. time (GAC).

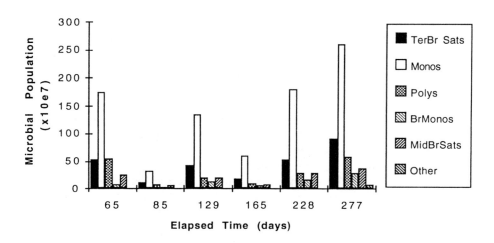

FIGURE 5. Community structure data vs. time (YWC).

been chosen for further studies in Phase 2. This choice has been based on media availability, cost, pollutant removal effectiveness, resistance to compaction, and an ability to host a thriving microbial population. Phase 2 will use bench- and pilot-scale studies to further optimize operational parameters for the design of a full-scale biofilter. Varying detention times, the effects of natural pH decline, compound retardation, and specific microorganism seeding of the medium will be evaluated.

REFERENCES

Enzien, M. V., F. Picardal, T. C. Hazen, R. G. Arnold, and C. B. Fliermans. 1994. "Reductive Dechlorination of Trichloroethylene and Tetrachloroethylene under Aerobic Conditions in a Sediment Column." *Applied and Environmental Microbiology 60*: 2200-2203.

Ergas, S. J., E. D. Schroeder, D.P.Y. Chang, and K. Scow. 1994. "Spatial Distribution of Microbial Populations in Biofilters." Presented at the 87th Annual Meeting & Exhibition of the Air & Waste Management Association, June 19-24.

Hodge, D. S., and J. S. Devinny. 1995. "Modeling Biofilter Performance." Accepted for publication in *Journal of Environmental Engineering*. American Society of Civil Engineers.

Leson, G., and A. M. Winer. 1994. "Biofiltration: An Innovative Air Pollution Control Technology for VOC Emissions." *Journal of Air & Waste Management Association 41*: 1045.

The Use of White-Rot Fungi as Active Biofilters

Annette Braun-Lüllemann, Christian Johannes,
Andrzej Majcherczyk, and Alois Hüttermann

ABSTRACT

White-rot fungi, growing on lignocellulosic substrates, have been successfully used as active organisms in biofilters. Filters using these fungi have a very high biological active surface area, allowing for high degrees of retention, a comparatively low pressure drop, and a high physical stability. The unspecific action of the extracellular enzymes of the white-rot fungi allows for the degradation of a wide variety of substances by the same organism. Degradation of several compounds in the gas phase by the white-rot fungi *Trametes versicolor, Pleurotus ostreatus, Bjerkandera adusta,* and *Phanerochaete chrysosporium* was tested. Among the aromatic solvents, styrene was the compound that was most readily degraded, followed by ethylbenzene, xylenes, and toluene. Tetrahydrofuran and dichloromethane were also degraded, whereas dioxane could not be attacked by fungi under the conditions used. Acrylonitrile and aniline were degraded very well, whereas pyridine was resistant to degradation. The process for removing styrene is now in the scaling-up stage.

INTRODUCTION

In recent years the problems of air pollution by organic compounds have attracted growing public interest. Emission of organic compounds from different industrial plants is the main source of this pollution (Bardtke et al. 1990). A special problem is the removal of low concentrations of organic compounds from waste airstreams of lacquering and processing plants of the plastics industry. Satisfactory physical-chemical cleaning processes are not available, are ineffective, or are not economical. In some cases, this problem could be solved using bacterial biofilters (Steinmüller et al. 1979, Gust et al. 1979, Bardtke & Fischer 1986, Ottengraf et al. 1986). Along with bacteria, pure cultures of white-rot fungi were investigated for their possible use (Majcherczyk et al. 1990, Braun-Lüllemann et al. 1992). White-rot fungi seem to be particularly suitable for biological air-cleaning for two reasons: first, they possess a very branched mycelium with a large surface area, and second, they produce unspecific enzymes permitting the degradation of lignin under natural conditions. This system also allows them to degrade other organic compounds (Sanglard et al. 1986, Bumpus 1989, and Fernando et al. 1990) and even chlorinated organic substances (Bumpus & Aust 1987). However, it has not been fully investigated whether white-rot fungi also can degrade xenobiotic organic compounds in the gas phase. In preliminary experiments, a large number of different strains of fungi were investigated for their ability to degrade several aromatic compounds. Based on degradation and biomass production, the four best fungi were selected and further tested for their ability to degrade other xenobiotics.

Four groups of compounds were investigated: simple aromatic hydrocarbons, nonaromatic solvents, and nitrogen- and sulfur-containing compounds. All compounds investigated are important air pollutants and are of toxicological relevance. Styrene, ethylbenzene, toluene, and xylenes were selected as model compounds for aromatic hydrocarbons. Tetrahydrofuran, dichloromethane, and dioxane were tested as commonly used nonaromatic solvents. Acrylonitrile, pyridine, and aniline were tested in the nitrogen-containing group of compounds. Thiophene, ethyl isothiocyanate, and methyl thiocyanate belong to the sulfur-containing group.

MATERIALS AND METHODS

Organisms and Cultures

Chopped winter wheat straw was used for all straw cultures. The experiments were performed in 10-mL screw bottles; 2.5 g straw was packed into the flasks, moistened with 5 mL distilled water, sterilized, and inoculated.

Fumigation of Cultures

Cultures and heat-killed controls were fumigated in triplicate. After 3 to 4 weeks of growth, bottles were tightly closed with screw-caps possessing two Teflon™ valves. Tested compounds were injected through one valve and cultures were incubated for 48 h at 22°C.

The following concentrations (g/m^3) of components were used:

- Aromatic hydrocarbons: ethylbenzene, 18.2; *m*-xylene, 18; toluene 15.6; styrene, 36.4
- Nonaromatic solvents: tetrahydrofuran, 30.9; dichloromethane, 31.5; dioxane, 32.2
- Nitrogen-containing compounds: acrylonitrile, 38.4; pyridine, 31.1; aniline, 15.4
- Sulfur-containing compounds: thiophene, 2.5; methyl thiocyanate, 23.5; ethyl isothiocyanate, 27.

Extraction and Sample Preparation

The straw cultures were treated with 60 mL 5% KOH solution in methanol added through one of the valves. The displaced air from the flasks was guided through the second valve and bubbled through the KOH solution, so that the remaining amounts of the gaseous compounds were also collected. For correct quantification, internal standards were added to the KOH solution: aromatic hydrocarbons, *o*-xylene; nonaromatic solvents, toluene; nitrogen- and sulfur-containing compounds, α-pinene. Subsequently, the flasks were mixed and incubated for 30 min. An aliquot of the solution was centrifuged, and 4 mL of the supernatant was mixed with 4 mL water and 1 mL of dichloromethane. After 30 min extraction and repeated centrifugation, ca. 0.5 mL of the dichloromethane phase was taken for analysis. Nitrogen-containing test compounds were extracted from cultures with pure methanol, and 4 mL 0.1 M phosphate buffer (pH 9.0) was used instead of water for extractions with dichloromethane.

GC/MS Analysis of the Different Compounds

All samples were analyzed by gas chromatography/mass spectrometry (GC/MS) detection (Hewlett Packard, USA) using a 5% phenyl methyl silicon column for the aromatic and a Carbowax column for the other compounds. Analysis was performed by single-ion monitoring; 1 μL of the sample was injected in the proportion 1:30.

RESULTS AND DISCUSSION

Aromatic Hydrocarbons

The best results for all of the aromatic hydrocarbons were obtained by the degradation of styrene. The concentration of this compound was reduced more than 95% by *Trametes versicolor* and *Bjerkandera adusta* (Figure 1). Until now degradation of styrene by biological systems had been demonstrated only for lower fungi (Cox et al. 1993). *Trametes versicolor* and *Bjerkandera adusta* reduced the concentration of ethylbenzene to 38 and 44% and *m*-xylene to 32 and 44%, respectively. The lowest degradation rate was in the case of toluene. Although Yadav and Reddy (1993) demonstrated degradation of benzene, toluene, ethylbenzene, and xylene mixtures using *Phanerochaete chrysosporium*, this fungus was not able to degrade any compound except styrene in the studied system. *Pleurotus ostreatus* could not attack toluene and *m*-xylene and showed less degradation of the other aromatic hydrocarbons than did *Trametes versicolor* and *Bjerkandera adusta*.

Nonaromatic Solvents

Phanerochaete chrysosporium could not degrade any of the components tested (Figure 2). All other fungi were able to reduce the concentration of dichloromethane from 7% (*Pleurotus ostreatus*) to 19% (*Bjerkandera adusta*). Tetrahydrofuran

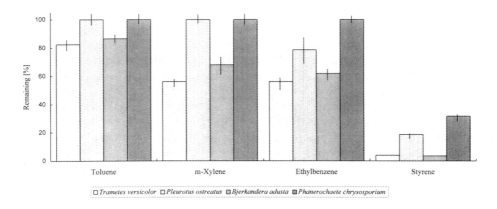

FIGURE 1. **Remaining contents of aromatic hydrocarbons after degradation by different white-rot fungi.**

was degraded only by *Trametes versicolor* to a remaining content of 85%. No reaction was obtained in the case of dioxane.

Nitrogen-Containing Compounds

For the nitrogen-containing compounds, very good results were obtained (Figure 3). Acrylonitrile was degraded nearly completely by *Trametes versicolor* and *Bjerkandera adusta*. The other fungi reduced the content of acrylonitrile to 50% of the controls. Aniline concentration was reduced by all fungi to a remaining

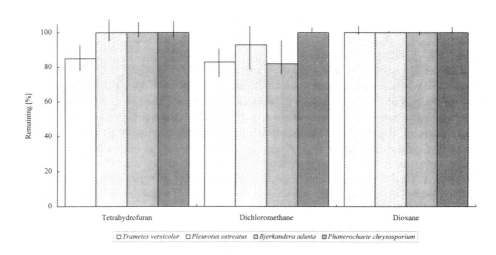

FIGURE 2. **Remaining contents of nonaromatic solvents after degradation by different white-rot fungi.**

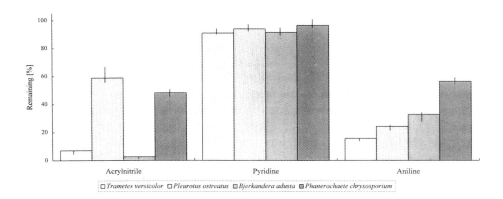

FIGURE 3. Remaining contents of nitrogen-containing compounds after degradation by different white-rot fungi.

content from 15% (*Trametes versicolor*) to 56% (*Phanerochaete chrysosporium*). The ability of all fungi to degrade pyridine was comparatively low; the concentration of this compound decreased only 4 to 10%. N-containing compounds in bacterial biofilters degraded in few cases (e.g., acrylonitrile: Gethke 1993). The ability of *Phanerochaete chrysosporium* to degrade N-containing compounds in liquid cultures has been shown by some authors (e.g., azo dyes: Capalash & Sharma 1992, and Spadaro et al. 1992).

Sulfur-Containing Compounds

For the investigated fungi, the best results were obtained for the degradation of thiophene (Figure 4). Except in the case of *Trametes versicolor*, which was nearly

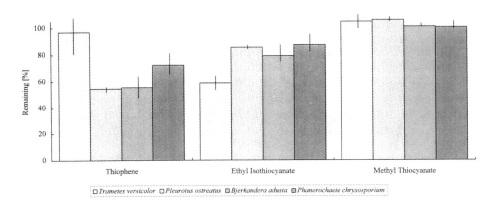

FIGURE 4. Remaining contents of sulfur-containing compounds after degradation by different white-rot fungi.

unable to degrade this compound, remaining thiophene contents ranged from 55 to 72%. However, *Trametes* could degrade ethyl isothiocyanate remarkably better than could the other fungi. No degradation was obtained for methyl thiocyanate.

In an approach to test the degradation potential of different fungi under applicable conditions of a possible biofilter, all tests were conducted under equal conditions, although conditions were not necessarily optimal for any specific fungus. Tested fungi were in stationary growth phase, but revealed different biomass productions. Enzyme systems responsible for degradation and the time of their expression differ significantly and did not allow for any direct comparisons.

The wide range of chemically different compounds tested demonstrated the large degradation potential of white-rot fungi. Among aromatic compounds, reactive side groups of styrene and aniline were responsible for their excellent degradation. Compounds with unpolar side chain or electron-deficient hetero-aromatics were degraded to a lesser extent. Aliphatic compounds in general were degraded less.

2-L Fumigation Installation

Fumigation with styrene was performed in a 2-L installation. Straw cultures (*Pleurotus ostreatus* and *Trametes versicolor*) were incubated for 3 weeks at 25°C prior to the experiments. The fumigation was conducted with an air flowrate of 1 to 2 L/min. For fumigations with high concentrations of styrene (1,245 and 3,300 mg/m^3), degradation rates of more than 90% and 80%, respectively, were obtained (Table 1). In prolonged fumigations with lower concentrations (553 mg/m^3), waste gas styrene was degraded more than 90%.

Based on these results, we are now conducting experiments using fumigation installations with 20-L filter elements. Furthermore, the experiments are being scaled-up for an installation with a filter area of 1 m^2, allowing airstreams up to 600 m^3/h.

ACKNOWLEDGMENT

The work was supported by the Deutsche Bundesstiftung Umwelt, Osnabrück, Germany, AZ 01205.

TABLE 1. Fumigation parameters of the experiments performed in the 2-L fumigation installation.

Remaining Concentration of Styrene (mg/m^3)	Flow (L/min)	Degradation Rate (%)
543	1.0	>90
1,245	1.25	>99
3,300	1.25	80

REFERENCES

Bardtke, D., and K. Fischer. 1986. *Untersuchungen zur Abbaubarkeit und Abbaukinetik verschiedener anorganischer und organischer Abluftinhaltsstoffe beim Biofilterverfahren.* Research report Ba 551/10-1, drawn up in the Institut für Siedlungswasserbau, Wassergüte- und Abfallwirtschaft, University of Stuttgart, Germany.

Bardtke, D., F. Sabo, and K. Fischer. 1990. "Systemstudie zur Erfassung und Bilanzierung organischer Schadstoffemissionen im Hinblick auf den Einsatz biologischer Abluftreinigungsverfahren." *Abschlußbericht zum Forschungsvorhaben* 01 VQ 86103.

Braun-Lüllemann, A., A. Majcherczyk, N. Tebbe, and A. Hüttermann. 1992. "Bioluftfilter auf Basis von Weißfäulepilzen." In A. J. Dragt and J. van Ham (Eds.), *Biotechniques for Air Pollution Abatement and Odour Control Policies,* pp. 91-95. Elsevier Science Publishers.

Bumpus, J. A. 1989. "Biodegradation of Polycyclic Aromatic Hydrocarbons by *Phanerochaete chrysosporium.*" *Appl. Environ. Microbiol.* 55: 154-158.

Bumpus, J. A., and S. D. Aust. 1987. "Biodegradation of Chlorinated Organic Compounds by *Phanerochaete chrysosporium,* a Wood-Rotting Fungus." *ACS Symp. Ser. (Solving Hazard Waste Probl.)* 338: 340-349.

Capalash, N., and P. Sharma. 1992. "Biodegradation of Textile Azo-Dyes by *Phanerochaete chrysosporium.*" *World J. Microbio. Biotechn.* 8: 309-312.

Cox, H.H.J., J.H.M. Houtman, H. J. Doddema, and W. Harder. 1993. "Enrichment of Fungi and Degradation of Styrene in Biofilters." *Biotechnology Letters* 15(7): 737-742.

Fernando, T., J. A. Bumpus, and S. D. Aust. 1990. "Biodegradation of TNT (2,4,6-Trinitrotoluene) by *Phanerochaete chrysosporium.*" *Appl. Environ. Microbiol.* 56: 1666-1671.

Gethke, H. G. 1993. "Positive und negative Erfahrungen mit großtechnischen Biofilteranlagen." In VDI Berichte 1034, *Fortschritte bei der thermischen, katalytischen, sorptiven und biologischen Abgasreinigung,* pp. 541-564. VDI-Verlag; Düsseldorf.

Gust, M., H. Grochowski, and S. Schirz. 1979. "Grundlagen der biologischen Abluftreinigung Teil V: Abgasreinigung durch Mikroorganismen mit Hilfe von Biofiltern." *Staub-Reinh. Luft* 39(11): 397-438.

Majcherczyk, A., A. Braun-Lüllemann, and A. Hüttermann. 1990. "Biofiltration of Polluted Air by a Complex Filter Based on White Rot Fungi Growing on Lignocellulosic Substrates." In M. P. Coughlan and M. T. Amaral Collaco (Eds.), *Advances of Biological Treatment of Lignocellulosic Materials,* pp. 311-322. London, Elsevier Applied Science.

Ottengraf, S.P.P., J.J.P. Musters, A.H.C. Van Den Oever, and H. R. Rozema. 1986. "Biological Elimination of Volatile Xenobiotic Compounds in Biofilters." *Bioprocess Engineering* 1: 61-69.

Sanglard, D., M.S.A. Leisola, and A. Fiechter. 1986. "Role of Extracellular Ligninases in Biodegradation of Benzo(a)pyrene by *Phanerochaete chrysosporium.*" *Enzyme Microbiol. Technol.* 8: 209-212.

Spadaro, J. T., M. H. Gold, and V. Renganathan. 1992. "Degradation of Azo Dyes by the Lignin-Degrading Fungus *Phanerochaete chrysosporium.*" *Appl. Environ. Microbiol.* 58: 2397-2401.

Steinmüller, W., G. Claus, and H.-J. Kutzner. 1979. "Grundlagen der biologischen Abluftreinigung Teil II: Mikrobiologischer Abbau von luftverunreinigenden Stoffen." *Staub-Reinh. Luft* 39(5): 149-152.

Yadav, J. S., and C. A. Reddy. 1993. "Degradation of Benzene, Toluene, Ethylbenzene, and Xylenes (BTEX) by the Lignin-Degrading Basidiomycete *Phanerochaete chrysosporium.*" *Appl. Environ. Microbiol.* 59(3): 756-762.

Reactor Switching: Proposed Biomass Control Strategy for the Biofiltration Process

Randall W. Farmer, Jian-Shin Chen,
David M. Kopchynski, and William J. Maier

ABSTRACT

A bench-scale biofiltration system was designed to study the elimination of methyl ethyl ketone (MEK) from a waste airstream and to investigate the effects of various parameters on rate of removal. The biofilter consisted of three ceramic-bed bioreactors connected in series, which allowed for changing the gas flow sequence. Results show that switching the sequence of flow has important process advantages. Biomass accumulation could be controlled and better distributed throughout the filter depths of the forward reactors, providing for longer operation with minimal pore plugging and essentially no increase in pressure drop across the reactors.

INTRODUCTION

Biofiltration is an emerging technology for the control of volatile organic compounds (VOCs) emitted in off-gases from manufacturing operations (Groenestijn and Hesselink 1993). Removal of VOCs from process air is accomplished by contact with an active microbial biofilm attached to a solid-phase packing. Biochemical oxidation results in complete mineralization of organic substrates to end products of water and carbon dioxide. Biofiltration is environmentally sound because it eliminates the emission of hydrocarbons to the atmosphere without the accumulation of harmful by-products and it replaces capital and energy-intensive combustion processes currently in use.

Basic design features of biofilters have been described in the literature (Leson & Winer 1991). However, the effects of operating variables on efficiency of removal and, specifically, the long-term effects of continuous loading have not been adequately described. Continuous operation is known to result in accumulation of biomass, which can cause removal efficiency to increase or decrease depending on whether accumulation results in thicker biofilms that have more biodegrading capacity or reduces available surface area for contact by clogging pore spaces. The latter eventually requires cleaning and repacking, which is costly and time-consuming (Sorial et al. 1994).

A series of pilot-plant tests have been carried out to examine the rate of biomass accumulation, its effect on process performance, and potential strategies for controlling biomass accumulation. This paper describes results from experiments aimed at developing strategies for controlling accumulation and distribution of biomass by changing the flow sequence of reactors operating in series.

EXPERIMENTAL PROCEDURES AND MATERIALS

The pilot plant consisted of three packed-bed reactors, 5-cm I.D. × 60-cm, connected to allow operating the reactors in series in any desired sequence. Quick-disconnect fittings were used to allow changing the flow sequence in a matter of minutes. The columns were packed with 13-mm-dia. ceramic Berl saddles to provide a readily available surface for biofilm formation. The packed beds were initially inoculated with cultures enriched from activated sludge obtained from the Minneapolis/St. Paul Sewage Treatment Plant. Berl saddles were used to minimize potential increases in pressure drop due to compaction. The porosity of the reactors without biofilm was 69% as measured via water displacement; the calculated surface area of the packing was 465-m^2/m^3.

House-compressed air was prehumidified to >95% relative humidity (RH) by sparging through an 8-L tank of water before adding the chemical to be tested—results obtained with methyl ethyl ketone (MEK) will be shown in this paper. MEK was prepared as an aqueous solution of 200-g/L and injected into the airstream using a syringe pump. Mininert® valves served as airstream sampling ports located at each column's inlet and at 20-cm and 40-cm depths within each column; a final port was placed at the biofilter outlet. Thus, the MEK-laden airstream passed through a cumulative bed depth of 180 cm when all three reactors were operating in series; and MEK concentrations in the airstream were measured at 20-cm intervals of depth. Air samples were measured using a gas chromatograph equipped with a 0.53-mm-I.D. × 15-m fused silica capillary column and a flame ionization detector. The detection limit was 0.12 ppmv MEK.

At these sampling ports CO_2 measurements were also performed on air samples to monitor mineralization of MEK and endogenous decay of biomass. Measurements were performed using a total carbon analyzer. Additionally, internal column pressure measurements were performed by inserting a needle connected to a digital pressure gauge.

Water containing nutrients (e.g., nitrate, phosphate, iron, and trace elements) was recycled through each column (Figure 1) at 5-mL/min to keep the biofilm moist and ensure that growth nutrients were available (Farmer 1994). However, 30 min prior to airstream sampling, recycle was discontinued and excess water was allowed to drain from the columns to eliminate the effects of water percolation on transport of MEK.

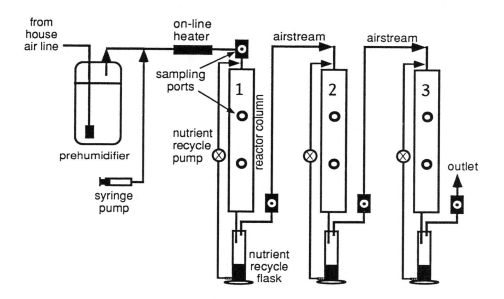

FIGURE 1. **Biofilter design: Three reactor columns connected in series.**

A set of experiments involving switching the flow sequence through the reactor columns was performed at various loading and flow conditions. Flow conditions were returned to base loading conditions after the sampling experiments were performed. Changes in biofilm volume and bed porosities were measured periodically by weighing the columns and then reweighing after filling with water to fill void spaces. Biomass density was also measured from samples of biofilm and found to be 1.003 g/L.

RESULTS

Initially, the reactors were operated for 27 days at 212-mg/h MEK loading and 213-L/h gas flowrate to build up an initial biomass layer on the packing material. Then the loading rate was cut back to 64-mg/h at a 71-L/h flowrate — known as the base loading conditions — where "complete" removal of MEK (i.e., removal beyond detection limits) was achieved prior to the outlet of the middle reactor. The total residence time was 3 min through the three reactors (i.e., approximately 1 min per reactor). One outcome of complete removal by the midpoint of the three reactors in series is that biomass production varied from reactor to reactor and eventually led to greater accumulations of biomass in the lead reactor as opposed to the final reactor.

Figure 2 shows the change in gas phase MEK concentration for the entire cumulative bed depth at a 361-mg/h MEK loading, with a 330-L/h air flowrate in a 1–2–3 flow sequence, and a 2–3–1 flow sequence through the columns, where

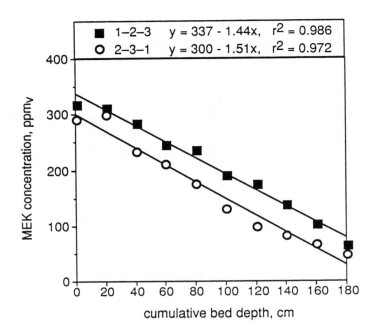

FIGURE 2. MEK gas-phase concentration vs. bed depth at a 361 mg/h loading with a 330 L/h gas flow in the 1-2-3 and 2-3-1 flow sequences.

the numerals 1, 2, and 3 refer to reactor columns 1, 2, and 3. Gas-phase MEK concentrations decreased linearly with bed depth in both flow sequences. The inlet and outlet concentrations were 320 and 65-ppmv in the 1–2–3 flow sequence and 293 and 49-ppmv in the 2–3–1 flow sequence. Total removal of MEK in the 1–2–3 and 2–3–1 flow sequences were 255 and 230-mg/h, respectively. Comparison of the overall slopes indicates that changing the reactor flow sequence did not have a significant effect on MEK removal.

The measured amounts of biofilm in reactors 1, 2, and 3 were 524, 383, and 264 g respectively. Switching reactor 2, which contained less biofilm than reactor 1, to the lead position did not significantly decrease removal. Nor did switching reactor 1, with the highest amount of biofilm, to the final position significantly affect removal.

Measurements of the pressure drop in the 1–2–3 flow sequence gave a 1.1-psig backpressure at the inlet port; most of this pressure drop occurred within the lead reactor, which contained the largest amount of biomass. Immediately after switching, there was no significant change in total pressure drop. It merely shifted to the final position with reactor 1. Seven days after switching the flow sequence to 2-3-1, the total pressure drop was only 0.2-psig due to loss of biomass in reactor 1.

Figure 3 shows the change in gas phase MEK concentration for the lead reactor vs. bed depth at a loading of 64-mg/h MEK, with a 71-L/h airflow (~base

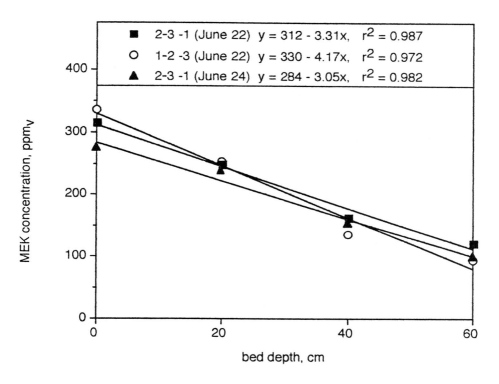

FIGURE 3. MEK gas-phase concentrations vs. bed depth in the lead reactor in the 2-3-1 and 1-2-3 flow sequences.

loading conditions) in both the 1–2–3 and 2–3–1 flow sequences 3 weeks after the flow sequence switching shown in Figure 2. Gas-phase MEK concentrations decreased linearly with bed depth in both flow sequences. The biomass measurements in reactors 1, 2, and 3 were 496, 572, and 363 g, respectively. Reactors 1 and 2 showed essentially the same amount of removal.

DISCUSSION

Using three columns in series — rather than one column of bed depth equal to the sum of these three — could be very beneficial in a couple of ways. First, there was no significant loss in overall removal efficiency by the biofilter outlet when the flow sequence was changed at the flow conditions shown in Figures 2 and 3.

Second, because complete removal of MEK is observed prior to the inlet of the reactor in the final position under base conditions, no net growth occurred in reactor 1 after it was moved to the final position. As a result, the biomass cells under went endogenous decay, and the actual amount of biofilm decreased, opening up pore spaces and decreasing the pressure drop across this column. Therefore, reactor 1 was able to partially regenerate itself in the final position.

Using a sequence of reactors avoids the problem of plugging at depths near the waste gas inlet, while the rest of the bed has sufficient void space for effective gas flow. In industrial applications, the reactors could be manifolded so the flow sequence could be changed at will. If the sequence were rotated often enough, biofilm growth might become more evenly distributed throughout the entire bed depth, again allowing for longer operation runs.

A final advantage, occurred when the biofilter was operated at base conditions. In this case, maintenance was performed on the final reactor section without unnecessary shutdown time. The final reactor could be taken off line without affecting the complete removal of MEK in the forward two reactors.

REFERENCES

Farmer, R. W. 1994. "Biofiltration: Process Variables and Optimization Studies." M.S. Thesis, University of Minnesota, Minneapolis, MN.

Leson, G., and A. M. Winer. 1991. "Biofiltration: An Innovative Air Pollution Control Technology for VOC Emissions." *J. Air Waste Manage. Assoc.* 41(8): 1045-1054.

Sorial, G. A., F. L. Smith, M. T. Suidan, P. Biswas, and R. C. Brenner. 1994. "Evaluation of the Performance of Trickle Bed Biofilters — Impact of Periodic Removal of Accumulated Biomass." 87th Annual Meeting and Exhibition of the Air and Waste Management Association (June 19-24, Cincinnati, Ohio), paper number 94-RA115A.05.

van Groenestijn, J. W., and P.G.M. Hesselink. 1993. "Biotechniques for Air Pollution Control." *Biodegradation* 4(4): 283-301.

Vapor-Phase Biofiltration: Laboratory and Field Experience

Patrick J. Evans, Katherine A. Bourbonais,
Lance E. Peterson, Jean H. Lee, and Gary L. Laakso

ABSTRACT

Application of vapor-phase bioreactors (VPBs) to petroleum hydro-carbons is complicated by the different mass transfer characteristics of aliphatics and aromatics. Laboratory- and pilot-scale VPB studies were conducted to evaluate treatment of soil vapor extraction (SVE) off-gas. A mixture of compost, perlite, and activated carbon was the selected medium based on pressure drop, microbial colonization, and adsorption properties. Two different pilot-scale reactors were built with a difference of 70:1 in scale. The smaller VPB's maximum effective elimination capacity (EC) was determined to be 7.2 g m^{-3} h^{-1}; the larger unit's EC was 70% to 80% of this value. Low EC values may be attributable to a combination of mass-transfer and kinetic limitations.

INTRODUCTION

VPBs have been used to treat gas streams contaminated with various chemi-cals. A potential application is SVE off-gases at sites contaminated with petro-leum hydrocarbons. Gasoline vapors typically contain both aromatic and aliphatic hydrocarbons. Oxygenates such as methyl t-butyl ether and ethanol also can be present.

Low-molecular-weight aliphatic and aromatic hydrocarbons are biodegradable by a variety of aerobic microorganisms. Readily biodegradable compounds have been removed by VPBs at ECs of 100 to 200 g-m^{-3}-h^{-1} (Ottengraf 1987). Aromatics such as benzene, toluene, ethylbenzene, and total xylenes (BTEX) have yielded ECs up to 100 g-m^{-3}-h^{-1} (Togna & Singh 1994). Aliphatics and gasoline, however, have yielded ECs < 10 g-m^{-3}-h^{-1} (Hodge et. al. 1991; Kampbell et al. 1987; Togna & Singh 1994). Aliphatics and aromatics have very different Henry's constants (H). For example, H at 20°C for benzene and *n*-hexane are 0.1879 and 36.71, respectively (Wilson & Clarke 1994). Aliphatics may have lower degradation rates than aromatics in VPBs due to different concentrations of actively degrading microorganisms, lower specific degradation rates, or mass transfer differences associated with H.

Biofiltration of dilute gas streams typically is more economical than other types of air pollution control, such as thermal incineration, catalytic incineration, activated carbon adsorption, chemical oxidation, condensation and recovery, and UV/ozonation (Diks 1992). If ECs for removal of target compounds with high H are low, larger and less economical VPBs will be required to meet design criteria. The objectives of this study were to select appropriate VPB media for treatment of gasoline SVE vapor and obtain design parameters from a pilot-scale VPB operating at an actual SVE site. These data would then be used to determine the design and economic feasibility of a full-scale VPB in series with SVE.

MATERIALS AND METHODS

Candidate VPB packing media tested for flow, physical, and microbiological characteristics included alder chips, compost, activated carbon, peat, perlite, and soil. These materials were purchased or collected from uncontaminated sites. Bench-scale testing involved packing media into glass columns 5 cm in diameter by 61 cm in length. Media were moistened equally to 20% (wt/wt) and pressure-flow relationships were measured. Moisture was maintained by periodically adding water to the media.

Soil, compost, and peat were tested for their ability to support growth of hydrocarbon-degrading microorganisms. Media (1 g) were placed in Erlenmeyer flasks with 100 mL of distilled water plus 25 μL of unleaded gasoline and shaken at room temperature (approximately 20°C) for 3 days. Aliquots of liquid were removed from the flasks and colony-forming units per gram (CFU/g) of media were determined on Bushnell-Haas mineral salts medium agar plates with either benzene or gasoline vapors.

Combinations of compost, perlite, activated carbon, alder, and peat were tested for their ability to remove gasoline vapors at bench scale. Gasoline vapor delivery was inconsistent, and after 25 days of operation, the gas influent stream was switched to benzene. Media samples were collected and analyzed for concentrations of gasoline- and benzene-degrading microorganisms.

Two pilot-scale biofilters were constructed: a 0.015 m^3 VPB (1.8 m height by 10 cm diameter) and a 1-m^3 VPB (1.24 m by 0.91 m by 0.91 m). Both were packed with a mixture of compost (70%), perlite (20%), and activated carbon (10%). Flow from an SVE system driven by a regenerative blower passed through a humidifier and upward through the VPBs in parallel. The 1-m^3 VPB contained a vapor distribution system beneath the media. Moisture was monitored and maintained between 40 and 60% (wt/wt). Inorganic nutrients were added as needed.

Hydrocarbon analysis of gases was performed with a flame ionization detector (FID) organic vapor analyzer (OVA), calibrated with a 100-ppmv methane standard, or by FID or photoionization detector (PID) capillary gas chromatography (GC).

BENCH-SCALE RESULTS

Pressure-flow relationships were measured for candidate VPB materials in packed columns. Figure 1 shows that alder chips had the least resistance to flow and peat the greatest.

Flask studies of soil, compost, and peat in suspension with liquid media containing gasoline demonstrated increased concentrations of gasoline-degrading organisms. Initial concentrations were 3.8×10^2, $< 1.0 \times 10^1$, and $< 1.0 \times 10^1$ CFU/g for soil, compost, and peat, respectively. After 3 days of incubation, respective concentrations had increased to 4.0×10^4, 2.5×10^4, and 4.0×10^4.

Columns packed with media mixtures supported gasoline-degrading microorganisms. Table 1 shows benzene removals and concentrations of gasoline- and benzene-degrading microorganisms. Compost mixtures supported greater concentrations of microorganisms than carbon or peat. All tested media were capable of removing benzene at the $0.14\text{-m}^3\text{-h}^{-1}$ flowrate and benzene loading rates ranging from $0.25 \text{ g-m}^{-3}\text{-h}^{-1}$ for peat/perlite to $1 \text{ g-m}^{-3}\text{-h}^{-1}$ for carbon. Removals may have been partially attributable to adsorption, especially in the case of carbon.

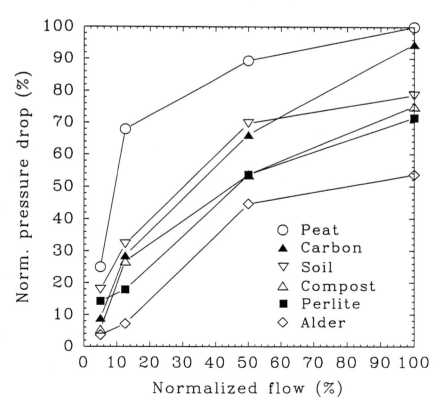

FIGURE 1. Normalized pressure-drop profiles as a function of normalized flowrate.

TABLE 1. Bench-scale VPB benzene removal and microbial concentrations.

Medium (percent composition)	Benzene			Microbial Concentration (CFU/g)	
	Influent (ppmv)	Effluent (ppmv)	% Removal	Gasoline[a]	Benzene[b]
50 peat/50 perlite	4.1	ND	100	2.0×10^1	2.0×10^1
50 compost/50 perlite	69	ND	100	8.5×10^2	4.0×10^4
75 compost/25 perlite	53	0.89	98	2.4×10^4	2.0×10^3
50 alder/50 compost	66	0.93	99	1.4×10^4	6.6×10^4
Compost	52	ND	100	4.0×10^4	7.0×10^4
Carbon	200	ND	100	$<1.0 \times 10^1$	6.0×10^3

(a) Concentration determined after 25 days of operation with gasoline vapor influent.
(b) Concentration determined after 24 days of operation with benzene vapor influent.
ND = Not detected.

PILOT-SCALE RESULTS

The pilot-scale 1-m³ VPB was initially designed for a residence time of 1 min, based on previous work by others. Removals of less than 50% were observed when the VPB was operated at a flowrate of 60 m³-h⁻¹, with influent hydrocarbon concentrations ranging from 50 to 500 ppmv. Increased residence time resulted in increased percent removals.

Flowrates and influent TPH concentrations were varied to characterize the biofilter's potential EC. Greater EC was observed with increasing TPH loading rate (data not shown). The maximum practical loading rate to obtain 60% removal of hydrocarbons from a 300-ppmv stream was 12 g-m⁻³-h⁻¹ in the 0.015 m³ VPB, and an EC of 7.2 g/m³h. This rate was observed over several months of operation without any apparent decline in activity. Comparison of the volumetric ECs of the 0.015-m³ and 1-m³ VPBs at a nominal influent concentration of 300 ppmv indicated the 1-m³ VPB was 70 to 90% as efficient as the 0.015-m³ VPB (data not shown).

The influent vapor was characterized by GC, and found to contain C₄ to C₈ straight- and branched-chain aliphatics and aromatics, including toluene and xylenes with traces of benzene and ethylbenzene (data not shown). Table 2 shows influent and effluent concentrations of total petroleum hydrocarbon (TPH) and BTEX at different percent removals.

Figure 2 shows TPH concentrations at different locations in the 0.015-m³ VPB under steady-state operation for different influent concentrations and flowrates. Data nonlinearity indicates mass-transfer limitations or first-order kinetics.

TABLE 2. Pilot-scale VPB removal of TPH and BTEX.

Analyte	53% Overall Removal[a]			84% Overall Removal[a]		
	Influent (ppmv)	Effluent (ppmv)	Removal (%)	Influent (ppmv)	Effluent (ppmv)	Removal (%)
TPH	480	210	56[b]	420	93	78[b]
Benzene	0.54	<0.5	>7.4	0.36	<0.2	>44
Toluene	2.1	0.47	78	1.6	<0.01	>99
Ethylbenzene	0.46	0.12	74	0.36	<0.01	>97
Xylenes	5.2	0.99	81	3.0	0.26	91

(a) Overall removals determined by OVA; TPH and BTEX concentrations determined by GC/FID/PID.
(b) Significantly less than percent removals of toluene, ethylbenzene, and xylenes at $p < 0.01$.

The concentration profile along the reactor length would be linear if the biofilm was saturated with substrate and the reaction in the biofilm was zero order (Ottengraf 1987).

The concentration of hexadecane degraders in a single media sample during operation was $>3 \times 10^6$/g wet compost. The specific microbial TPH degradation rate was calculated to be 0.07 min^{-1} based on wet compost bulk density of 0.5 g-mL^{-1}, single microorganism dry weight of 10^{-12} g (Bailey & Ollis 1977), EC of 7.2 g-m^{-3}-h^{-1}, and microbial concentration.

DISCUSSION

VPB development from laboratory to pilot-scale involved selection of a medium based on flow characteristics and microbial colonization. Peat and carbon supported lower concentrations of microorganisms than compost. No significant differences were observed between the different compost mixtures. A mixture of compost and perlite (80:20 vol/vol) was determined to be an optimal compromise based on pressure drop and volumetric surface area. Benzene removal was shown to be feasible in the bench-scale VPB, but system operation was not optimized.

A pilot-scale 1-m^3 VPB was constructed. Activated carbon (10% vol/vol) was added to the compost/perlite mixture to increase the VPB's ability to moderate influent contaminant spikes. The VPB was connected to an SVE system at a gasoline-contaminated site. The vapor contained light-end gasoline components, primarily C$_4$ to C$_8$ aliphatics and some aromatics. A 1-min residence time resulted in poor performance.

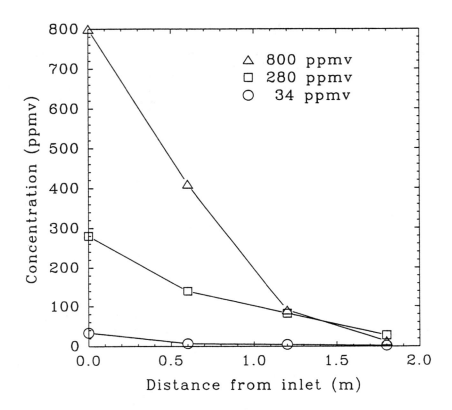

FIGURE 2. Vapor TPH concentration profiles along the length of the 0.015-m³ VPB for varying influent concentrations.

The 0.015-m³ column VPB was constructed and operated on the same influent stream as the pilot-scale VPB to ascertain the significance of mass transfer limitations. Aromatics were selectively removed; individual compound percent removals were 1.2 to 1.4 times greater than that of TPH, even though their concentrations were less than 1% of the TPH concentration (Table 2). This may have been attributable in part to the effect of H and mass transfer limitations (Figure 2) leading to lower aqueous-phase concentrations of aliphatics at a given gas-phase concentration.

The specific microbial degradation rate, 0.07 min^{-1}, was very high considering these limitations and low substrate concentrations. The microbial concentration may have been underestimated; nevertheless, kinetic as well as mass transfer limitations in the VPB may have existed.

VPB EC was dependent on TPH loading rate. Based on a minimum practical percent removal of 60%, the maximum loading rate was approximately 12 g-m^{-3}-h^{-1}. Adjusting for decreased efficiency of the pilot-scale unit, the loading rate was approximately 8 to 10 g-m^{-3}-h^{-1}. The economics of VPB application to treatment of gasoline vapors are thus not as favorable as those of compounds with lower H. Further analysis is needed to determine whether the VPB is competitive with other air pollution control technologies for gasoline vapor treatment. Further VPB development will focus on increasing the concentration of hydrocarbon degraders in the VPB in order to overcome potential kinetic limitations and on enhancing VPB mass transfer characteristics.

REFERENCES

Bailey, J. E., and D. F. Ollis. 1977. *Biochemical Engineering Fundamentals*. McGraw-Hill, New York, NY.

Diks, R.M.M. 1992. "The Removal of Dichloromethane from Waste Gases in a Biological Trickling Filter." Ph.D. Thesis. Eindhoven University of Technology, Eindhoven, The Netherlands.

Hodge, D. S., V. F. Medina, R. L. Islander and J. S. Devinny. 1991. "Treatment of Hydrocarbon Fuel Vapors in Biofilters." *Environmental Technology* 12: 655-662.

Kampbell, D. H., J. T. Wilson, H. W. Read, and T. T. Stocksdale. 1987. "Removal of Volatile Aliphatic Hydrocarbons in a Soil Bioreactor." *J. Air Pollution Control Association* 37(10): 1236-1240.

Ottengraf, S.P.P. 1987. "Biological Systems for Waste Gas Elimination." *Trends in Biotechnology* 5: 132-136.

Togna, A. P., and M. Singh. 1994. "Biological Vapor-Phase Treatment Using Biofilter and Biotrickling Filter Reactors: Practical Operating Regimes." *Environmental Progress* 13(2): 94-97.

Wilson, D. J., and A. N. Clarke. 1994. "Soil Vapor Stripping." In: D. J. Wilson and A. N. Clarke (Eds.), *Hazardous Waste Site Soil Remediation: Theory and Application of Innovative Technologies*. Marcel Dekker, New York, NY. pp. 171-242.

Biofiltration of Toluene Vapors in a Carbon-Medium Biofilter

Victor F. Medina, Joseph S. Devinny, and Mukund Ramaratnam

ABSTRACT

Treatment of toluene vapors in a biofilter with a packed bed of activated carbon pellets was tested. The flowrate was 0.14 m^3/min, the volume of the bed was 0.25 m^3, and the bed depth was 1 m. The reactor was capable of removing 70% of the toluene at an input concentration of 2,700 µg/L with an empty bed detention time of 1.8 min. Total removal was 64 g/m^3/h. Organic carbon removed from the air in a biofilter may be oxidized or incorporated into the biomass. Measurements of heat production and carbon dioxide production indicated, respectively, that 29% or 38% of the removed toluene was oxidized, while the remainder, 62% to 71%, was incorporated into biomass by cell synthesis. The rapid growth of the biomass indicates clogging may occur in biofilters operated at these rates.

INTRODUCTION

In biofiltration, contaminated air is treated by passing it through a damp porous medium that supports a vigorous culture of microorganisms. Contaminants dissolve in the water, are adsorbed on the solids, and are degraded by the microorganisms. Biodegradation continuously removes contaminants from the water, keeping concentrations below the saturation value there and on the surface of the solids. This in turn causes more transfer from the air phase. Although biofilters have been successfully deployed in pilot- and full-scale applications in Europe, Japan, and the United States (Leson and Winer 1991), the majority of published studies of biofiltration describe bench-scale experiments.

A key component of any biofilter is the packing material, or support medium. Natural materials, such as soil and compost, are most commonly used in commercial application. However, these materials may have significant inhomogeneities, making them difficult to study. Further, organic material found in soil and compost is available for microbial degradation, which can complicate studies of microbial activity associated with contaminant destruction.

Plastic media have been used, and have the advantage that shape and size can be precisely controlled to maximize porosity and surface area. High porosities will reduce head loss during operation, and higher surface areas will increase the amount of biofilm present. Plastics do not degrade or compact with time. However, they have little adsorptive capacity, and biofilters with plastic media may be more susceptible to shock loads.

Granular activated carbon (GAC) is a possible alternative support medium. GAC provides a reasonably benign environment for microorganisms, although care must be taken to adjust pH and provide nutrients. Water-treatment systems using biologically active carbon have shown that microorganisms grow well on GAC (Pirbazari et al. 1990). Although the adsorptive capacity of GAC is reduced by the presence of water and microorganisms, it remains more adsorptive than soil and compost, making the biofilter more resistant to shock loading. Biodegradation rates in biofilters often increase with increasing contaminant concentration (Kampbell et al. 1987; Hodge et al. 1991b; Baltzis and Shareefdeen 1993; Hodge and Devinny 1995). Adsorption of the contaminant on carbon provides a region of high concentration at the surface of the particles where the microorganisms are active, increasing the overall rate of degradation. Organic media, such as compost, are degraded as the biofilter is used. This results in compaction, increased pressure drop, increased operating costs, channelling, and poorer performance. GAC does not compact and can be used indefinitely. Recent work has demonstrated that GAC biofilters are effective at treating gasoline vapors (Medina et al. 1992), diesel vapors (Hodge et al. 1991a), ethanol vapors (Hodge et al. 1992), and various other compounds (Govind et al. 1992).

Studies have indicated some problems with GAC-based biofilters. Clogging from vigorous biological growth has been reported for heavily loaded systems (Govind et al. 1992; Hodge and Devinny 1995). The greatest drawback to the use of GAC in a commercial biofilter is that it is 5 or 27 times more expensive per unit volume than compost-based mixes or yard waste compost, respectively (Leson [personal communication] 1992; Ramaratnam [personal communication] 1992). However, the high initial cost of the medium may be acceptable if greater efficiency of GAC allows construction of a smaller biofilter for a given wastestream.

This study used a pilot-scale GAC biofilter for treating an artificial wastestream of toluene vapors in air. Toluene is ranked 27th in chemical production in the United States (Sax and Lewis 1987). It is used as aviation gasoline, as high-octane blending stock, and as a solvent for manufacturing plastics and explosives. Liu et al. (1994) studied biofiltration of toluene for relatively low input concentrations and detention time (0.041 mg/L, 2 s). This study evaluated the effectiveness of carbon biofilters for much higher concentrations and examined alternative methods of evaluating treatment phenomena.

METHODS

Experimental Apparatus

The pilot-scale reactor was constructed by Westates Carbon of Los Angeles, California (Figure 1), was 0.35 m in diameter, and contained a carbon bed 1 m

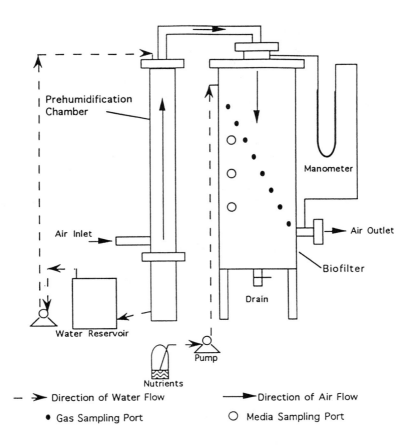

FIGURE 1. Schematic diagram of experimental apparatus.

deep. The airflow was humidified in a countercurrent prehumidification chamber. A sprayer was installed to directly dampen the biofilter medium. The irrigation solution consisted of water with 1 drop/L of Schultz Instant Plant Food
and was added at the rate of 2.2 L/day. Three ports were used for removing
medium samples from the reactor, and three more allowed the introduction of
thermocouples. A water manometer was used to measure pressure drop across
the biofilter. The carbon medium was supported on a 0.3-m layer of plastic balls
covered by a polypropylene screen. These were intended to prevent the carbon
at the bottom of the column from clogging the output port and to reduce the
tendency of the flow to channel near the ports. The plastic balls were not thought
to contribute to treatment, because the pores between them were large.

Biofilter Medium

The medium was Type KP 601 activated carbon (Westates Carbon, Los
Angeles, California), consisting of cylindrical pellets about 0.5 cm long by 0.3 cm

in diameter. The reactor and medium had been previously used for treating gasoline vapors, but had been inactive for 6 months. The medium presumably included dormant hydrocarbon-degrading microorganisms and biofilm polysaccharides, and much of the adsorptive capacity was exhausted. To ensure activity, it was reseeded with a variety of inocula. The following materials were mixed for 15 min in a concrete mixer: 250 L GAC; 4,400 g soil from an active petroleum refinery landfarm; 3 L sewage treatment sludge; 250 g commercial seed (Solmar L-103); 260 g commercial seed (Solmar I-107); 450 g commercial seed (Microlite Technologies Munox 501); 5.45 kg Grace-Sierra Max Bac nutrient pellets; 361 drops Schultz Instant Plant Food; and 123 L water (64 L of fluid drained from the medium after mixing).

The seeding procedure increased the total-volatile-solids content of the medium from 9.4% to 19.0%, reflecting the organic matter in the soil and sludge. Plate counts of viable bacteria grown on a general medium known to support petroleum-degrading organisms were determined, and expressed as numbers per gram of wet biofilter medium (Marshall and Devinny, 1986). These rose from $1.7 \times 10^6/g$ to $5.8 \times 10^8/g$, and counts of fungi rose from $3.0 \times 10^5/g$ to $1.3 \times 10^7/g$ during the course of the experiment.

Sampling and Analysis

Vapor samples were collected in Teflon™ bags and transported to the laboratory for analysis within 1.5 hours. After use, the bags were filled and purged three times using nitrogen gas. Analyses included total hydrocarbons using a flame ionization detector (FID) and carbon dioxide using a methanizer catalyst and an FID. Medium samples were collected from within the packed bed using a grain thief, sealed in autoclaved jars, and transported to the laboratory for determination of water content, pH, and plate counts (Table 1).

Operating Conditions

Records were kept on flowrates, and pollutant concentrations, and used to calculate approach velocity, interstitial velocity, and detention times (Table 2).

TABLE 1. Analyses of support medium.

Analysis	Method
Water content	Oven drying at 103°C
pH	Equilibrium in deionized water, pH meter
Plate counts of bacteria and fungi	Growth on general medium with fungicide or bactericide (Marshall and Devinny 1986)

TABLE 2. Operating conditions.

Input flow	0.14 m³/min
Approach velocity	49 cm/min
Interstitial velocity	163 cm/min
Detention time (empty bed)	1.8 min
Detention time (packed bed)	0.53 min
Pollutant load	90 g/m³/h

RESULTS

Treatment Success

Cumulative mass input and output curves were prepared by multiplying the length of time between samples by the flowrate and the average concentrations during the period. Input and output concentrations were calculated as the average of the determinations at the beginning and end of the period. These were summed to show cumulative input and output as a function of time (Figure 2). Some variability in input concentration occurred because of the difficulty of producing a uniform airstream in the laboratory.

Percent cumulative removal was also calculated, and reached a steady-state value of about 70%. This did not change appreciably as the input concentration changed, consistent with an assumption of first-order kinetics. The overall removal rate averaged 64 $g/m^3/h$, which compares favorably with the results of other studies. Liu et al. (1994) found overall removal rates of 19 $g/m^3/h$, and Morales et al. (1994) found 25 $g/m^3/h$, but these investigators were treating a much lower concentration of toluene. Biofilters used in actual applications could be made with a somewhat longer detention time, and achieve higher removals without excessive cost. Such a system also could be used in advance of activated carbon adsorbers, substantially reducing the amount of carbon regeneration required.

Organic carbon removed from the air in a biofilter may be oxidized or incorporated into the biomass. If incorporation is dominant, the total amount of biomass present will grow rapidly and eventually clog the biofilter. The amount of toluene oxidized in this experiment was estimated in two ways: (1) from the amount of heat generated and (2) from the amount of carbon dioxide released.

Heat Balance

Oxidation of the contaminants in the biofilter produces heat. The amount of heat produced can therefore be used as an indicator of the fraction of the

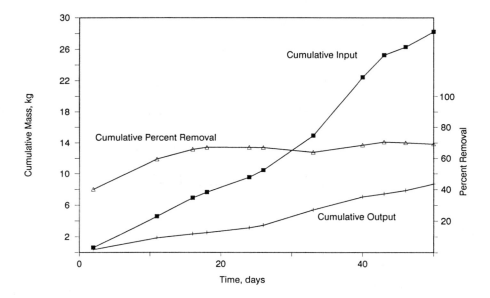

FIGURE 2. Cumulative input, output, and percent removal calculated from input and output concentrations.

contaminant that is being biodegraded. Collection of data on temperature, water input and output, and medium water content allows calculation of an approximate heat balance for the biofilter. Input concentrations of toluene averaged 2,700 µg/L, of which 70% (or 1,895 µg/L) were removed. The heat of combustion of toluene is 934.2 kcal/g, so complete oxidation of the removed toluene would generate 20 cal of heat/L of air. This energy is seen in two forms: (1) as increased temperature of the discharged air stream and (2) as the latent heat of vaporization of water evaporated in the biofilter.

The temperature of the air passing through the biofilter increased by an average of 7°C. The heat capacity of air is 0.31 cal/L/°C, so heating of the air accounts for 2.17 cal/L. The amount of water evaporated can be estimated in two ways. If it is assumed that both the input air and output air were saturated, then the input partial pressure of water vapor was 0.023 atm (20°C), and the output partial pressure was 0.035 atm (27°C). This corresponds to the evaporation of 6.4×10^{-3} g water/L of air, requiring a heat of vaporization of 3.5 cal/L of air. A second estimate of water evaporation can be made by noting that total input to the biofilter during the experiment, minus total leachate, was 0.85 L/day. Despite this addition, the water content of the medium declined from 40% to 25%, corresponding to a further loss of 0.76 L/day, for a total evaporative loss of 1.61 L/day. This is 7.9×10^{-3} g water/L of air, requiring a heat of vaporization of 4.2 cal/L of air.

Thus, the estimated total energy production was either 5.63 or 6.37 cal/L of air treated, or 29% or 33% of the energy production expected for complete oxidation of the removed toluene. It is presumed that the remainder of the removed toluene was incorporated into biomass.

Carbon Balance

The carbon which enters the biofilter as toluene may either be oxidized or incorporated into microorganisms. Incorporation can be measured in terms of new biomass generated. Oxidation of toluene can be estimated on the basis of CO_2 production. Input and output concentrations of CO_2 were measured on two occasions, and values increased by 1,161 ppmv and 1,297 ppmv as the air passed through the biofilter. The average value of 1,229 ppmv implies that 55×10^{-6} mols of CO_2/L of air were produced. Complete oxidation of the toluene removed would have produced 144×10^{-6} moles of CO_2/L of air. This suggests that 38% of the toluene removed was oxidized.

Incorporation

Incorporation of toluene into biomass was assumed equal to total toluene removal minus oxidation estimated from heat or CO_2 balances (Table 3). Assuming that all of the carbon and hydrogen in the toluene appeared as biomass, the amounts calculated correspond to about 9% of the mass of the medium. Efforts to measure the biomass in the medium as total volatile solids before and after the experiment were not successful, probably because a large, unknown amount of organic seed material was washed out of the reactor during the experiment. The relatively large fraction of toluene that was incorporated rather than oxidized to CO_2 indicates a rapidly growing biomass. Continued growth at this rate could cause clogging problems in biofilters operated for a longer period of time than the 50 days of this experiment. However, as the biomass matures, populations of predators may be established, increasing mineralization rates. Sloughing of biomass was not important because flowrates of water through the biofilter were minimal.

TABLE 3. Fate of toluene treated, as percent of input.

Method of Estimation	Input	Output	Oxidation	Incorporation
Heat balance, from water content of input and output air	100	30	20	50
Heat balance, from medium water loss	100	30	23	47
CO_2 production	100	30	27	43

Pressure Drop and pH

The measured pressure drop was modest throughout the experiment (always less than 0.2 cm of water), indicating no problems with clogging during this short run. The pH declined moderately immediately after the beginning of the experiment, then stabilized. The pH averages for the inlet, middle, and outlet portions were 5.6, 6.2, and 6.5, respectively. Presumably the declines resulted from the production of organic acid intermediates.

REFERENCES

Baltzis, B. C., and Z. Shareefdeen. 1993. "Modeling and Preliminary Design Criteria for Packed Bed Biofilters." Preprint. Presented at the 86th Annual Meeting and Exhibition of the Air and Waste Management Association, June 13-18, Denver, CO.

Govind, R., V. Utgikar, Y. Shan, W. Zhao, D. F. Bishop, S. I. Safferman, and G. D. Sayles. 1992. "Fundamental Studies on the Treatment of VOCs in a Biofilter." In *Bioremediation of Hazardous Wastes*. United States Environmental Protection Agency Biosystems Technology Development Program, Ada, OK.

Hodge, D. S., and J. S. Devinny. 1995. "Modeling Removal of Air Contaminants by Biofiltration." *Journal of Environmental Engineering, American Society of Civil Engineers*, in press.

Hodge, D. S., V. F. Medina, R. L. Islander, and J. S. Devinny. 1991a. "Treatment of Hydrocarbon Vapors in Biofilters." *Environmental Technology* 12: 655-662.

Hodge, D. S., V. F. Medina, and J. S. Devinny. 1991b. "Biological Methods for Air Decontamination." In P. A. Krenkel (Ed.), *Environmental Engineering. Proceedings of the 1991 Specialty Conference*, pp. 352-357. American Society of Civil Engineers, New York, NY.

Hodge, D. S., V. F. Medina, Y. Wang, and J. S. Devinny. 1992. "Biofiltration: Application for VOC Control." *Proceedings of the 47th Annual Purdue Industrial Waste Conference*. Purdue University, West Lafayette, IN.

Kampbell, D. H., J. T. Wilson, H. W. Read, and T. T. Stocksdale. 1987. "Removal of Volatile Aliphatic Hydrocarbons in a Soil Bioreactor." *Journal of the Air Pollution Control Federation* 37(10): 1236-1240.

Leson, G., and A. M. Winer. 1991. "Biofiltration: An Innovative Air Pollution Control Technology for VOC Emissions." *Journal of the Air and Waste Management Association* 41(8): 1045-1054.

Liu, P.K.T., R. L. Gregg, H. K. Sabol, and N. Barkley. 1994. "Engineered Biofilter for Removing Organic Contaminants in Air." *Journal of the Air and Waste Management Association* 44: 299-303.

Marshall, T., and J. S. Devinny. 1986. "Methods for Enumerating Microorganisms in Petroleum Land Treatment Soils." *Hazardous Waste and Hazardous Materials* 3(2): 175-182.

Medina, V. F., T. Webster, M. Ramaratnam, D. S. Hodge, and J. S. Devinny. 1992. "Treatment of Soil Vapor Extraction Off Gases by GAC Based Biological Filtration: Bench and Pilot Studies." *Emerging Technologies for Hazardous Waste Management*, American Chemical Society I&EC Division Special Symposium, Atlanta, GA, September 21-23.

Morales, M., F. Perez, R. Auria, and S. Revah. 1994. "Toluene Removal from Air Streams by Biofiltration." In E. Galindo and O. T. Ramirez (Eds.), *Advances in Bioprocess Engineering*, pp. 405-411. Kluwer Academic Publishers, The Netherlands.

Pirbazari, M., T. C. Voice, and W. J. Weber. 1990. "Evaluation of Biofilm Development on Various Natural and Synthetic Media." *Hazardous Waste and Hazardous Materials* 7(3): 239-250.

Sax, N. J., and R. J. Lewis. 1987. *Hawleys Condensed Chemical Dictionary*, 11th ed. Van Nostrand Reinhold and Company, New York, NY.

Removal of Gasoline Volatile Organic Compounds Via Air Biofiltration

R. Scott Miller, Amireh G. Saberiyan, Peter DeSantis, Jeffrey S. Andrilenas, and Charles T. Esler

ABSTRACT

Volatile organic compounds (VOCs) generated by vapor extraction and air-stripping systems can be biologically treated in an air biofiltration unit. An air biofilter consists of one or more beds of packing material inoculated with heterotrophic microorganisms capable of degrading the organic contaminant of concern. Waste gases and oxygen are passed through the inoculated packing material, where the microorganisms will degrade the contaminant and release CO_2 + H_2O. Based on data obtained from a treatability study, a full-scale unit was designed and constructed to be used for treating gasoline vapors generated by a vapor-extraction and groundwater-treatment system at a site in California. The unit is composed of two cylindrical reactors with a total packing volume of 3 m^3. Both reactors are packed with sphagnum moss and inoculated with hydrocarbon-degrading microorganisms of *Pseudomonas* and *Arthrobacter* spp. The two reactors are connected in series for air-flow passage. Parallel lines are used for injection of water, nutrients, and buffer to each reactor. Data collected during the startup program have demonstrated an air biofiltration unit with high organic-vapor-removal efficiency.

INTRODUCTION

Volatile organic compounds (VOCs) typically are generated as off-gases by two widely used remedial processes: vapor extraction and air stripping (Vasconcelos et al. 1991). Both operations entail media transfer of the contaminants from the soil or groundwater to an airstream without any contaminant destruction.

Air biofiltration is a relatively new technology that uses microorganisms to remove VOCs from the airstream. An air biofilter consists of biologically active packing material. Waste gases and oxygen are passed through the packing material, where microbial destruction takes place (Douglass 1991). Microorganisms grow on the surface area of the packing in a wet biolayer known as the biofilm. The microbes use the VOCs as the food source in the presence of oxygen and nutrients. The by-products of this usage are carbon dioxide, water, and biomass. The two basic processes that govern the success of the biofilter process are (1) diffusion of the contaminant into the biofilm, and (2) the metabolism of the contaminant by the microorganism (Diks and Ottengraf 1991).

PILOT-SCALE STUDY

Because of the lack of practical and theoretical design information regarding the application of this technology, a pilot-scale study was performed prior to full-scale system design (Saberiyan et al. 1992). The objective of the study was to determine the maximum loading rate and maximum concentration of gasoline vapors per unit volume of packing material. This study was performed to assess the practicality of treating gasoline-impacted air via air biofiltration. The primary goal of the study was to determine the basic engineering design requirement for remediation of gases generated by the vapor extraction system at a gasoline-impacted site. The pilot-scale study was designed to provide data suitable for the design of a field-scale unit and to simulate operating conditions that might be limiting factors in the operation of the field-scale unit.

Three treatment cells for packing material were used, each with a different volume of packing material. The pilot-scale biofilters were constructed containing 1 ft^3, 2 ft^3, and 4 ft^3 (0.028 m^3, 0.057 m^3, and 0.113 m^3). After 1 month of acclimation, various flowrates and contaminant loading rates were investigated. Based on calculated retention times and the degradation obtained from this study, the maximum possible gasoline-vapor loading rate was determined for 90% removal efficiency. The approximate value obtained was 16 g/h of gasoline vapors per 1 ft^3 (0.028 m^3) of packing material.

FIELD-SCALE DESIGN AND FABRICATION

The field-scale design of the biofiltration unit was based on the pilot-scale study and the anticipated contaminant loading as determined by tests performed at the project site. Site-specific feasibility testing and the remediation design indicated that an anticipated total air flowrate of 350 cfm (10 m^3/min) would be achieved. There are three sources of air emissions: soil vapor extraction, air stripping of extracted groundwater, and off-gases from the vacuum enhancement of the groundwater recovery wells. Based on the anticipated loading rate and the achievable degradation rate of the packing material, a total of 3 m^3 of packing material was determined to be required for the unit.

Two primary requirements were set for the field-scale design of the air biofiltration unit: (1) general duplication of the environment conducive to the biological degradation of hydrocarbons as demonstrated by the pilot study, and

(2) ease of field operation and maintenance of the unit. The design of the unit also had to consider the transportation and handling of the units. Using two 1.5-m³ units in place of one large unit allowed the units to be moved using a standard forklift. The double unit design provides additional flexibility for the future; when the concentrations in the vapors begin to drop, one of the two units may simply be removed for use at a different site.

The units, each 1.5 m in diameter and over 1.5 m tall, are fabricated of carbon steel. A unit consists of a removable lid for access, expanded-steel-grate subfloor for an internal sump, and steel tubing at the base for the forklift. Each unit also contains standard plumbing fittings to connect the air and water piping (Figure 1).

Sphagnum mass was used as packing material for the units because its large surface area supports effective microbial distribution, it retains moisture, and its high porosity minimizes airflow resistance. In addition, sphagnum moss was used in the pilot study that provided the design hydrocarbon-consumption rate.

As demonstrated in Figure 1, off-gases enter the top of the units, pass through the packing material, and exit near the base of the unit (the two units are in series). The moisture of the packing material within the units is maintained by four spray nozzles located at the top of each unit. Water is recirculated through each unit using the collection sump at the base and an external pump.

TESTING AND OPERATION

Field-scale units are currently undergoing startup and testing in Portland, Oregon. Startup of the units consists of neutralizing the sphagnum moss,

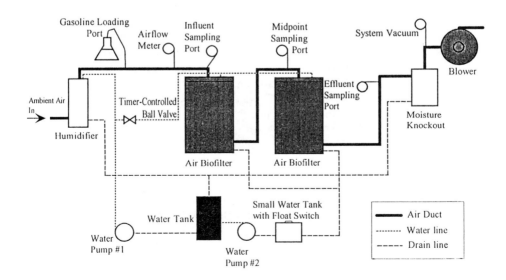

FIGURE 1. Schematic diagram of air biofilter units.

introducing the nutrient solution and bacterial culture to the sphagnum moss, loading the gasoline fuel in the influent air stream, and performing operational monitoring.

The feed system for the introduction of hydrocarbon during the startup period consists of side-stream gases that are drawn through a trap filled with gasoline. The gases that are drawn from the headspace above the trap are saturated with gasoline vapors. The rate of fuel introduction is controlled by a needle valve.

A pH range of 6 to 8 is known to be suitable for biological processes. Sphagnum moss contains high levels of humic acids. At high moisture levels pH can drop to as low as 4. The pH of the packing material was adjusted to 7 prior to contaminant loading using sodium bicarbonate. After stabilization of pH, the packing material was inoculated with indigenous petroleum-degrading bacteria. These bacteria were isolated from groundwater samples collected from the site, and were characterized as *Pseudomonas* spp. In addition, a consortium of available commercial bacteria were added to the moss through the water recirculation system. To enhance and support biological activity, an inorganic nutrient solution containing ammonium nitrate and ammonia phosphate was added to the biofilter. Water within the system can leave the biofilter through evaporation into the airstream, which passes through the biofilter. To address this problem, a humidifier was used to provide more than 90% humidity in the influent airstream.

Multiple parameters are monitored in conjunction with startup of this unit. Oxygen concentration, volatile organic gas concentration, and carbon dioxide are monitored at the influent to the air filters, between the two filters, and at the effluent of the filters; the ambient levels within the room are also monitored. Oxygen concentration is monitored with an oxygen sensor/transmitter for standalone applications. The concentration of the volatile organic gases is monitored with the MSA/Baseline 1015C total organic vapor monitor. Carbon dioxide is measured with the infrared gas monitor with an operation range of 0 to 0.2% carbon dioxide. These parameters are monitored at the aforementioned locations using a combination of solenoid valves. Each of the sampling locations are monitored for 1 h. The signals from each of the sensors are recorded with a Unidata Model 7000 macro data logger. Sensor values are recorded at 2-min intervals during the first month of operation. Data collected during subsequent months were recorded at 15-min intervals.

RESULTS

At the end of the first week of operation (equilibrium period), under a low contaminant loading rate, 50% removal efficiency was achieved. Laboratory analysis of air samples collected 5 months after the start of the equilibrium period showed a 73% reduction in gasoline and 100% reduction in benzene, toluene, ethylbenzene, and xylenes (BTEX) (U.S. EPA Methods 5030/8020). These results are lower than the reduction rate predicted by the results of the biotreatability pilot-scale study, which achieved a removal efficiency as high as 90%. The reduction in gasoline and benzene concentrations was measured at the inlet to the units and at the midpoint between the two units. Sample analysis indicates no significant reduction is currently provided by the second unit.

The contaminant and carbon dioxide concentration-gradient profiles at steady-state condition (temperature = 25°C, pH ~7, flowrate ranging from 195 to 225 cfm [5.5 to 6.4 m³/min], constant moisture) are shown in Figure 2. Both laboratory analysis of the filter units and continuous monitoring (with photoionization detector [PID]) are presented in Figures 2 and 3, respectively. Carbon dioxide and VOCs are shown for only the first months of operation. Laboratory analysis indicates that aromatic components of the gasoline are removed at a higher rate compared to alkane and alkene chains. Diffusion of the contaminant to the wet layer (biofilm) is the limiting factor in contaminant removal; hence, aromatic hydrocarbons (BTEX) with higher solubility than other gasoline components maintain higher degradation rates.

DISCUSSION

Air biofiltration is a relatively new technology, which can be used in advance of an activated-carbon system, reducing the contaminant load and frequency of

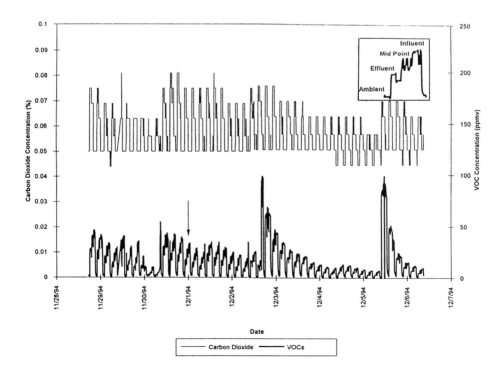

FIGURE 2. CO₂ and VOC monitoring data.

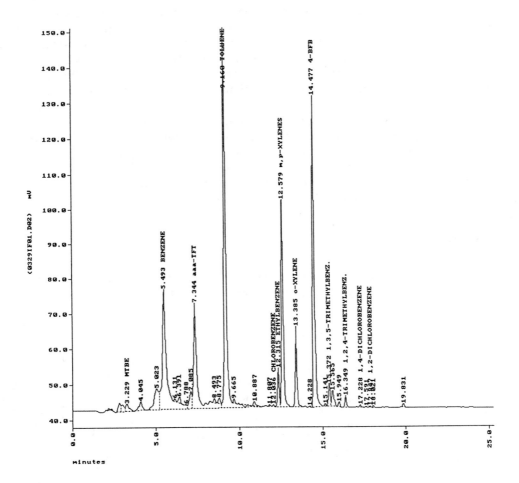

FIGURE 3. PID chromatographic traces for volatile aromatic hydrocarbons.

carbon regeneration. Two biofilter units were designed in series (Figure 1) and will be used as the secondary air treatment units at a gasoline-impacted site. The units are currently under acclimation and equilibrium. After 2 weeks of operation, 50% removal efficiency was achieved, followed by an increase in reduction to over 70% after 5 months.

During the startup period, biofilters have a zero-order kinetic and the rate of the diffusion is the controlling removal mechanism. As the contaminant concentration increases, the zero-order kinetic would change to first order. At this time, the rate of removal depends on the rate of biodegradation and contaminant concentration. The CO_2 data and sample analysis indicates that with the current loading rates, most of the contaminant destruction happens in the first unit.

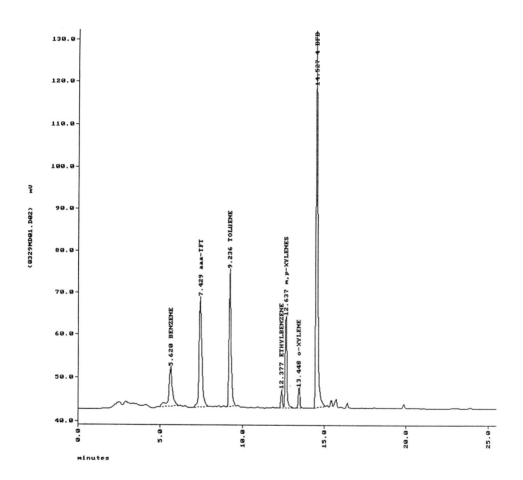

FIGURE 3. (continued).

To confirm the biological activity within the system, the rate of CO_2 production to contaminant reduction was calculated and compared with the theoretical ratio. Gasoline vapors are a complex mixture of various hydrocarbons. An average chemical formula of C_7H_{12} was selected. Following is an assumption of the bioremediation stoichiometric relationship (McCarty 1988):

$$C_7H_{12} + 5CO_2 + NH_3 \rightarrow 2CO_2 + 4H_2O + C_5H_7O_2N$$

$C_5H_7O_2N$ represents the bacterial cell and C_7H_{12} represents the gasoline vapors. The theoretical ratio of CO_2 production to contaminant reduction is 0.92. The empirical data show a ratio of 1.25. The higher empirical ratio can be due

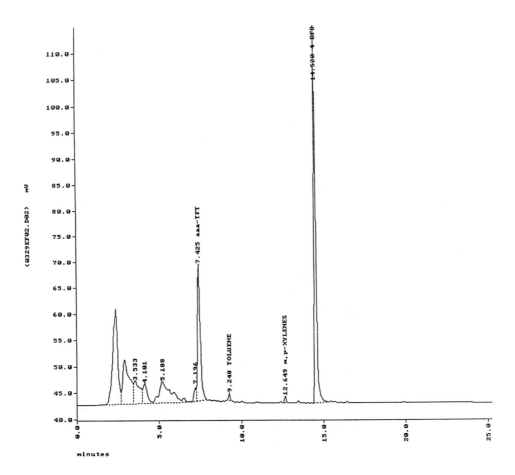

FIGURE 3. (continued)

to either moss respiration or partial mineralization of the contaminant. Based on preliminary data collected and the previous pilot study, the field-scale air biofiltration unit is achieving the predicted reduction in gasoline vapors. It is anticipated that the continuing growth of biofilm and an increase in the hydrocarbon loading will produce greater contaminant degradation.

REFERENCES

Diks, R.M.M., and S.P.P. Ottengraf. 1991. "A Biological Treatment System for the Purification of Waste Gases Containing Xenobiotic Compounds." In R. E. Hinchee and R. F. Olfenbuttel

(Eds.), *On-Site Bioreclamation: Processes for Xenobiotic and Hydrocarbon Treatment*, pp. 452-463. Butterworth-Heinemann, Stoneham, MA.

Douglass, R. H. 1991. "Biofiltration Shows Potential as Air Pollution Control Technology." *The Air Pollution Consultant* (November/December): 1.1-1.2.

McCarty, P. L. 1988. "Bioengineering Issues Related to In Situ Remediation of Contaminated Soil and Groundwater." *Environmental Biotechnology*, pp. 128-147.

Saberiyan, A. G., M. S. Wilson, E. O. Roe, J. S. Andrilenas, C. T. Esler, G. H. Kise, and P. E. Reith. 1992. "Removal of Gasoline Volatile Organic Compounds Via Air Biofiltration: A Technique for Treating Secondary Air Emissions From Vapor-Extraction and Air-Stripping Systems." In R. E. Hinchee, B. C. Alleman, R. E. Hoeppel, and R. N. Miller (Eds.), *Hydrocarbon Bioremediation*, pp. 1-12. Lewis Publishers, Boca Raton, FL.

Vasconcelos, J. J., L.Y.C. Leong, and J. E. Smith. 1991. "VOC Emissions and Associated Health Risks." *Water Environment & Technology*, (May): 47-50.

Biological Treatment of Waste Gas Containing Volatile Hydrocarbons

Jiyu Lei, Denis Lord, Rolf Arneberg,
Denis Rho, Charles Greer, and Benoit Cyr

ABSTRACT

A biological system to treat volatile hydrocarbon-contaminated gases generated during in situ bioventing and air sparging of subsurfaces contaminated with gasoline was field-tested. The system consisted of an air/water separator, a trickling filter, and a biofilter in series. During the field test, extensive monitoring was carried out to evaluate system performance, including the measurement of physical, chemical, biochemical, and microbiological parameters. Degradation and mineralization of volatile hydrocarbons such as benzene and toluene were demonstrated by gene probing and mineralization assays. Data collected showed an average removal of 90% of the BTX (benzene, toluene, and xylenes) and 72% for total hydrocarbons.

INTRODUCTION

In situ biological treatment (bioventing and air sparging) is a promising technology for remediation of soil and groundwater contaminated with hydrocarbons. The treatment involves vapor extraction from the contaminated zone with or without air and nutrient injections. The vapor extraction phase contains volatile hydrocarbons and requires purification. Technologies suitable for volatile hydrocarbon control include carbon adsorption and thermal or catalytic oxidation (Reisinger et al. 1994). Recently, biological gas cleaning has been considered as an attractive alternative because of its low cost, technical simplicity, and the production of little or no secondary waste. Our biological system has been field-tested to treat off-gas extracted during an in situ soil treatment operation. The test was carried out within the framework of the federal-provincial Development and Demonstration of Site Remedial Technologies (DESRT) Program.

MATERIAL AND METHODS

Site Description and Contaminant Characteristics

Field tests were conducted at a gasoline service station located on the out-skirts of Quebec City. During work to replace storage tanks, volatile hydrocarbon contamination due to gasoline leakage from an underground tank was discovered. The contaminated area measured approximately 250 m² and the volume of soil to be treated was estimated at 400 m³. The soil contamination was located at an average depth of 3 m. In situ treatment (bioventing and air sparging) was carried out at the remediation site. Wells were bored to inject air into the contam-inated areas and to recover the gas products. The waste gas recovered contained volatile hydrocarbons including benzene, toluene, and xylenes. The treatment system for these gases had to be highly efficient due to the close proximity of commercial activity (restaurants, boutiques, and a bus stop).

Description of the Gas Cleaning System

The system, presented schematically in Figure 1, was made up of three units: an air/water separator, a trickling filter reactor, and a biofilter. Waste gas, extracted from the subsurface of the contaminated area at a flowrate of 140 to 210 m³/h, was first drawn in through the air/water separator used to remove nongas-phase impurities in the air, such as water droplets, solid particles, etc. The effluent then travelled through the trickling filter reactor packed with 3 m³ of a porous synthetic polymer. The density of the polymer was 29 to 32 kg/m³, and the porosity was about 85%. In the trickling filter, gas and liquid streams flowed countercurrently. The liquid was recirculated continuously from the bottom to the top of the packing. The use of a polymer as packing material enhanced mass transfer between the liquid and the gas phases. Moreover, the polymer could fix a large concentration of specific bacteria helpful in degrading the contaminants found in the liquid phase. Following treatment in the trickling

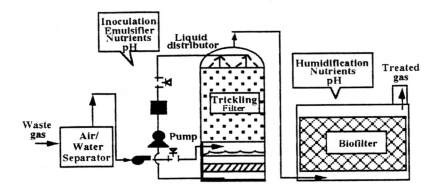

FIGURE 1. Biological treatment system of waste gases.

filter reactor, the waste gas was channelled to a 4-m^3 compost-based biofilter where residual contaminants were removed.

Operation and Monitoring

Field testing was carried out at the site for 5 months. The system was first used to treat bioventing off-gas for 60 days. It was then shut down for 40 days and subsequently used again during air sparging. The system was used on site from early March to August, during which the ambient temperature varied from −21.6 to 25.6°C.

During the startup phase, a liquid containing mineral nutrients, including NH_4NO_3 and K_2HPO_4, an emulsifying agent, and hydrocarbon-degrading bacteria was introduced into the trickling filter unit. The bacteria were previously isolated from a gasoline-contaminated soil and subcultured for 8 months in a mineral salts medium containing gasoline as sole carbon source. A biologically active compost-based material was used as the biofilter bed. No nutrient supplementation in the biofilter was required because of a high initial level of N and P in the compost bed. However, during the field test, occasional irrigation of the filter bed was used to keep the moisture level at 50 to 60% (W/W).

Analytical follow-up was conducted to evaluate system performance. In addition to physical/chemical and microbiological analyses including pH, temperature, nutrients, and the enumeration of total aerobic heterotrophic bacteria and hydrocarbon-degrading bacteria (Lei et al. 1994), the biochemical characterization of microorganisms was carried out using a mineralization assay (Rho et al. 1994) and gene probing (Greer et al. 1993). To assess contaminant elimination, gas samples were analyzed for BTXs and total volatile hydrocarbons.

Analytical Methods

Mineralization Test. The microcosms used consisted of 100-mL glass serum bottles containing a liquid sample collected from the trickling filter or a compost sample taken from the biofilter, with ^{14}C-benzene or ^{14}C-toluene added to the microcosms. A tube containing 0.5 mL of 1 M KOH was introduced into each microcosm to trap CO_2. Mineralization rates were calculated as a percentage of the initial amount of radioactive compound spiked into the microcosms.

DNA Hybridization Analysis. Isolated bacterial colonies on an agar medium, obtained by sample plate counts, were transferred directly onto nylon membranes for gene probing (Greer et al. 1993). Three gene probes were used: *alk* B, *xyl* E, and *tod* E. The *xyl* E gene probe is a fragment of the *xyl* E gene, encoding the enzyme catechol / 2,3-dioxygenase. The enzyme is a key enzyme involved in the bacterial biodegradation of compounds such as toluene, xylenes, and naphthalene. The *tod* E gene probe is a fragment from the internal coding region of the 3-methycatechol 2,3-dioxygenase gene of the *tod* operon from *Pseudomonas putida*. The gene is involved in the degradation of toluene. The third probe, *alk* B, is a fragment of the alkane hydroxylase gene of the OCT plasmid of *P. oleovorans*.

The enzyme is involved in the bacterial degradation of straight-chain alkanes with from 6 to 12 carbons.

BTX Analysis. Gas samples were collected in 250-mL glass cells (chromatographic specialties, cal # C72805) and analyzed by gas chromatography/mass spectrometry (GC/MS). Samples (50 to 200 µL) were injected into the chromatographic columns with a gastight syringe. GC/MS was calibrated with a standard gas sample containing 20 ± 1 ppm (v/v) of the following chemicals: benzene, toluene, *o*-xylene, *p*-xylene, and *m*-xylene.

Total Hydrocarbon Analysis. Total concentrations of volatile hydrocarbons were determined using a Hydrocarbon Surveyor, model 1238 (GasTech, Inc.). The instrument was calibrated with hexane, and the concentration of total hydrocarbons was expressed in hexane equivalents.

RESULTS AND DISCUSSION

In the trickling filter unit, due to the difficulty of desorbing all immobilized microorganisms from the packing, it was inappropriate to enumerate viable microorganisms in the packing material. Therefore, microbiological analysis was done in the recirculating liquid of the trickling filter. The total heterotrophic microorganisms recovered on low nutrient agar medium ranged from 10^6 to 10^8 colony forming units per milliliter of liquid (CFU/mL) throughout the monitoring period (Table 1). At the beginning of the operation of the trickling filter, the population level was around 3.7×10^7 CFU/mL. The following sampling showed a reduction in population by one order of magnitude. Then the next two samplings showed an increase in total population to above 10^8 CFU/mL after 42 days of operation which was followed by another decrease to 10^6 CFU/mL. Because microorganisms inoculated in the trickling liquid at the beginning are expected to attach onto the packing material, it may be normal to observe a decrease in the liquid microbial population during the startup period. The variations observed for the total heterotrophic population after that period could be due to variations in the organic loading rate (volatile hydrocarbons). Gene probing showed the presence of all three genotypes (*alk* B, *xyl* E, and *tod* E) in the recirculating liquid of the trickling filter (Table 1). At time zero, the levels of all three probe-positive populations in the liquid were very high (around 35% for *alk* B, 43% for *xyl* E, and 20% for *tod* E). This demonstrates that the choice of microorganisms used to inoculate the recirculating liquid was judicious.

The population levels of heterotrophic microorganisms observed in the biofilter medium were much higher than in the liquid of the trickling filter (Table 1). Screening with gene probes showed that all three probe-positive populations were present at high concentrations. This represents a very substantial population of potential volatile hydrocarbon degraders. However, the populations were not fully active during the startup phase since the hydrocarbon-degrading population,

TABLE. 1 **Microbiological characterization of the trickling filter (liquid phase) and the biofilter (solid phase).**

Phase	Time (d)	THM[a]	Probe-positive		
			alk B	*xyl* E	*tod* E
Trickling filter (CFU/mL)	0	3.7×10^7	1.3×10^7	1.6×10^7	7.3×10^6
	14	4.3×10^6	8.6×10^5	7.3×10^5	NA[b]
	28	2.5×10^7	8.8×10^5	NA	4.9×10^5
	42	1.3×10^8	NA	1.7×10^7	4.2×10^6
	56	4.0×10^6	5.1×10^3	3.2×10^5	NA
	119	1.8×10^6	1.0×10^3	8.1×10^4	NA
	147	2.3×10^6	NA	4.0×10^5	2.1×10^5
Biofilter (CFU/g)	0	1.6×10^9	4.1×10^6	5.1×10^7	4.0×10^7
	14	7.9×10^8	2.9×10^6	4.4×10^7	NA
	28	1.1×10^9	2.0×10^6	NA	1.8×10^7
	42	2.1×10^9	NA	1.4×10^8	4.5×10^7
	56	1.5×10^9	6.0×10^6	6.8×10^7	NA
	119	2.9×10^8	1.5×10^6	6.2×10^6	NA
	147	2.5×10^8	NA	5.0×10^7	1.5×10^7

(a) Total heterotrophic microorganisms.
(b) Not analyzed.

enumerated with the most-probable number method, was only 10^3 CFU/g (data not shown). The population reached 10^6 CFU/g after 4 weeks of operation (Lei et al. 1994). The mineralization potential of the biofilter samples for ^{14}C-benzene and ^{14}C-toluene was assayed. Results showed that both chemicals could be converted to CO_2 (Rho et al. 1995).

Figure 2 presents the concentrations of BTX and total volatile hydrocarbons in inlet and outlet gases of the treatment system. As expected, the concentrations of the contaminants were high at the beginning of the field test, and they gradually decreased later (Figure 2). High percentages of BTX removal (approximately 90%) were observed even with an inlet concentration as high as 377 mg/m^3 (Figure 2). The removal of total volatile hydrocarbons ranged from 47% to 75% at an influent concentration of more than 400 mg/m^3. At BTX concentrations less than 200 mg/m^3, the treatment system showed ≥95% removal of the total volatile hydrocarbons.

Data were gathered to evaluate the performance of the trickling filter and the biofilter separately. Figures 3 and 4 show the elimination capacity of BTX and of total hydrocarbons plotted against the contaminant loads. In terms of

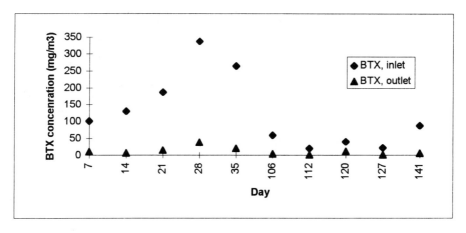

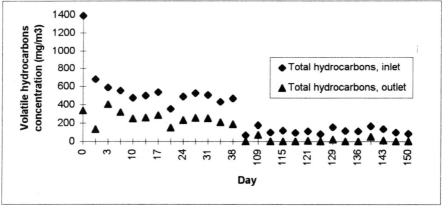

FIGURE 2. BTX and total volatile hydrocarbons concentrations in inlet and outlet gases.

BTX removal, the biofilter unit demonstrated a higher elimination capacity than the trickling filter at a similar load (Figure 3). However, the latter was more efficient in removing total hydrocarbons (Figure 4). For example, at a loading rate of 10 g total hydrocarbon/m^3 h, the elimination capacity of the trickling filter was 5.5 g/m^3 h as compared to 2.5 g/m^3 h for the biofilter.

CONCLUSION

Field-testing results presented here and in a previous paper (Lei et al. 1994) show that the biological gas cleaning system can be effectively used to eliminate volatile hydrocarbons, including BTX, from waste gases. The system, made up of mobile units, is particularly suitable in the treatment of soil remediation off-gases.

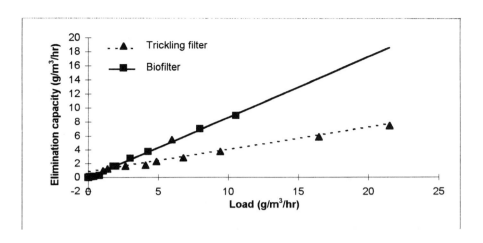

FIGURE 3. BTX elimination capacity as a function of organic load.

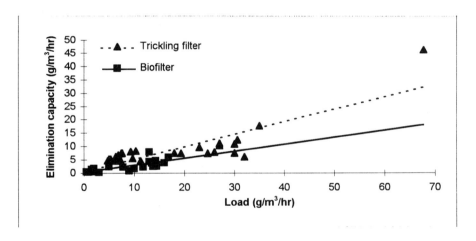

FIGURE 4. Total hydrocarbons elimination capacity as a function of organic load.

The use of an emulsifier in the trickling filter unit minimizes the startup time and efficiently eliminates a large variety of volatile compounds, including those that are only slightly water soluble.

ACKNOWLEDGMENTS

Funding from the DESRT program is gratefully acknowledged. Special acknowledgment is given to Pierre Mercier, Jean-François Jetté, Danielle Beaumier,

Marc Péloquin, Stéphane Richard, and Sonia Gignac for their contribution to the project. The authors would like to extend their gratitude to Y. Lefebvre and M. Audet-Lapointe for useful discussions.

REFERENCES

Greer, C. W., L. Masson, Y. Comeau, R. Brousseau, and R. Samson. 1993. "Application of Molecular Biology Techniques for Isolating and Monitoring Pollutant Degrading Bacteria." *Water Pollut. Res. J. Can.* 28: 275-287.

Lei, J., D. Lord, R. Arneberg, D. Rho, and B. Cyr. 1994. "Design, Field Operation and Performance of a Biological System to Remove Volatile Hydrocarbons from Waste Gases." In *4th Annual Symposium on Groundwater & Soil Remediation.* Environment Canada, Calgary, Alberta, Canada. pp. 691-696.

Reisinger, H.-J., E. F. Johnstone, and P. Hubbard. 1994. "Cost Effectiveness and Feasibility Comparison of Bioventing vs. Conventional Soil Venting." In R. E. Hinchee, B. C. Alleman, R. E. Hoeppel, and R. N. Miller (Eds.), *Hydrocarbon Bioremediation.* Lewis Publishers, Boca Raton, FL. pp. 40-57.

Rho, D., P. Mercier, J. F. Jetté, R. Samson, J. Lei, and B. Cyr. 1994. "Performances of a 30-L Biofilter for the Treatment of Contaminated Air with Toluene and Gasoline Vapors: Filter Media Selection, Microbial Activity and Substrate Interactions." In *4th Annual Symposium on Groundwater & Soil Remediation.* Environment Canada, Calgary, Alberta, Canada. pp. 671-682.

Rho, D., P. Mercier, J.F. Jetté, R. Samson, J. Lei, and B. Cyr. 1995. "Respirometric Oxygen Demand Determinations of Laboratory- and Field-Scale Biofilters." In R. E. Hinchee, G. D. Sayles, and R. S. Skeen (Eds.), *Biological Unit Processes for Hazardous Waste Treatment.* Battelle Press, Columbus, OH. pp. 211-218.

Bioremediation of TNT-Contaminated Soil by the TERRANOX® System

Holger Stolpmann, Hiltrud Lenke, Juergen Warrelmann,
Elisabeth Heuermann, Antje Fruechtnicht, Gregor Daun,
and Hans-Joachim Knackmuss

ABSTRACT

A method for biological on-site treatment of 2,4,6-trinitrotoluene (TNT)-contaminated soil is presented. Laboratory and large-scale pilot experiments provide new data on how TNT and its bacterial metabolites (such as aminoaromatic chemicals) can be detoxified. The pilot project is performed in a controlled biological anaerobic/aerobic treatment process using a bioreactor called TERRANOX®. Experiences with TNT-contaminated soil from several former TNT production sites in Germany are based on extensive laboratory-scale experiments indicating that TNT disappears completely under strictly anaerobic conditions. A stable polymeric substance that is not able to leak out of the soil anymore is generated in the second step of the process from 2,4,6-triaminotoluene (TAT) by aerobic treatment. The careful analysis of this immobilization process, long-term behavior of the product, and ecotoxicological data are currently being investigated.

INTRODUCTION

Because of its potential impact on human health, 2,4,6-trinitrotoluene (TNT) has generated strong public attention as a contaminant of existing and former ammunition facilities (Gaebell 1992; Schneider & Koenig 1987; Wolff 1989). Although TNT is a xenobiotic substance, it is subject to biotransformation, yielding compounds claimed to be less toxic than TNT (Kaplan 1992, Naumova et al. 1992, Funk et al. 1993). We learned from present investigations that bioremediation of TNT-contaminated soil can be carried out by a combined anaerobic/aerobic treatment process that reduces TNT and its cocontaminants, the aminonitrotoluenes, completely in the first step (Lenke et al. 1994, Rieger & Knackmuss 1995). During this anaerobic and subsequent aerobic treatment (second step) the resulting 2,4,6-triaminotoluene (TAT) is bound irreversibly to the soil.

SCIENTIFIC POINT OF VIEW

Due to the strong electron-withdrawing character of the nitro groups, TNT is a highly electron-deficient compound. Thus, TNT seems to be unsusceptible to oxygenation by aerobic microbial populations. In contrast, TNT is readily reduced by anaerobic consortia to yield TAT if strict anaerobic conditions (Eh $\leq$ –200 mV) are provided (Lenke et al. 1994). This process is based on gratuitous transformation activities that are ubiquitously found in the autochthonous microflora. The process requires only the presence of an electron donor such as glucose (Lenke et al. 1993). The main steps of the degradation process are illustrated in Figure 1.

TAT is yielded from complete reduction of TNT and is irreversibly bound to the soil without any participation of oxygen (Lenke et al. 1994). Initial adsorption of TAT is based on its intercalation into an ionic binding to clay minerals. Under the influence of oxygen during subsequent aerobic treatment, the process of interlamellar adsorption must be followed by oxidative polymerization. In the presence of O_2 and heavy metals (e.g., Mn^{2+}), TAT was found to undergo rapid polymerization (Preuss et al. 1993, Lenke et al. 1994). Dark, high-molecular-weight products are generated.

Apart from the intercalation and cationic binding capacities of the clay soil with a high content of humic substances, this soil also exhibits additional cation exchange through its polyanionic structure. Its phenolic and quinoide structures are also good candidates for covalent binding of TAT through radical-induced carbon-carbon and carbon-oxygen bonds and Schiff-base formation (Rieger & Knackmuss 1995). In general, aromatic amines are very reactive and appear to bind easily to organic matter in soil (Parris 1980, Bollag & Loll 1983). Diverse mechanisms explain the enormous capacity of the soil matrix and the humic compounds for irreversible binding of TAT.

EXPERIMENTAL PROCEDURES

Laboratory-Scale Anaerobic Reduction of TNT in a Soil Slurry Contaminated with TNT and Aminonitrotoluenes

TNT-contaminated soil (500 g) was incubated with an anaerobic sludge in mineral salts medium (5 L in a bioreactor) supplemented with glucose (20 mM) and ammonia. Glucose was added daily at a rate of 10 mM per day. The pH value was adjusted to 7.4. Concentrations of TNT, aminodinitrotoluene (ADNT), and diaminonitrotoluene (DANT) were determined by high-performance liquid chromatography (HPLC) in the supernatant. The amounts of sorbed TNT, ADNT and DANT were determined in methanolic extracts of the soil samples.

The investigated soil was obtained from a former TNT production plant (Hessisch Lichtenau – Hirschhagen near Kassel, Germany) and has the following essential properties: 4.4% organic carbon, 9% clay, and pH = 7.2 (0.01 M $CaCl_2$). For the experiment, the air-dried soil material was passed through a sieve (2 mm) and homogenized by treatment with mortar and pestle.

Technical Scale of the Anaerobic/Aerobic
Biotreatment Process of TNT-Contaminated Soil

Before scaling up any process from laboratory scale to technical scale, a process should be established among these scales. The reactor chosen for the biotreatment process should be able to operate under almost the same conditions as in the laboratory but provide experiments with larger amounts of soil as well.

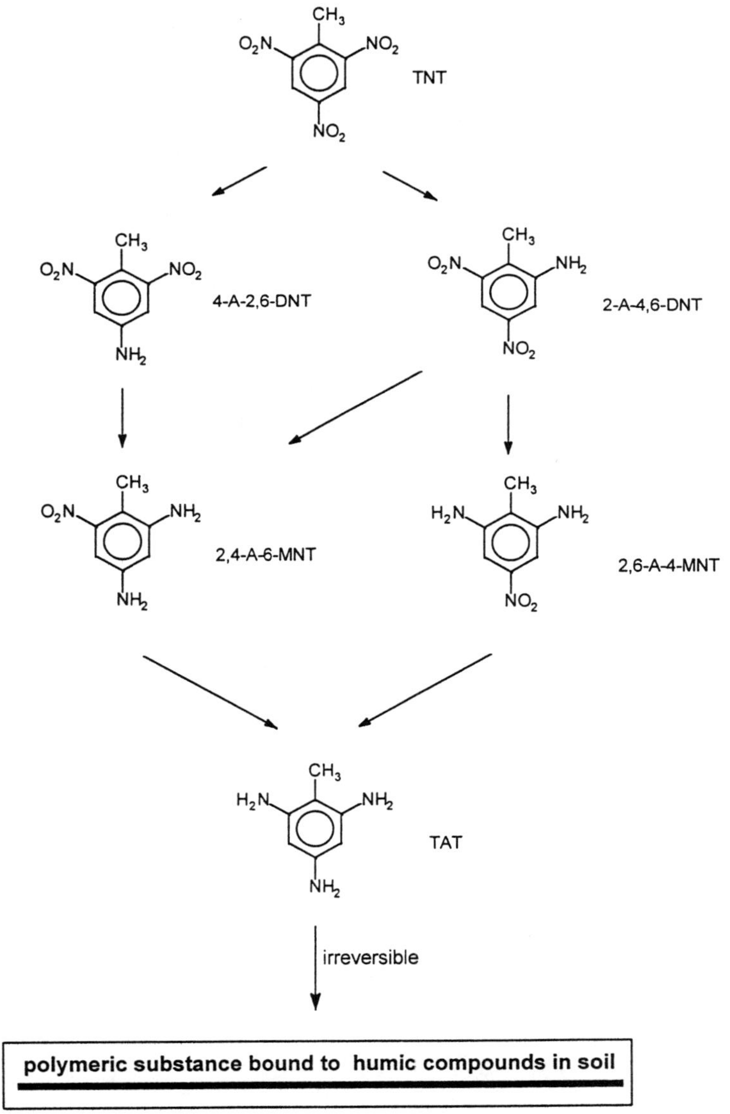

FIGURE 1. Breakdown and detoxification of trinitrotoluene (TNT) in soil.

The TERRANOX® reactor, developed for anaerobic biotreatment processes of soil, consists of a trough that is built in segments to a length up to 50 m (164 ft) with a total content of 100 tons of soil. The entire reactor can be closed with an aluminum-coated plastic foil to minimize the exchange of gas between the contents of the reactor and the atmosphere. The reactor is equipped with a stirring device that turns the soil upside down to mix the soil and the added nutrients.

In the actual pilot project, 30 tons of contaminated soil from a former TNT production plant (Hessisch Lichtenau – Hirschhagen; risk assessment and management: Hessische Industriemüll GmbH - ASG, Wiesbaden, Germany) are treated by a two-step anaerobic/aerobic biological process. In the first step of the biotreatment, TNT and aminonitrotoluenes are reduced in the TERRANOX® reactor by adding nutrients for the autochthonous anaerobic soil bacteria. The final aerobic treatment process provides irreversible binding of TAT to the soil fraction and mineralization of the fermentation products from the anaerobic biotreatment.

RESULTS

As illustrated in Figure 2, TNT and aminonitrotoluenes as contaminants in the soil were reduced completely under anaerobic conditions in the soil slurry. After 4 days of incubation TNT and the aminonitrotoluenes disappeared completely from the aqueous phase, whereas considerable amounts of contaminants

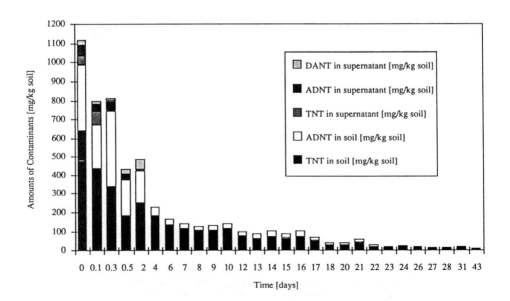

FIGURE 2. Breakdown data for TNT and metabolites.

were still present in soil. Obviously, at lower contaminant levels, desorption of the soil becomes rate limiting.

During the first days of the experiment, the redox potential achieved the required value of Eh < −200 mV, indicating that TNT must be reduced completely to TAT. The TAT could not be detected either in the aqueous supernatant or in the methanolic soil extract due to the irreversible binding of TAT to the soil fraction. At the beginning of the experiment, 350 mg TNT/kg and 240 mg ADNT/kg were found. After 28 days of biological treatment, the concentrations have decreased to 7 mg TNT/kg and 4 mg ADNT/kg and below 1 mg DANT/kg. This means that 98.9% of TNT was removed.

DISCUSSION

The direct anaerobic/aerobic biotreatment process of TNT-contaminated soil leads to a complete reduction of TNT and aminonitrotoluenes to TAT. The TAT was found to be irreversibly bound to the soil, especially under aerobic conditions. The irreversible binding of TAT prevents further microbial degradation of TAT under anaerobic as well as under aerobic conditions. Instead of complete degradation, the biotreatment process results in humification of the contaminants in the soil which is a very cost-effective process for cleaning up TNT-contaminated sites.

The direct anaerobic/aerobic biological treatment of TNT-contaminated soil leading to humification of the contaminants requires a careful analysis of the treated soil. First, the long-term behavior of the humified TAT is tested by the use of ^{14}C-labeled TNT. In addition, the product of oxidative polymerization with and without soil is analyzed by pyrolysis gas chromatography/mass spectroscopy (GC/MS) and compared with the structure of naturally occurring humic substances. Finally, the ecotoxicological quality of the treated soil is assessed by exposing aquatic organisms to aqueous soil extracts. Thus, tests with *Daphnia magna*, *Photobacterium phosphoreum*, and algae will provide information about any poisonous material left in the treated soil. These tests are the subject of accompanying investigations during the actual pilot project of anaerobic/aerobic biotreatment in the TERRANOX® reactor of TNT-contaminated soil from a former TNT production plant.

ACKNOWLEDGMENTS

Part of the work has been supported by the Hessische Industriemuell-ASG and the German Ministry for Research and Development (BMFT). Laboratory experiments from Umweltschutz Nord were performed in cooperation with Daimler-Benz Aerospace-Raumfahrt Infrastruktur (DASA-RI).

REFERENCES

Bollag, J.-M. and M. J. Loll. 1983. "Incorporation of xenobiotics into soil humus." *Experientia* 39: 1221-1231.

Funk, S. B., D. J. Roberts, D. L. Crawford, and R. L. Crawford. 1993. "Initial-phase optimization for bioremediation of munition compound contaminated soils." *Appl. Environ. Microbiol.* 59: 2171-2177.

Gaebell, H. C. 1992. "Ruestungsaltlast Hallschlag." In H. Pfaff-Schley and L. Schimmelpfeng (Eds.), *Ruestungsaltlasten '92:* pp. 69-88. Erich Schmidt Verlag, Berlin.

Kaplan, D. L. 1992. "Biological degradation of explosives and chemical agents." In *Current Opinion in Biotechnology* 3: 253-260.

Lenke, H., B. Wagener, G. Daun, and H. J. Knackmuss. 1994. "TNT-contaminated soil: A sequential anaerobic/aerobic process for bioremediation." *Abst. O-383, Abstr. 94th Ann. Meet. Am. Soc. Microbiol.* 1994: 456.

Lenke, H., G. Daun, D. Bryniok, and H. J. Knackmuss. 1993. "Biologische Sanierung von Ruestungsaltlasten." *Spektrum der Wissenschaft* 10: 106-108.

Naumova, R. P., L. A. Golovleva, R. M. Aliyeva, and P. I. Gvosdyak. 1992. "Microbial bioconversion of pollutants." *Reviews of Environmental Contamination and Toxicology* 124: pp. 41-78.

Parris, G. E. 1980. "Environmental and metabolic transformations of primary aromatic amines and related compounds." *Residue Rev.* 76: 1-30.

Preuss, A., J. Fimpel, and G. Diekert. 1993. "Anaerobic transformation of 2,4,6-trinitrotoluene (TNT)." *Arch. Microbiol.* 139: 345-353.

Rieger, P.-G. and H.-J. Knackmuss. 1995. "Basic knowledge and perspectives on biodegradation of 2,4,6-trinitrotoluene and related nitroaromatic compounds in contaminated soil." In J. C. Spain (Ed.), *Biodegradation of Nitroaromatic Compounds*, pp. 1-8. Plenum Press, New York, NY, in press.

Schneider, U. and W. Koenig. 1987. *Sprengstoff aus Hirschhagen*, 2nd ed., Verlag Gesamthochschulbibliothek Kassel, Kassel, Germany.

Wolff, H. J. 1989. *Die Allendorfer Sprengstoffwerke DAG und WASAG.* Magistrat der Stadt Stadtallendorf (Ed.), 2nd ed., Stadtallendorf, Germany.

Periodically Operated Bioreactors for the Treatment of Soils and Leachates

Robert L. Irvine and Daniel P. Cassidy

ABSTRACT

Limited contaminant bioavailability at concentrations above the required cleanup level reduces biodegradation rates and renders solid-phase bioremediation more cost effective than complete treatment in a bioslurry reactor. Slurrying followed by solid-phase bioremediation combines the advantages and minimizes the weaknesses of each treatment method when used alone. Periodic aeration during solid-phase bioremediation has the potential to lower treatment costs relative to continuous aeration. A biological treatment system consisting of slurrying followed by periodic aeration in solid-phase sequencing batch reactors (SP-SBRs) was developed and tested in the laboratory using a silty loam contaminated predominantly with the plasticizer *bis*(2-ethylhexyl)phthalate (BEHP or DEHP) and a silty clay loam contaminated with diesel fuel. The first experiment evaluated the effect of water content and mixing time during slurrying on subsequent treatment in continuously aerated solid-phase bioreactors. The second experiment compared treatment of slurried soil in SP-SBRs using three different periodic aeration strategies with continuous aeration. Results showed that slurrying for 1.5 h markedly increased the rate and extent of contaminant biodegradation achieved in solid-phase bioreactors. Slurrying the soil at or above its saturation moisture content resulted in lengthy dewatering times which retarded biological treatment in the solid-phase bioreactors. The results also showed that an SP-SBR can provide a greater ratio of biological to abiotic contaminant removal than a continuously aerated bioreactor.

INTRODUCTION

Bioreactors should be designed to enrich and maintain a microbial consortium that will degrade organic contaminants present in leachates and/or soils. Typically, some combination of water, nutrients, electron acceptors, cosubstrates, surfactants, and, sometimes, biomass is added to encourage the degradation of contaminants. Treatment is carried out either in situ (i.e., with the "ground" as the reactor) or ex situ (i.e., in bioslurry reactors or solid-phase systems including landfarming and composting). For in situ bioremediation, the amendments needed are delivered to the contaminated source, eliminating the need for soil or groundwater removal. The phrase "pump-and-treat" usually describes above-ground reactors that are used for the ex situ bioremediation of recovered contaminated groundwaters and leachates.

Early research and development of periodically operated bioreactors were directed at the suspended growth, sequencing batch reactor (SBR) (Irvine and Busch 1979; Irvine and Ketchum 1989). The SBR has been employed successfully to achieve nutrient removal (Alleman and Irvine 1980a; Alleman and Irvine 1980b; Ketchum et al. 1987; Irvine and Busch 1979; Manning and Irvine 1985; Palis and Irvine 1985), control of bulking sludge (Brenner et al. 1992; Chiesa and Irvine 1985; Chiesa et al. 1985; Dennis and Irvine 1979), and the removal of hazardous organic compounds found in industrial wastes and leachates from landfills (Brenner et al. 1992; Chozick and Irvine 1991, Herzbrun et al. 1985; Irvine et al. 1993b; Irvine and Wilderer 1988; Smith and Wilderer 1987; Ying et al. 1987). The soil slurry-SBR (SS-SBR) has also been applied successfully to treat excavated contaminated soils (Irvine et al. 1993a). Periodically operated reactor systems used for treating hazardous and nonhazardous wastes present in leachates and soils are described herein, with a focus on soil bioremediation using a combination of slurry-phase and solid-phase techniques.

Ex situ soil bioremediation consists of treatment in either slurry-phase or solid-phase reactors. Slurry-phase bioremediation is more expensive than solid-phase treatment because of the energy required to keep the soil particles in suspension. However, mixing in bioslurry reactors lowers mass transfer resistances by dispersing isolated pockets of contaminant and disrupting aggregates of clay and silt particles to provide intimate contact between added nutrients, contaminants, microorganisms, and oxygen. The result is a greater rate, extent, and uniformity of contaminant biodegradation than is typically realized in solid-phase systems (Torpy et al. 1989; Ross 1990; Mueller et al. 1991a and b). Solid-phase bioremediation (sometimes called biopiling or soil heaping) includes composting, and is one of the least expensive ex situ treatment methods for contaminated soil. The energy consumed by forced aeration (usually done continuously) is small, making solid-phase bioremediation cost effective even when relatively long treatment times are required. The effectiveness of solid-phase bioremediation is often limited by heterogeneous ("patchy") contaminant removal, and low rates and extent of biodegradation. Many authors have attributed this to a lack of soil homogenization and biological reactant distribution (Ross 1990; Mueller et al. 1991b; Stegmann and Franzius 1991; Blume 1992; Garakani 1992; Battaglia et al. 1993; Schulz-Berendt 1993). Periodic aeration has the potential to reduce solid-phase treatment costs by lowering the energy consumed by pumping air, and by lowering or eliminating the costs associated with treating volatile emissions.

Slurry-phase and solid-phase studies show that rates of contaminant biodegradation in soils often decrease markedly as the contaminant bioavailability becomes rate limiting (Ryan, 1986; Mueller et al. 1991a and b; Heely et al. 1992;

Riser-Roberts 1992; van Afferden et al. 1992; Lewis 1993; Lynch and Genes 1993; Irvine et al. 1993a). Limited contaminant bioavailability at concentrations above the required cleanup level can lengthen treatment times considerably, rendering solid-phase bioremediation more cost effective than complete treatment in a bio-slurry reactor. Slurrying followed by solid-phase bioremediation combines the advantages and minimizes the drawbacks of each treatment method when used alone.

COMBINED SLURRY-PHASE AND SOLID-PHASE STUDIES

Two soils were used in these studies: a silty loam contaminated with *bis*-(2-ethylhexyl) phthalate (BEHP or DEHP) and petroleum hydrocarbons having between 25 and 30 carbon atoms (12% clay-sized particles); and a silty clay loam contaminated with diesel fuel (38% clay-sized particles). Stable, fine-grained aggregates comprised approximately 25% by weight of the silty loam and 20% of the silty clay loam. Table 1 lists the important properties for the bulk soil and the aggregates of the silty loam. As can be seen from Table 1, the content of clay and of native soil organic matter (SOM) in the silty loam is markedly higher in the aggregates than in the bulk soil. The contaminant concentrations are approximately twice as high in the aggregates as in the bulk soil, indicating that a disaggregation step would enhance biological treatment. TPH concentrations in the silty clay loam were also two times greater in the aggregates than in the bulk soil.

The SP-SBR employs a periodically operated aeration strategy for treatment of contaminated soils. The laboratory SP-SBRs consisted of 1.7-L sealed glass Imhoff

TABLE 1. Properties of the silty loam contaminated with BEHP and petroleum hydrocarbons.

Parameter	Bulk Soil	Aggregates
Bulk density (g/cm^3)	1.65	1.9
% Clay (<2 μm)	12.0	30.1
Carbonate content (%)	28	12
Cation exchange capacity (meq/100 g)	10.3	15.0
Total organic content (%)	5.2	14.6
Soil organic matter content (%)	3.1	7.0
BEHP (g/kg soil sieved to 2 mm)	42.6	92.3
TPH (g/kg soil sieved to 2 mm)	17.7	34.4

cones with approximately 1.5 L of soil. A controlled flow of air (0.5 L/min) was moved through the SP-SBR using a vacuum pump and valves. Periodic aeration was achieved by controlling the pumps with timers. The SP-SBRs had monitoring ports for gas analysis, a water trap to collect leachate, and activated carbon traps to collect volatile organics in the effluent air (Figure 1). The control valves were closed daily to measure changes in the headspace oxygen and carbon dioxide concentrations in the SP-SBRs. Soil samples were taken weekly for quantification of contaminant concentrations, dehydrogenase activity, and moisture content.

An experiment was conducted with the BEHP-contaminated soil to evaluate the effect of water content and mixing time during slurrying on subsequent treatment in five continuously aerated solid-phase bioreactors. One reactor contained soil centrifuged from a slurry having a 40% solids concentration (w/v), which had been provided with nutrients and active BEHP and TPH degraders, and allowed to react for 5 days. Two reactors contained soil slurried with nutrients for 1.5 h in an 80-L cement mixer at a volumetric moisture content of 75% of the saturation capacity. Another contained unmixed soil packed into the column in layers of 300 g, with nutrients added to the top of each layer. The final reactor was an autoclaved control. A bioslurry reactor with a 10% solids concentration was operated for 100 days to compare slurry-phase with solid-phase biological treatment.

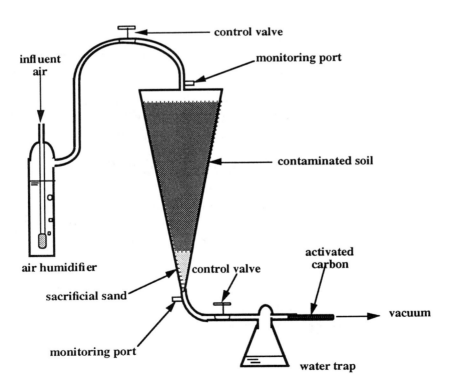

FIGURE 1. Laboratory SP-SBRs used in the studies.

Three different SP-SBRs were operated with the diesel fuel-contaminated silty clay loam to compare the performance of periodic aeration with continuous aeration. The soil in all reactors was slurried for 1.5 h in a cement mixer with a moisture content of 60% saturation. A biologically active solid-phase reactor and an autoclaved control were aerated continuously. The three periodic aeration strategies used in the SP-SBRs were (1) 1 h of aeration followed by 1 h without aeration; (2) 6 h of aeration followed by 6 h without aeration; (3) 24 h of aeration followed by 24 h without aeration.

Soil samples were extracted with CS_2 according to the method described by Irvine et al. (1993a) and Yocum (1994). Quantification of BEHP and TPH in the CS_2 extract was performed using GC/FID. Recovery of the internal standard (diethyl phthalate or DEP) was greater than 87%. Dehydrogenase activity was measured using the INT method described by Coho (1990). Oxygen and carbon dioxide were measured in the SP-SBR headspace using an Illinois Instruments model 3600 O_2/CO_2 analyzer.

RESULTS AND DISCUSSION

Results of BEHP removal, oxygen consumption rates, and dehydrogenase activity (expressed as µg INT-formazan produced/g soil-day) in the continuously aerated bioreactors are shown in Figures 2 through 4. Less than 1% of the BEHP and TPH was removed from the reactors in the effluent air stream and the leachate, which is consistent with the low volatility and solubility of the contaminants. BEHP removal (Figure 2) was due to biological degradation, as has been

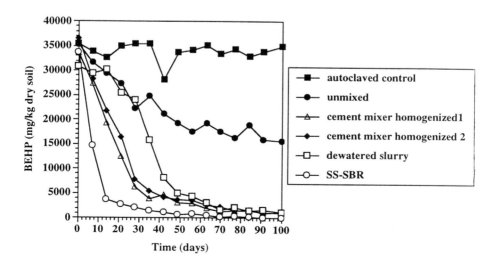

FIGURE 2. Concentration of BEHP remaining in the solid-phase reactors with time.

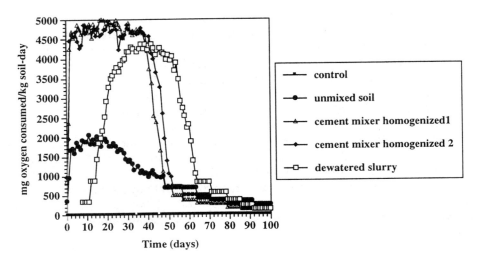

FIGURE 3. Rates of oxygen consumption in the solid-phase reactors with time.

established in previous studies (Maag and Lokke 1990; Irvine et al. 1993a). The relative oxygen consumption rates (Figure 3) and dehydrogenase activity (Figure 4) in the reactors show the same trends as BEHP removal, further indicating that BEHP removal is biological. Trends of carbon dioxide production rates (not shown here) were identical to oxygen consumption rates.

Figure 2 shows that complete treatment in a slurry reactor does provide a greater extent of BEHP removal than in solid-phase reactors. As can be seen in Figures 2 through 4, the solid-phase reactors containing slurried soil achieved approximately twice the BEHP removal as unslurried soil during the 100-day experiment. The reactor containing soil slurried with a 40% solids concentration and the two reactors containing soil slurried with a moisture content of 75% of the saturation capacity achieved approximately the same maximum rates and extent of BEHP removal during the 100-day experiment. However, it took approximately 30 days for the reactor containing soil slurried with a 40% solids concentration to achieve maximum rates of biological activity. The high pore water content limited the passage of air through the pore spaces in the soil and reduced oxygen transfer to the microorganisms. Other studies (not discussed here) indicated that dewatering times were much longer for soils having a higher clay content. The final water content in all solid-phase reactors was approximately 48% of saturation. On day 49, the soil in the second solid-phase reactor containing soil slurried in the cement mixer was remixed for 1.5 h with nutrients at a moisture content of 75% saturation. BEHP removal was not noticeably enhanced by remixing and nutrient addition, indicating that contaminant bioavailability was limiting rates of biodegradation below 5,000 mg BEHP/kg soil.

Table 2 compares the percent of total TPH removed due to biodegradation and volatilization achieved after 100 days in the solid-phase bioreactors containing

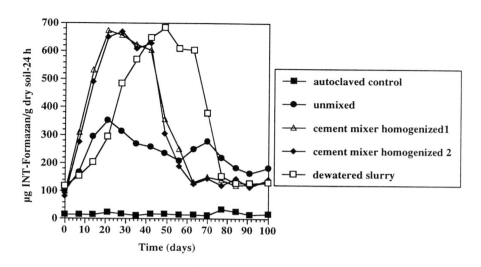

FIGURE 4. Rates of INT-formazan production in the solid-phase reactors with time.

the slurried silty clay loam contaminated with diesel fuel. No leachate was collected from any of the reactors in this study because the moisture content of the soil was at field capacity. The initial TPH concentration in all reactors was approximately 18,000 mg/kg. As can be seen in Table 2, 84% of the TPH was removed in the continuously aerated solid-phase reactor, compared with 76%, 59%, and 54% in the SP-SBRs periodically aerated every hour, every 6 hours, and every 24 hours, respectively. TPH removal in the autoclaved control was 12% of the total, and 100% volatile. Biological removal (of the total TPH removed)

TABLE 2. Total, volatile fraction, and biological fraction of TPH removed from the diesel fuel-contaminated silty clay loam for SP-SBRs and continuously aerated solid-phase reactors.

Reactor	Total TPH removed (%)	Volatile fraction removed (%)	Biological fraction removed (%)
Autoclaved control	12	100	0
Continuously aerated	84	10.1	89.9
Periodically aerated: 1 hour	76	5.7	94.3
Periodically aerated: 6 hours	59	15.5	84.5
Periodically aerated: 24 hours	54	16.5	83.5

in the SP-SBR aerated hourly was approximately 94%, which is somewhat greater than the 90% biological removal achieved in the continuously aerated reactor. The percent of total TPH removed biologically in the SP-SBRs aerated every 6 and 24 hours was less than 85%.

CONCLUSIONS

The results from these investigations show that (1) complete treatment in a bioslurry reactor provided greater rates and extent of contaminant removal than did treatment in solid-phase systems, both with and without a slurry pretreatment step; (2) slurrying at a moisture content less than saturation, for as little as 1.5 h, markedly enhanced the rates and extent of biodegradation of BEHP and petroleum hydrocarbons in the solid-phase reactors relative to using no slurry pretreatment; (3) slurrying at or above the saturation soil moisture content caused lengthy dewatering times, thereby increasing the overall treatment time in the solid-phase system; and (4) periodic aeration can increase the ratio of biological to abiotic contaminant removal. The potential for a SP-SBR to maximize biodegradation relative to volatile contaminant removal is greater for contaminants more volatile than diesel fuel, such as gasoline and jet fuel 4.

REFERENCES

Alleman, J. E., and R. L. Irvine. 1980a. "Storage-Induced Denitrification using Sequencing Batch Reactor Operation." *Water Res.* 14:1483.

Alleman, J. E., and R. L. Irvine. 1980b. "Nitrification in the Sequencing Batch Reactor." *J. Water Pollut. Control Fed.* 52:2747.

Battaglia, A., D. J. Morgan, D. G. Linz, and T. D. Hayes. 1993. "Application of the GRI Accelerated Protocol to Contaminated Soils From Manufactured Gas Plant Sites." In P. T. Kostecki, E. J. Calabrese, and M. Bonazountas (Eds.), *Hydrocarbon Contaminated Soils*. Lewis Publishers, Chelsea, MI.

Blume, H. P. 1992. *Handbuch des Bodenschutzes: Bodenöcologie und -Belastung; Vorbügende und Abwehrende Schutzmassnahmen*. Auflage 2, Ecomed Verlaggesellschaft, mbH, Landsberg, Germany.

Brenner, A., R. Chozick, and R. L. Irvine. 1992. "Treatment of a High Strength Mixed Phenolic Waste in an SBR." *Water Environ. Res.* 64:128.

Chiesa, S. C., R. L. Irvine, and J. F. Manning, Jr. 1985. "Feast/Famine Growth Environments and Activated Sludge Population Selection." *Biotechnol. and Bioeng.* 27:562.

Chiesa, S. C., and R. L. Irvine. 1985. "Growth and Control of Filamentous Microbes in Activated Sludge: An Integrated Hypothesis." *Water Res.* 19:471.

Chozick, R., and R. L. Irvine. 1991. "Preliminary Studies on the Granular Activated Carbon-Sequencing Batch Biofilm Reactor." *Environ. Prog.* 10:282-289.

Coho, J. W. 1990. "Biodegradation of Jet Fuel in Vented Columns of Water-Unsaturated Sandy Soil." Master's Thesis, University of Florida.

Dennis, R. W., and R. L. Irvine. 1979. "Effect of Fill to React Ratio on Sequencing Batch Reactors." *J. Water Pollut. Control Fed.* 51:255.

Garakani, M. 1992. "Remediation of Petroleum-Contaminated Soil." *Hazwaste Fall 1992*:7-17.

Heely, D. A., E. S. Werk, and R. G. Kowalski. 1992. "Bioremediation and Reuse of Soils Containing No. 5 Fuel Oil in New England Using an Aboveground Treatment Cell: A Case Study." In P. T. Kostecki, E. J. Calabrese, and M. Bonazountas (Eds.), *Hydrocarbon Contaminated Soils.* pp. 243-255. Lewis Publishers, Chelsea, MI.

Herzbrun, P. A., R. L. Irvine, and K. C. Malinowski. 1985. "Biological Treatment of Hazardous Waste in the SBR." *J. Water Pollut. Control Fed.* 57:1163-1167.

Irvine, R. L., and A. W. Busch. 1979. "Sequencing Batch Biological Reactors — An Overview." *J. Water Pollut. Control Fed.* 51:235-243.

Irvine, R. L., J. P. Earley, G. J. Kehrberger, and B. T. Delaney. 1993a. "Bioremediation of Soils Contaminated with *bis*-(2-ethylhexyl) Phthalate (BEHP) in a Soil Slurry-Sequencing Batch Reactor." *Environmental Progress* 12:39-44.

Irvine, R. L., and L. H. Ketchum, Jr. 1989. "Sequencing Batch Reactor for Biological Wastewater Treatment." *CRC Crit. Rev. Environ. Control* 18:255.

Irvine, R. L., and P. A. Wilderer. 1988. "Aerobic Processes." In H. Freeman (Ed.), *Standard Handbook on Hazardous Waste Control Technology.* McGraw-Hill, New York, NY, 9.3.

Irvine, R. L., P. S. Yocum, J. P. Earley, and R. Chozick. 1993b. "Periodic Processes for In Situ and On-Site Bioremediation of Leachates and Soils." *Wat. Sci. Tech.* 27(7-8):97-104.

Ketchum, Jr., L. H., R. L. Irvine, R. E. Breyfogle, and J. F. Manning, Jr. 1987. "A Comparison of Biological and Chemical Phosphorus Removals in Continuous and Sequencing Batch Reactors." *J. Water Pollut. Control Fed.* 59:13.

Lewis, R. F. 1993. "SITE Demonstration of Slurry Phase Biodegradation of PAH Contaminated Soil." *Air & Waste* 43:503-510.

Lynch, J., and B. Genes. 1993. "Land Treatment of Hydrocarbon Contaminated Soils." In E. J. Calabrese and P. T. Kostecki (Eds.), *Principles and Practices for Petroleum Contaminated Soils.* pp. 409-422. Lewis Publishers, Chelsea, MI.

Maag, J., and H. Lokke. 1990. "Landfarming of DEHP Contaminated Soil." In F. Arendt, M. Hinsenveld, and W. J. van den Brink (Eds.), *Contaminated Soil '90.* pp. 975-982. Kluwer Academic Publishers, Netherlands.

Manning, Jr., J. F. and R. L. Irvine. 1985. "The Biological Removal of Phosphorus in a Sequencing Batch Reactor." *J. Water Pollut. Control Fed.* 57:87.

Mueller, J. G., P. J. Chapman, B. O. Blattman, and P. H. Pritchard. 1990. "Isolation and Characterization of a Fluoranthene-Utilizing Strain of *Pseudomonas paucimobilis*." *Appl. Environ. Microbiol.* 56:1079-1086.

Mueller, J. G., S. E. Lantz, B. O. Blattman, and P. J. Chapman. 1991a. "Bench-Scale Evaluation of Alternative Biological Treatment Processes for the Remediation of Pentachlorophenol- and Creosote-Contaminated Materials: Slurry Phase Bioremediation." *Environ. Sci. Technol.* 25:1045-1061.

Mueller, J. G., S. E. Lantz, B. O. Blattman, and P. J. Chapman. 1991b. "Bench-Scale Evaluation of Alternative Biological Treatment Processes for the Remediation of Pentachlorophenol- and Creosote-Contaminated Materials: Solid Phase Bioremediation." *Environ. Sci. Technol.* 25:1055.

Palis, J. C., and R. L. Irvine. 1985. "Nitrogen Removal in a Low-Loaded Single Tank Sequencing Batch Reactor." *J. Water Pollut. Control Fed.* 57:82-86.

Riser-Roberts, E. 1992. *Bioremediation of Petroleum Contaminated Sites.* CRC Press, Inc., Boca Raton, FL.

Ross, D. 1990. "Slurry-Phase Bioremediation: Case Studies and Cost Comparisons." *Remediation* 1:61-74.

Ryan, J. 1986. "The Land Treatability of Appendix VIII Organics Present in Petroleum Industry Wastes." In Loehr et al. (Eds.), *Land Treatment: A Hazardous Waste*, pp. 347-367.

Schulz-Berendt, V. V. 1993. "Biologische Reinigung Oelkontaminierter Böden." *Spektrum der Wissenschaft* (Oktober):93-97.

Smith, R. G., and P. A. Wilderer. 1987. "Treatability of Hazardous Landfill Leachate using Sequencing Batch Reactors with Silicone Membrane Oxygenation." *41st Purdue Indust. Waste Conf.*, Purdue University. Lewis Publishers, Chelsea, MI, p. 272.

Stegmann, R., and V. Franzius. 1991. "Übersicht über die Verfuegbaren Verfahren zur Bodenreinigung." In 15. Muelltechnisches Seminar, "Sanierung Kontaminierten Böden," *Berichte aus Wasserguete- und Abfallwirtschaft.* Berichtsheft Nummer 108. pp. 39-52. Technische Universitat München.

Torpy, M. F., H. F. Stroo, and G. Brubaker. 1989. "Biological Treatment of Hazardous Waste." *Pollution Engineering* 21:80-86.

van Afferden, M., M. Beyer, J. Klein, and D. Essen. 1992. "Significance of Bioavailability for the Microbial Remediation of PAH-Contaminated Soil." *Preprints from the International Symposium on Soil Decontamination Using Biological Processes*, pp. 605-610. Karlsruhe, DECHEMA, Frankfurt am Main, Germany.

Wilderer, P. A., J. Brautigam, and I. Sekoulov. 1985. "Application of Gas Permeable Membranes for Auxiliary Oxygenation of Sequencing Batch Reactors." *Conservation & Recycling* 8:181.

Ying, W., R. R. Bonk, and S. A. Sojka. 1987. "Treatment of a Landfill Leachate in Powdered Activated Carbon Enhanced Sequencing Batch Bioreactors." *Environ. Prog.* 6:1.

Yocum, P. 1994. "Bioremediation of Leachate and Soil Contaminated with Petroleum Products." Doctoral Dissertation, University of Notre Dame, Notre Dame, IN.

Degradation of Chlorinated Organic Compounds by Microbial Mats

Judith Bender, Peter Phillips, Richard Lee,
Susana Rodríguez-Eaton, Gautam Saha,
Bommanna Loganathan, and Lucinda Sonnenberg

ABSTRACT

An algal/bacterial consortium, or microbial mat, offers a unique combination for sequential reductive dechlorination and aerobic ring cleavage because of the close proximity of oxic and anoxic zones. Mineralization of octachlorocyclopentadine (chlordane), 2,2'-4,4'-5,5'-hexachlorobiphenyl (PCB), and trichloroethylene (TCE) by a microbial mat was demonstrated, as well as reduction in adsorbable organochlorine compounds (AOX) from pulp- and paper-mill effluent. Chlordane, in water, was 91% mineralized after 21 days and no parent compound remained. Most ^{14}C was detected as large macromolecules, such as cellular protein. In the same time period, 17% of PCB was mineralized to $^{14}CO_2$. Chlordane, in sediments, was degraded more slowly. After 15 days, 119 µg TCE/kg of microbial mat was mineralized (dark conditions) compared to 147 µg/kg in the presence of 20 mg/L zinc. AOX removal was 78% by day 1 and >90% by day 7. Bioremediation systems designed for treatment of these recalcitrant compounds ideally should present a consortium of microbes that spontaneously generates heterogeneous microzones in close proximity, degrading microbes that can penetrate soils and sediments, and mechanisms that transport contaminated materials from the sediment into the region of high microbial density. The microbial mat biological system potentially offers these attributes.

INTRODUCTION

Chlorinated hydrocarbons are recalcitrant contaminants that are widely distributed in soil, sediment, and water. The difficulty for a single microbial species to perform both dechlorination and ring cleavage of chloroaromatics probably accounts for the persistence of these contaminants in the environment (Quensen et al. 1988). Mixed populations of aerobes and anaerobes working in concert may be required for complete degradation of highly chlorinated aromatics, but the environment in which these contaminants are deposited limits their exposure to these microbes. Partitioning of most sediment and soil environments into discrete oxic/anoxic zones (with their characteristic microbial populations) requires transport of contaminants to the various zones. Because chlorinated hydrocarbons often are bound in the soil or sediment matrix, such transport is unlikely.

Reductive dechlorination of structurally dissimilar chlorinated organics has been demonstrated with sediment bacterial consortia (Rhee et al. 1993; Suflita et al. 1983; Genthner et al. 1989) as well as with novel recombinant bacteria (McCullar et al. 1994; Adams et al. 1992; Mokross et al. 1990). When white-rot wood fungus, *Phanerochaete chrysosporium*, was cultured under nutritionally limiting conditions, obligating the production of lignase, it was able to degrade chlordane (Kennedy et al. 1990; Bumpus 1988).

Most of the studies carried out to date suggest that aerobic degradation of chlorinated biphenyls is primarily restricted to the less chlorinated biphenyls, i.e., monochlorinated and dechlorinated congeners. For example, the half-lives of monochlorobiphenyls at concentrations of 1 µg/L in river water were 2 to 3 days (Bailey et al. 1983). However, the degradation of tetrachlorobiphenyls did not take place. Bacteria able to oxidize monochlorobiphenyls have been isolated from river water, soil, and sewage (Ahmed and Focht 1973; Fava et al. 1994). The first reaction is the cleavage of the aromatic ring. Few studies have shown aerobic degradation of highly chlorinated biphenyls in natural sediments of waters. Kohler et al. (1988) found that two species of *Actinobacter* isolated from soil were able to degrade aerobically many of the highly chlorinated biphenyls of Aroclor 1254, including several tetrachlorobiphenyls and pentachlorobiphenyls. Some strains of pseudomonads (e.g., *Pseudomonas* sp. strain LB400) are capable of degrading aerobically many PCB congeners, including some hexachlorobiphenyls. Biphenyl dioxygenase, a multicomponent enzyme, is an important enzyme system which initiates the degradation of PCBs, followed by ring cleavage (Erickson and Mondello 1993). Intermediates in this degradative pathway include chlorobenzoic acids (Harkness et al. 1993). The role that these related bacteria play in the natural environment is unclear because highly chlorinated biphenyls are very persistent.

Rapid and complete degradation of highly chlorinated recalcitrant aromatics by environmental consortia or constructed microbial systems has been difficult to demonstrate. The constructed microbial consortium reported here offers a unique combination of remediation mechanisms for degradation of these contaminants.

Bioremediation systems, designed for treatment of these recalcitrant compounds ideally should present several characteristics: (1) a consortium of microbes that spontaneously generates heterogeneous microzones (oxic/anoxic) in close proximity, (2) degrading microbes that can penetrate soils and sediments, and (3) mechanisms that transport contaminated materials from the sediment into the region of high microbial density. This research describes a biological system, the microbial mat, which potentially offers these attributes.

The goals of this project were to investigate several aspects of removal and degradation of chlorohydrocarbons. Specific research questions were selected because they represent central issues associated with this class of environmental contaminants. Problems such as rate of degradation, fate of contaminants and their metabolites, industrial mixed contaminant samples, and difficulties associated with the bioavailability of sediment-bound chlorohydrocarbons were investigated. This research project targets several recalcitrant chlorohydrocarbons: (1) aromatic (chlordane and PCB), (2) aliphatic (TCE), and (3) mixed industrial samples (pulp- and paper-mill effluent). Degradation studies using model recalcitrant compounds (chlordane and PCB) were performed to determine metabolic products and degree of mineralization. Additionally, since much of the environmental contamination of aromatic chlorohydrocarbons occurs in the sediment regions, this research investigated the sequester and degradation of chlordane mixed in estuarine sediments.

EXPERIMENTAL PROCEDURES AND MATERIALS

Microbial Mat

The cyanobacteria (blue-green algae) and bacteria consortium, or microbial mat, is a proprietary entity and process developed by J. Bender and P. Phillips (inventors, patent pending). These mats are developed as a free-floating biomass in quiescent laboratory ponds, or immobilized to a variety of inert substrates. The method for their development is discussed in Bender and Phillips (1994). The microbial mat method is now being used in field remediation of coal-mine drainage (Phillips et al. 1994). Microbial mats generally are acclimated to the contaminant before the experiment is performed.

Octachlorocyclopentadine (Chlordane)

Water-Based Experiments. The initial chlordane degradation by microbial mat in water was reported by Bender et al. (1994a). A second series of experiments examined chlordane uptake and degradation by rates of removal from the water column, generation of metabolites, and mineralization to CO_2 in 21 days. Although chlordane is not readily soluble in water, a globule of chlordane was sequestered by a floating microbial mat by means of contractile biofilms produced by the mat. Initial addition of 10 µg of ^{14}C-chlordane (specific activity, 5.91 mCi/mM) was made to 75 mL of water containing 1.1 g of microbial mat. The experiment was conducted under 12L:12D (light:dark) conditions. Controls were heat-sterilized mats with chlordane. For ^{14}C mass balance experiments, KOH traps were used to collect $^{14}CO_2$, and specific extraction methods were used to separate chlordane and its metabolites from other macromolecules assumed to be components of mat structure. In these procedures, mats were extracted initially with chloroform and methanol to remove chlordane and its metabolites. Radioactivity in the mat after this extraction was assumed to be associated with cellular macromolecules, such as protein. The extracted mats were also treated with 10% trichloroacetic acid to precipitate proteins. This was followed by eight washes with mixtures of methanol, ethanol, and ethylacetate to remove all remaining residues of chlordane and metabolites. Resulting extracts and precipitates were counted for radioactivity.

Metabolic intermediates for chlordane and PCB were analyzed after 21-day treatment by GC/MS (EPA Method 8270). Analytical parameters were column: 30 m × 0.25 mm, 1 μm film thickness, silicone-coated fused-silica column from J & W Scientific [DB-5]; mass range: 35 to 500 amu; initial temperature: 40°C hold time 4 min; temperature program: 40 to 270°C at 10°C/min; final temperature: 270°C; sample volume: 1 μL; carrier gas: hydrogen at 50 cm/s or helium at 30 cm/s.

Estuarine Sediment Experiments. Estuarine sediment (250 mL, approximately equivalent to 300 g) from the tidal flats at the Skidaway Institute of Oceanography was added to seawater to create a 400-mL volume; 5 mL slurry was added per glass flask. An aliquot of blended microbial mat slurry and 10 μg of ^{14}C-chlordane, dissolved in 10 μL of ethanol, was added to each flask. Controls (sediment samples without mats) were compared to (1) mat + sediment and (2) mat + sterilized sediment. This experimental structure allows a comparison of the specific enhancement of mats over the microbial components within the sediment. Light/dark exposures were either 12L:12D or 24D cycle. At the end of each incubation period, $^{14}CO_2$ was measured and the floating mat was removed and analyzed for ^{14}C.

Extraction: Chlordane and its metabolites were extracted by the procedure of Bligh and Dyer (1959). After centrifugation the lower organic phase was taken to near dryness with a rotary evaporator and taken up in a small volume of chloroform. An aliquot of the organic phase was counted. The radioactivity of precipitated macromolecules in the aqueous phase was determined by passing the aqueous phase through a glass-fiber filter, and the filter was placed in scintillation fluid and counted.

Thin-Layer Chromatography and Radioautography: The chloroform extracts were applied to silicic acid thin-layer plates. The solvent system used was petroleum ether:ethyl ether (70:30 v/v). Using x-ray film (Xomat AR film, Eastman Kodak Co.), autoradiography of the thin-layer plates was carried out to locate the position of chlordane and metabolites. Radiolabeled spots were scraped from the thin-layer plates and added to scintillation fluid. Radioactivity was determined with a scintillation counter (Tri-Carb 300C, Packard).

Collection of $^{14}CO_2$: After incubation for various times, 2.0 mL of 2M H_2SO_4 were added to the sediment slurry, and the respired $^{14}CO_2$ was collected in a center well containing a piece of Whatman No. 1 filter paper soaked with 0.4 mL of 2M NaOH. After absorption of the $^{14}CO_2$ by the NaOH for 3 h, the soaked papers were transferred to a second set of flasks equipped with CO_2 traps containing phenethylamine (Eastman Kodak), and 2 mL of 2M H_2SO_4 was added to each flask. After 3 h the $^{14}CO_2$ in the phenethylamine-soaked paper was transferred to scintillation vials and counted in the liquid scintillation counter. The two steps for collection of $^{14}CO_2$ were necessary because small amounts of ^{14}C-labeled organic compounds volatilized and were collected in the NaOH traps.

2,2'-4,4'-5,5'-Hexachlorobiphenyl (PCB)

All experiments were conducted in water. PCB presents a solubility problem in water similar to that of chlordane. The initial specific activity was 4.37 mCi/mM. The experiment lasted 21 days. Separate experiments using nonradiolabeled PCB were conducted to determine the presence of chlorinated metabolites by GC/MS analyses. Controls were heat-sterilized mats with PCB.

Trichloroethylene (TCE)

The TCE experiments included the substrate alone and in combination with 20 mg/L Zn. Experiments were initiated with a TCE-specific activity of 14.2 mCi/mM. Degradation values are based on mineralization of ^{14}C-TCE to CO_2. Five mineralization trails were performed in a triplicate light or dark series. Appropriate controls were included, and $^{14}CO_2$ was collected in KOH. Sealed test tubes were held at 20 to 22°C. Lighted tubes were maintained under continuous incandescent and fluorescent lighting. Degradation, as mineralization to CO_2 of the originally spiked μg/L of substrate, was calculated at the end of the 15-day experimental period based on the micrograms of TCE substrate/mat dry weight. Controls were heat-sterilized mats.

AOX From Pulp- and Paper-Mill Effluent

In bench-scale experiments, three 2-L experimental tanks with acclimated microbial mats grown over limestone rocks and control tanks consisting of limestone rocks were used. Control tanks contained limestone without mat. The effluents were sampled from a bleached hardwood and softwood kraft mill. The amount of adsorbable organochlorine compounds (AOX) and the color (absorbance at 510 nm, unbuffered) of wastewaters were measured.

RESULTS

Chlordane

Water-Based Experiments. A totally dechlorinated chlordane product, 4,7-methano indene, was detected by GC/MS analysis after 21 days of mat treatment. Budget analysis of ^{14}C-chlordane showed that it was 91% degraded after 21 days, with no parent compound remaining (Table 1). Of the 91%, 13% of the label

TABLE 1. Mineralization of ^{14}C-chlordane, ^{14}C-PCB, ^{14}C-TCE, and ^{14}C-TCE + Zn in water.

Contaminant and Exposure Time	Light (L): Dark (D) Cycle	Polar Metabolites (%)	Large Macromolecules in Aqueous Phase (%)	Carbon Dioxide (%)
Chlordane, 21 d	12L/12D	2	78	13
Control		0	0	1
Hexachloro-PCB, 21 d	12L/12D	NA[(a)]	NA	17
Control		NA	NA	0
Trichloroethylene, 15 d	24D	NA	NA	21
Control		NA	NA	0.4
Trichloroethylene + Zn, 15 d	24D	NA	NA	23
Control		NA	NA	0.5

(a) NA = Not analyzed.

was detected as ^{14}CO$_2$ and 78% was detected as large cellular macromolecules. Because the extraction procedure excluded chlordane and its metabolites (see Methods), most of the ^{14}C is assumed to be in cellular material, such as protein. This phenomenon is to be expected because the algal component of the mats probably would sequester the ^{14}CO$_2$ produced during degradation for photosynthesis, thereby excluding it from the KOH trap. Additionally, in contrast to heterotrophic bacterial systems, mats produce large amounts of structural material. Carbon resources (chlordane and PCB) might be converted to small molecules such as acetate, which are subsequently used for biomass construction rather than energy production. Such activities would account for the large percent of the ^{14}C being found in cellular macromolecules of the mats, rather than in the CO$_2$ traps.

Estuarine Sediment Experiments. Recovered radioactivity was between 85 and 102% of the total added. The results presented in Table 2 show an increasing pattern of mineralization of chlordane over time. Controls (sediment without mat) show little chlordane removal. Comparisons of the sterile/nonsterile sediment plus mat suggest some augmentation of the sediment component by the mats. However, further experiments are needed to confirm this effect.

PCB

As seen in Table 1, 17% of the PCB was mineralized to ^{14}CO$_2$. No analysis of labeled cellular macromolecules was conducted. Separate experiments using nonradiolabeled PCB were conducted to determine the presence of chlorinated metabolites by GC/MS analyses. The results show trichloro- and tetrachloro-PCB production from hexachloro-PCB, suggesting progressive dechlorination by the microbial mat.

TABLE 2. Distribution of ^{14}C-radioactivity in different fractions after addition of ^{14}C-chlordane to sediment and microbial mats.

Analysis	Day	Parent Compound (%)			Polar Metabolites (%)			Large Macromolecules in Aqueous Phase (%)			Carbon Dioxide (fraction) (%)		
		7	14	21	7	14	21	7	14	21	7	14	21
Sediment (Control)	L/D	97	90	81	0	1	2	1	5	7	2	3	3
	D	95	89	NA	1	2	NA	2	2	NA	2	3	NA
Sediment + Mat	L/D	91	85	68	3	5	5	7	8	11	4	12	13
	D	91	93	NA	1	3	NA	2	1	NA	2	1	NA
Mat + Autoclaved	L/D	86	74	49	4	8	11	6	14	27	9	15	12
Sediment	D	97	94	NA	0	0	NA	0	0	NA	1	2	NA

(a) Ten μg of ^{14}C-chlordane was added to each flask. Experimental materials were exposed to a 12L:12D (light, dark) cycle or 24D. The amount is the percent of recovered radioactivity from the total radioactivity added.

TCE

TCE was mineralized to CO_2 alone and in the presence of 20 mg/L zinc (Table 1). During 15 days, under dark conditions, 119 µg TCE/kg microbial mat was mineralized, whereas 147 µg/kg was mineralized in the presence of 20 mg/L Zn. Mineralization rates in the lighted series probably were inaccurate because the photosynthesizing mat was using the $^{14}CO_2$ produced.

AOX

The reductions in AOX and color are shown in Figure 1, which shows 78% of the AOX removal occurring by the end of the first day. After 7 days, approximately 15% more AOX was removed, resulting in total AOX removal, >90% for the effluents. Overall removal rate, based on first day decreases, was 670 mg AOX removed per m^2 of mat per day. Color removal occurred more slowly; however, after 7 days there was a reduction in absorbance of 72%. No studies of molecular mechanisms or fate of the AOX compounds were made.

DISCUSSION

PCB and Chlordane in Water

The rapid and complete dechlorination/mineralization process achieved by microbial mats with both chlordane and hexachloro-PCB indicates the action of unique mechanisms associated with this consortium. We suggest that the

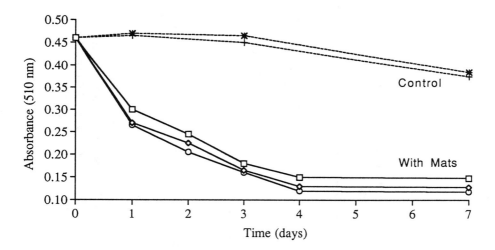

FIGURE 1. Treatment of AOX and color (unbuffered absorbance at 510 nm) in whole pulp- and paper-mill effluents with microbial mats in 2-L tanks with limestone-rock substrate.

fluctuating and simultaneous existence of oxic/anoxic zones in close proximity within the mat (Bender et al. 1994b, Revsbech et al. 1989) probably accounts for this efficiency. Anaerobic consortia, present in the anoxic zones, probably dechlorinate the hydrocarbon producing the metabolite 4,7-methano-1H indene, from chlordane. Ring cleavage of this totally dechlorinated product by the aerobic bacteria residing in the adjacent oxic zones of the microbial mat might be expected. Oxic or anoxic zones are enhanced during the light or dark periods, respectively. Thus, the transport of metabolic intermediates is unnecessary and does not present an impossible technical challenge for this bioremediation system.

Chlordane in Sediments

Although microbial isolates have been demonstrated to dechlorinate highly chlorinated aromatics, the role that these related bacteria play in the natural environment is unclear. Our results show that chlordane deposited in sediment is progressively mineralized by the mat consortium over time, and rates surpass those observed in the natural environment. Additionally, comparison with the estuarine-sediment microbial consortium shows that a much more efficient remediation system exists in the mats. Stimulation and enhancement of microbial mats in the estuarine ecosystem may represent a noninvasive method of treating these regions.

TCE

Similar TCE mineralization occurred with and without the presence of zinc. Microbial mats release negatively charged bioflocculants (Bender et al. 1994b), which may bind the heavy metals before they can cause toxic effects or deactivate cellular enzymes responsible for TCE degradation. Goodroad et al. (1994) have shown that mats can mineralize organic contaminants in mixture with heavy metals (TNT + Pb, chrysene + Zn). In all cases the metals were sequestered simultaneously with the degradation processes. This capacity of microbial mats to remediate mixed organic/inorganic contaminants is relatively unusual for microbial remediation systems.

Pulp- and Paper-Mill Effluents; Mixtures of Chlorinated Organics

Treatment of mixed waste from pulp- and paper-mill samples (Figure 1) shows only that AOX was reduced and that most of the reduction occurred during the first day. The process of AOX reduction may be simple adsorption to mat surfaces. Because model compounds were not a part of this study, no assumptions on the ultimate fate of the chlorohydrocarbons can be made. However, the mineralization of chlordane and PCB described above suggests that degradation can take place over time. Mats showed no toxic effects from long-term exposure to the mill effluents. Because on-site reduction of contaminants in pulp- and paper-mill effluents is an important issue in terms of environmental compliance, AOX reduction, metabolic products, and the final fate of the sequestered contaminants merit further investigation.

ACKNOWLEDGMENTS

Major support for this research was provided by U.S. Environmental Protection Agency Assistance ID No. CR8168901. Partial support for chlordane degradation analysis was provided by the U.S. Agency for International Development, under Grant Nos. DAN-5053-G-001050-00 and PCE-5053-G-00-4020-00. Battelle Marine Sciences Laboratory is acknowledged as a collaborative partner in the trichloroethylene research.

REFERENCES

Adams, R. H., C. M. Huang, F. K. Higson, V. Brenner, and D. D. Focht. 1992. "Construction of a 3-Chlorobiphenyl-Utilizing Recombinant From an Intergeneric Mating." *Applied and Environmental Microbiology* 58: 647-654.

Ahmed, M., and D. D. Focht. 1973. "Degradation of Polychlorinated Biphenyls by Two Species of *Achromobacter*." *Canadian Journal Microbiology* 19: 47-52.

Bailey, R. E., S. J. Gonslor, and W. L. Rhinehart. 1983. "Biodegradation of the Monochlorobiphenyls and Biphenyl in River Water." *Environmental Science and Technology* 17: 617-621.

Bender, J., and P. Phillips. 1994. "Implementation of Microbial Mats for Bioremediation." In J. L. Means and R. E. Hinchee (Eds.), *Emerging Technology for Bioremediation of Metals*, pp. 85-98. Lewis Publishers, Boca Raton, FL.

Bender, J., R. Murray, and P. Phillips. 1994a. "Microbial Mat Degradation of Chlordane." In J. L. Means and R. E. Hinchee (Eds.), *Emerging Technology for Bioremediation of Metals*, pp. 135-139. Lewis Publishers, Boca Raton, FL.

Bender, J., J. R. Washington, B. Graves, P. Phillips, and G. Abotsi. 1994b. "Deposit of Zinc and Manganese in an Aqueous Environment Mediated by Microbial Mats." *Water, Air and Soil Pollution* 75: 195-204.

Bligh, E. G., and W. J. Dyer. 1959. "A Rapid Method of Total Lipid Extraction and Purification." *Canadian Journal of Biochemistry and Physiology* 37: 911-917.

Bumpus, J. A. 1988. "Biodegradation of Polycyclic Aromatic Hydrocarbons by *Phanerochaete chrysosporium*." *Applied and Environmental Microbiology* 55(1): 154-158.

Erickson, B. D., and F. J. Mondello. 1993. "Enhanced Biodegradation of Polychlorinated Biphenyls After Site-Direct Mutagenesis of a Biphenyl Dioxygenase Gene." *Applied and Environmental Microbiology* 59: 2858-3862.

Fava, F., D. Digioia, S. Cinti, L. Marchetti, and G. Quattroni. 1994. "Degradation and Dechlorination of Low-Chlorinated Biphenyls by a 3-Membered Bacterial Coculture." *Applied Microbiology and Biotechnology* 41: 117-123.

Genthner, B.R.S., W. A. Price II, and P. H. Prichard. 1989. "Anaerobic Degradation of Chloroaromatic Compounds in Aquatic Sediments Under a Variety of Enrichment Conditions." *Applied and Environmental Microbiology* 55: 1466-1471.

Goodroad, L., J. Bender, P. Phillips, J. Gould, G. Saha, S. Rodríguez-Eaton, Y. Vatcharapijarn, R. Lee, and J. Word. 1994. "Potential for Bioremediating Using Constructed Mixed Microbial Mats." Superfund XV, Hazardous Materials Control Resources Institute, November 29-December 1, 1994, Washington, DC.

Harkness, M. R., J. B. McDermott, D. A. Abramowicz, J. J. Salvo, W. P. Flanagan, M. L. Stephens, F. J. Mondello, R. J. May, J. H. Lobos, J. M. Carroll, M. J. Brennan, A. A. Bracco, K. M. Fish, G. L. Warner, P. R. Wilson, D. K. Deitrich, D. T. Lin, C. B. Morgan, and W. L. Gately. 1993. "*In Situ* Stimulation of Aerobic PCB Biodegradation in Hudson River Sediments." *Science 259*: 503-507.

Kennedy, D. W., S. D. Aust, and J. A. Bumpus. 1990. "Comparative Biodegradation of Alkyl Halide Insecticide by the White Rot Fungus *Phanerochaete chrysosporium* (BKM-F-1767)." *Applied and Environmental Biology 56*(8): 2347-2353.

Kohler, H. P., E. D. Kohler-Staub, and D. D. Focht. 1988. "Cometabolism of Polychlorinated Biphenyls: Enhanced Transformation of Aroclor 1254 by Growing Bacterial Cells." *Applied and Environmental Microbiology 54*: 1940-1945.

McCullar, M. V., V. Brenner, R. H. Adams, and D. D. Focht. 1994. "Construction of a Novel Polychlorinated Biphenyl-Degrading Bacterium: Utilization of 3,4'-Dichlorobiphenyl by *Pseudomonas acidovorans* M3GY." *Applied and Environmental Microbiology 60*: 3833-3839.

Mokross, H., E. Schmidt, and W. Reineke. 1990. "Degradation of 3-Chlorobiphenyl by *In Vivo* Constructed Hybrid Pseudomonads." *FEMS Microbiology Letters 71*: 179-186.

Phillips, P., J. Bender, R. Simms, S. Rodriguez-Eaton, and Cynthia Britt. 1994. "Use of Microbial Mat and Green Algae for Manganese and Iron Removal from Coal Mine Drainage." Presented at International Land Reclamation and Mine Drainage Conference and the Third International Conference on Abatement of Acidic Drainage, April 24-29, 1994, Pittsburgh, PA.

Quensen, J. F., J. M. Tiedje, and S. A. Boyd. 1988. "Reductive Dechlorination of Polychlorinated Biphenyls by Anaerobic Microorganisms From Sediments." *Science 242*: 752-754.

Revsbech, N. P., P. B. Christensen, and L. P. Nielsen. 1989. "Microelectrode Analysis of Photosynthetic and Respiratory Processes in Microbial Mats." In Y. Cohen and E. Rosenberg (Eds.), *Microbial Mats, Physiological Ecology of Benthic Microbial Communities*, pp. 153-162. American Society for Microbiology, Washington, DC.

Rhee, F. Y., B. Bush, C. M. Bethony, A. DeNucci, H. M Oh, and R. C. Sokol. 1993. "Anaerobic Dechlorination of Aroclor 1242 as Affected by Some Environmental Conditions." *Environmental Toxicology and Chemistry 12*: 1033-1039.

Suflita, J. M., J. A. Robinson, and J. M. Tiedje. 1983. "Kinetics of Microbial Dehalogenation of Haloaromatic Substrates in Methanogenic Environments." *Applied and Environmental Microbiology 45*:1466-1473.

Fluidized-Bed Reactor and In Situ Hydrogen Peroxide-Enhanced Bioremediation

John P. Gandee, Joshua L. Jenkins, and Patrick N. McGuire

ABSTRACT

Both in situ and ex situ bioremediation technologies can be favorable alternatives for the cleanup of hydrocarbon-contaminated sites. This case study presents 12 months of operating data on a bioremediation system consisting of groundwater recovery with nutrient and hydrogen peroxide (H_2O_2) reinjection designed to stimulate biodegradation of contaminated soils and groundwater. A granular activated carbon fluidized-bed reactor was used to treat extracted groundwater prior to reinjection.

TECHNOLOGY SELECTION

Bioremediation of organic contaminants in soils and groundwater is an established means of hydrocarbon remediation. The uniform mixing of microorganisms supplemented with nutrients and electron acceptors (e.g., oxygen, nitrate, and sulfate) has been demonstrated to achieve degradation of various contaminants in a controlled environment.

A biological remedial program was selected in the evaluation of treatment systems based on the extent of contamination and the remedial potential of the constituents. Air sparging and vapor extraction technology also had been considered, but were eliminated because of the extensive area of impact requiring a significant number of wells and the constraints of remediating an active manufacturing and production area.

CASE STUDY

Site History

A remedial investigation/feasibility study (RI/FS) was prepared to identify site characteristics and the extent of hydrocarbon impacts, and to review remedial technologies for a former underground storage tank (UST) site. Initial soil samples were obtained to assess the chemical and biological parameters of the site, and to conduct a bench-scale treatment analysis of the soils. The baseline data obtained from the initial sampling are presented in Table 1. A leachate test was conducted on the soils using H_2O_2 and nutrients, and the test determined that naturally occurring bacteria could be stimulated and total petroleum hydrocarbons (TPH) could be biologically degraded. However, the reaction of H_2O_2 within the native soils indicated that a decrease in soil permeability of up to two orders of magnitude could occur. Results of the permeability test are presented in Table 2.

Additional hydrocarbon sampling was conducted prior to implementing the remediation system (2 years later) to identify baseline data for performance evaluation. This sampling event showed higher TPH soil concentrations of approximately 2,000 to 7,000 mg/kg. Groundwater sampling showed that the total concentration of benzene, toluene, ethylbenzene, and xylenes (BTEX) in one downgradient well was higher than expected (i.e., 300 mg/L). Based on the more recent contaminant assessment, the quantity of the release may have been more substantial than previously estimated.

TABLE 1. Baseline data and results of remediation system operation.

Parameter	Baseline Data	First 6 Months of Operation	12 Months of Operation
Total Heterotrophs[a]	5-61 × 10⁶ CFU/g	1-20 × 10⁶ MPN	NA
Total BTEX Degraders	0.5-19 × 10⁶ CFU/g	0.005-1 × 10⁶ MPN	NA
TPH Degraders	NA	0.005-0.2 × 10⁶ MPN	NA
TPH in Soils Tank Pit Excavation at startup:	80-2,400 mg/kg	NA	
12-14 ft (3.6-4.2 m)	6,700 mg/kg	< 5 mg/kg	< 10 mg/kg
14-16 ft (4.2-4.9 m)	2,000 mg/kg	490 mg/kg	150 mg/kg
BTEX in Groundwater	0.01-300 mg/L	0.01-200 mg/L	0.009-90 mg/L
Oxygen Concentrations[b]	0.2-2 mg/L	0.2-5 mg/L	1.03-5 mg/L
Phosphorus Concentrations	0-0.6 mg/L	0.4-3.9 mg/L	0.8-2.0 mg/L
Ammonia Concentrations	0-0.052 mg/L	0.1-2.7 mg/L	0.7-1.7 mg/L

(a) CFU/g = colony-forming units/gram. MPN = most probable number.
(b) Oxygen concentrations within a 15-ft (4.6-m) radius of the reinjection wells were observed at approximately 5 mg/L. Areas further downgradient showed oxygen concentrations less than 3.0 mg/L.

TABLE 2. Treatability study test results.

Pore Volumes	Hydraulic Conductivity	H_2O_2[a]
0	1.9×10^{-3} cm/s	0 mg/L
4	5.2×10^{-5} cm/s	0 mg/L
6	7.2×10^{-5} cm/s	30 mg/L
8	1.1×10^{-5} cm/s	20 mg/L
14	1.1×10^{-5} cm/s	20 mg/L

(a) Initial H_2O_2 concentration was 300 mg/L.

System Design

A treatment scheme was selected to remediate BTEX and TPH contaminants in groundwater and soils using a groundwater recovery and reinjection system. The system was designed to meet the requirements of three state regulatory mandates: (1) provide maximum reinjection of recovered groundwater to minimize discharge to the local publicly owned treatment works (POTW); (2) demonstrate plume control via groundwater modeling in design and via groundwater monitoring during operation; and (3) obtain drinking water maximum contaminant level (MCL) concentrations for treated groundwater.

First, 12 reinjection points were selected for oxygen addition across the site. Most of the reinjection wells were limited to a distance of 100 ft (30.5 m) from the recovery wells. Each well included air sparging and vapor extraction points as a contingency for future remediation.

The reinjection rates varied between 2 and 6 gpm (0.1 and 0.4 L/s) per well to obtain effective coverage and maximum reinjection. Two recovery wells were required to maintain plume control at a combined groundwater recovery rate of 70 gpm (4.4 L/s).

Reinjection water was amended with H_2O_2, nitrogen, and phosphorus nutrients. Treatability testing determined that adverse effects could occur on naturally occurring bacteria using H_2O_2. Due to the destructive nature of high concentrations of H_2O_2, the injection concentration was limited to less than 500 mg/L initially, and was gradually increased. The reinjectate contained 5 mg/L of nitrogen and 1 mg/L of phosphorus.

Several alternatives were considered for treatment of the recovered groundwater. High groundwater BTEX concentrations were expected for treatment based on the site characterization data. A fluidized-bed biological reactor was selected during the RI/FS due to its cost effectiveness compared to conventional activated carbon treatment. The fluidized bed consists of activated carbon for adsorption of BTEX that is recirculated with oxygen, and nutrient addition to stimulate biological growth on the carbon surface. The fluidized bed is designed to maintain an effluent dissolved oxygen concentration of 2.0 mg/L and maintain effluent water quality near MCL concentrations. A portion of the water discharged from the reactor is reinjected, and the remaining portion is discharged to the POTW.

Treated groundwater used for reinjection contains 10 to 15 mg/L of suspended solids that are filtered by multimedia sand filters and bag filters. These units minimize mechanical blockage by suspended solids entering the reinjection wells. Backwash water from the sand filters is sent to a holding tank for solids removal, and supernatant is returned to the treatment process.

System Monitoring and Performance

A groundwater monitoring program was implemented on a monthly basis to monitor BTEX and biological parameters. Groundwater elevation data are collected from monitoring wells and entered into a software program to evaluate movement of the dissolved-phase plume.

Problems encountered with the remediation system include reinjection well plugging and biological fouling of process instruments and equipment. Well plugging was determined to be due to either oxidation of native minerals or heavy biological growth in the area of the wells. An acid washing program was implemented to restore the reinjection wells and has been effective in maintaining reinjection. Process equipment including flowmeters and well pumps required more frequent maintenance after about 6 months of operation due to heavy accumulation of biological growth. Biological growth continues to be a problem for the entire system.

Minor problems were also encountered with the fluidized bed, including biological fouling. The fluidized bed has been able to operate under varying conditions and has responded favorably to system shutdowns.

RESULTS

Oxygen and nutrient concentrations showed marginal concentration increases in the system monitoring wells. Breakthrough concentrations for oxygen and nutrients at the recovery wells have not been observed to date. Dissolved-phase BTEX concentrations increased in two monitoring wells over the first 6 months due to changes in groundwater movement and decreased over the next 6 months. TPH analysis of the soil in the remediation area collected 6 months after startup showed up to 100% reduction in TPH compared to samples collected at startup. TPH analysis 1 year after startup showed similar results. Bacteria enumeration shows a high presence of hydrocarbon-utilizing bacteria in the impact area. A comparison of bacteria enumeration with baseline data is not conclusive due to the difference in sampling techniques of previous sampling events and the lack of sufficient data. Results of oxygen and BTEX monitoring are presented in Figures 1 and 2 for two downgradient monitoring wells. Table 1 presents a comparison of baseline data with performance sampling results.

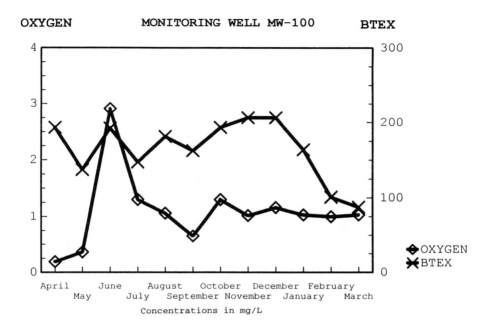

FIGURE 1. Groundwater remediation system monitoring well MW-100.

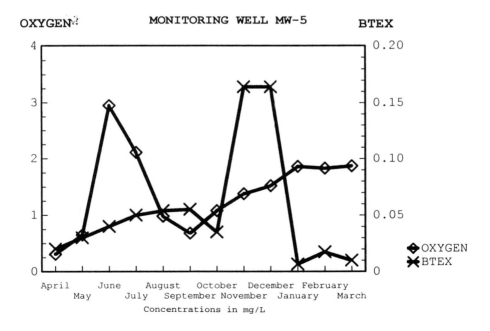

FIGURE 2. Groundwater remediation system monitoring well MW-5.

Reinjection has been maximized to the extent possible and has been slightly inhibited by reinjection well plugging. A decrease in oxygen concentrations occurred as a result of plugged reinjection wells. The reinjection problems were resolved and the oxygen concentration began increasing again.

Recovered groundwater has decreased in BTEX concentrations. The amount of BTEX initially recovered averaged approximately 100 lb (45.4 kg) per month and has decreased to approximately 5 lb (13.6 kg) per month. Oxygen addition averages approximately 600 to 1,000 lb (272 to 454 kg) per month, which equates to degradation of approximately 200 to 333 lb (91 to 151 kg) of hydrocarbons, assuming no loss in H_2O_2 efficiency, and 1 lb (0.45 kg) of hydrocarbon degradation for 3 lb (1.36 kg) of oxygen consumed. The recovery component of the system appears to be minor in terms of remediation, and in situ degradation is the primary means of remediation.

The fluidized bed has maintained a consistent effluent quality significantly below the drinking water MCL, and an effluent dissolved oxygen concentration of 2.0 mg/L.

Groundwater recovery with reinjection has been an effective transport mechanism for oxygen transfer to the subsurface contamination. H_2O_2 stability and its effects on soil permeability appear to be manageable. Soil and groundwater contamination and oxygen delivery will be evaluated periodically to direct changes in reinjection rates to ensure optimum oxygen delivery.

Biodegradation of Volatile Chlorinated Solvents in a Methanotrophic Biofilm Reactor

Eric H. Marsman, Wiecher W. van Veen,
and Leon G.C.M. Urlings

ABSTRACT

Two full-scale methanotrophic biofilm reactors (13.5 m^3) were tested for the treatment of groundwater (27 m^3/h) contaminated with dichloroethylene (DCE) and monochlorobenzene (MCB). The two plugflow BIOPUR® reactors were built on site, and each contained five compartments in series. Natural gas was injected into the second compartment resulting in methanotrophic conditions in compartments 2 through 5 to stimulate cometabolic degradation of DCE. Compartment 1 was intended for the biodegradation of MCB. For DCE, removal efficiencies of 60 to 80% were achieved. Stripping amounted to between 5 and 20%. A significant simultaneous removal of DCE and MCB was observed in the first compartments. The methanotrophic biofilm reactor operated safely and at low costs; however, a one-step, economically attractive process does not seem to be feasible for the removal of DCE from contaminated groundwater down to the low levels required by Dutch legislation.

INTRODUCTION

Currently, biological treatment of groundwater contaminated with organic compounds has a proven track record. Rotating biological contactors and submerged filters are being used for groundwater treatment. Aerated biofilm reactors have been shown to remove contaminants such as fuel compounds; benzene, toluene, ethylbenzene, and xylenes (BTEX); total petroleum hydrocarbons; phenolic compounds; chlorinated aromatics; and the lighter polycyclic aromatic hydrocarbons (Marsman et al. 1994).

However, little is known about the use of microorganisms in full-scale bioreactors for treating groundwater contaminated with volatile chlorinated solvents such as tetrachloroethylene (TeCE), trichloroethylene (TCE), dichloroethylene

(DCE), and 1,1,1-trichloroethane (TCA). Although extensive research has been carried at both bench and pilot scale, almost no results are available for full-scale applications. The aim of this project was to test a full-scale methanotrophic biofilm reactor for removing DCE from groundwater. The results of this project and an economic evaluation are presented.

BACKGROUND

Biological treatment of groundwater contaminated with chlorinated solvents has several important advantages compared to air stripping and activated carbon filtration. The adsorption capacity of activated carbon is limited, especially for vinyl chloride (VC), and destruction of loaded activated carbon by means of thermal incineration can result in the production of dioxines. In the literature, much attention is paid to cometabolic biodegradation of chlorinated solvents by means of microorganisms that grow on methane (methanotrophic bacteria) and bacteria able to use toluene or phenol as a growth substrate (Oldenhuis et al. 1989, Folsom & Chapman 1991, Speitel et al. 1994, Dolfing et al. 1993). Microorganisms are isolated that can degrade even VC (Hartmans & de Bont 1992). Anaerobic dechlorination also is possible; however, this can result in the formation of the carcinogenic compound VC (De Bruin et al. 1992).

Over a 6-month period, we investigated the applicability of these three process conditions on a pilot-plant scale before selecting one technology for field application. The volume of each biofilm reactor was 100 L, and the biomass was attached to reticulated polyurethane. The preconditions for the investigation were (1) a relatively short, economically attractive hydraulic retention time (1 h); (2) good removal efficiencies (>90%); and (3) safe and simple handling of the biofilm reactor. The influent concentrations of TCA, TeCE, and TCE ranged from 300 to 600 µg/L.

The laboratory tests showed the best results for the methanotrophic biofilm reactor in which removal efficiencies of 70 to 95% were obtained for TCA, TeCE, and TCE. In the anaerobic bioreactor, all three compounds were also removed to a certain extent, but harmful intermediates, including DCE and VC, were formed. The removal efficiencies of the biomass grown on toluene were low compared to those for the methanotrophic bioreactor. Moreover, the carrier medium became clogged during the experiment. Although adsorption of the chlorinated compounds took place, experiments showed that the adsorption equilibrium had been reached within a few days.

One interesting result was the removal of TeCE in the aerobic methanotrophic biofilm reactor; this may have been caused by direct cometabolism by the methanotrophic biomass, or by reductive dechlorination of TeCE in anaerobic zones of the biofilm.

Our conclusion was that methanotrophic biomass can be used to treat water contaminated with TeCE, TCE, and TCA. Considerable removal efficiencies were achieved, while stripping of the compounds was less than 5%.

FULL-SCALE BIOFILM REACTOR ON SITE

The contractor, Broerius Bodemsanering, built two biofilm reactors (volume 13.5 m^3 each) for treating groundwater contaminated with DCE (800 to 1,200 µg/L), MCB (400 to 600 µg/L), minor concentrations of TeCE and TCA (40 to 60 µg/L), and methylisobutylketone (200 µg MIBK/L). Incidentally, VC was measured (<50 µg/L). The total flux amounted to 27 m^3/h. The configuration of the groundwater treatment plant is shown in Figure 1.

Groundwater passes a buffer (volume 10 m^3) and is pumped through two biofilm reactors in parallel. Each bioreactor consists of five compartments in series, and each compartment is aerated (20 m^3/h) with the off-gas from the previous compartment. Water and air are passed cocurrently through the compartments. Both bioreactors are connected to natural gas mains (80% v/v methane) with automatic valves. Methane is injected into the second compartment to create methanotrophic conditions in compartments 2 through 5, whereas the first compartment is intended for the biodegradation of MCB and MIBK. Methane is injected at a concentration of 3% v/v and 5 lower explosion level (LEL) meters prevent the methane concentration rising above 80% of the LEL. The groundwater remediation started in the spring of 1993, without methane injection (phase 1, 70 days). From June onward, methane injection took place (phase 2, 20 days), and in the beginning of July the bioreactor was inoculated with *Methylosinus trichosporium* OB3b (phase 3, 45 days).

Whenever the effluent contains higher concentrations of DCE than the maximal allowable value of 50 µg/L, it is possible to aerate the buffer and the fifth compartments of both bioreactors as a security measure to remove the volatile compounds by stripping. The off-gas from the buffer and the biofilm reactors is treated by means of activated carbon filtration.

RESULTS

Dichloroethylene

Figure 2 shows the concentration of DCE in the groundwater and in the effluent from the biofilm reactors. As can be seen in Figure 2, in phase 1 (without methane addition) about 45% of the DCE was removed. After starting the injection of methane (phase 2), the removal efficiency appears to increase to an average 49%. Addition of a suspension of *M. trichosporium* OB3b (phase 3) resulted in a further increase in the removal of DCE (59%). The maximum removal efficiency was 80%, which was observed at day 109. Mass balances showed that the percentage of stripping ranged between 5 and 25%, depending on the concentration of DCE in the groundwater.

However, at day 110, the concentration of DCE in the effluent once again started to rise. At that time, the water authorities demanded that further steps be taken and we started aerating the fifth compartments in both bioreactors

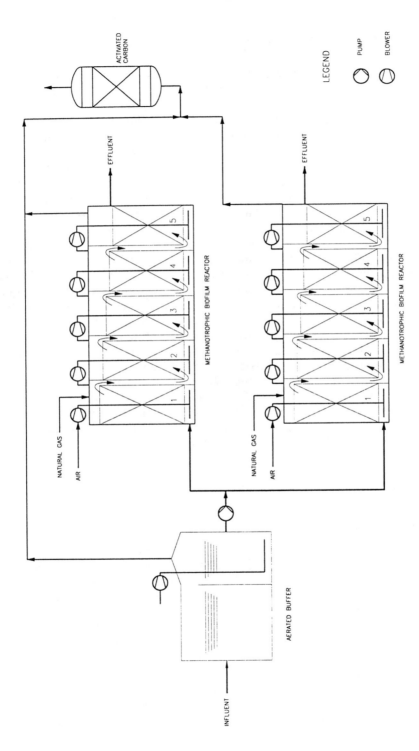

FIGURE 1. Configuration of the groundwater treatment plant.

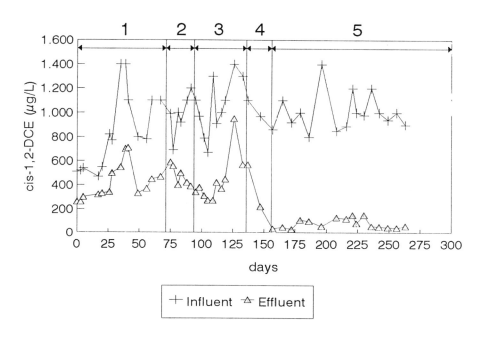

FIGURE 2. Concentration of dichloroethylene in groundwater and effluent of the biofilm reactors. Phase 1: without methane injection; phase 2: methane injection; phase 3: inoculation with *Methylosinus trichosporium* OB3b; phase 4: additional aeration of compartment 5; phase 5: additional aeration of the influent buffer.

(phase 4). Additional aeration of the buffer started on 21 August 1993 (phase 5); and from then on, the effluent concentration was more or less in the range of the required value, i.e., 50 µg/L.

To assess whether or not the methanotrophic biomass really could biodegrade DCE, biomass samples were taken for use in in vitro experiments. The biomass was placed in small flasks to which DCE, formate (10 mM, flask A), or methane and formate (respectively 5% v/v, 10 mM, flask B) were added. Formate and methane served as an exogenous source of reduction equivalents to stimulate the cometabolism of DCE. No reduction equivalents were added to flask C. Two flasks functioned as a blank to which no biomass (flask D) or an additional 1 mL of HCl was added (flask E). The concentration of DCE at the beginning of the experiment was 2.88 mg/L. The results are shown in Figure 3.

Figure 3 shows that the cultivated biomass can indeed cometabolize DCE, although low DCE concentrations were not reached. The highest initial activity was obtained in the flask where only formate was added. The in vitro experiments indicate that approximately 5% of the biomass consisted of the desired microorganisms based on specific oxidation rates reported by Oldenhuis et al. (1989).

To improve the DCE removal efficiency, a two-step biological pilot plant (200 L) was built on the site. The first reactor was intended for the biodegradation of MCB; the second was a methanotrophic bioreactor. A chemostat, in which *M. trichosporium* OB3b was grown continuously, was connected to the second bioreactor to feed it with the specific microorganism. However, the removal efficiency did not improve. Furthermore, the cultivation of *M. trichosporium* OB3b in a chemostat, under field conditions, proved to be a difficult and time-consuming process.

Monochlorobenzene

MCB was biodegraded with a removal efficiency of more than 95% in the first compartments of both bioreactors. The overall removal efficiency was >99%. Stripping of MCB was less than 1%. One interesting result was the combined removal of DCE and MCB in the first compartment. Mass balances showed a mean removal efficiency of 30 to 35% for DCE of which about 10% was stripped. This indicates that microorganisms growing on (chlorinated) aromatics can also cometabolize DCE to a certain degree (Shields et al. 1991).

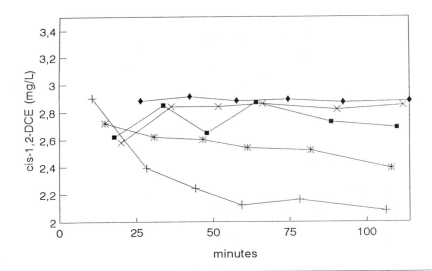

FIGURE 3. **In vitro experiments to test the presence of DCE-degrading bacteria in the methanotrophic biomass, with different sources of reduction equivalents: flask A with formate (10 mM), flask B with formate (10 mM) and methane (5% v/v), flask C (no addition of methane and formate), flask D (with 1 mL HCl), flask E (no sludge added). The experiments were carried out by S. Hartmans of the Wageningen Agricultural University.**

DISCUSSION

Removal Efficiency

The results show that considerable amounts of DCE were removed. However, removal to a very low concentration in one biological step in an economically attractive hydraulic retention time (approx. 1 h) was not achieved. The following explanations can be given for this result. First, it is difficult to cultivate a specific methanotrophic microorganism under field conditions. Other methane-consuming microorganisms in the pumped groundwater also will start to grow in the biofilm reactor, and, unfortunately, these organisms lack the ability to cometabolize chlorinated solvents in substantial amounts. As a result, *M. trichosporium* OB3b is outgrown by other microorganisms. Furthermore, the affinity of *M. trichosporium* OB3b for DCE is relatively low, with an apparent Michaelis-Menten constant of 3 mg DCE/L. The groundwater contains DCE-concentrations of only 1 mg/L, and the desired effluent value is 50 µg/L, which means that the activity of the biomass also is very low. In contrast with the high removal efficiency of BTEX and gasoline in aerobic bioreactors, the removal efficiency of DCE in the methanotrophic reactor was much less.

Economic and Technological Evaluation

The groundwater treatment plant has been in operation since 1993 without any major technological problems. Methane injection was carried out safely at a flowrate of 0.8 m³/h. In an economic evaluation, the groundwater treatment system was compared with stripping and activated carbon filtration. The evaluation showed that the costs of treating groundwater by means of stripping are less than those of a two-step process involving a methanotrophic biofilm reactor followed by stripping (0.3 $/m³). If, however, the groundwater is also contaminated with other biodegradable compounds, it may be possible to achieve a reduction of chlorinated solvents. In that case, due to cometabolic degradation by microorganisms growing on other contaminants, only a small stripper for posttreatment and minor amounts of activated carbon would be required, and a two-step process may become economically attractive.

CONCLUSION

The experiments in our laboratory with artificial groundwater showed that removal efficiencies of approximately 90% could be achieved under methanotrophic process conditions within hydraulic retention times of 1 h. The methanotrophic biofilm reactor has proved itself to be a technologically stable installation, with low operational costs. The full-scale biofilm reactor was capable of removing 60 to 80% of the DCE and more than 99% of the present MCB. About 20 to 25% of DCE was removed in the first compartment (without methane injection) and about 55 to 60% was removed in the methanotrophic part. Stripping the DCE amounted to 10 to 25%. However, the desired value of 50 µg/L was not reached, and additional aeration of the installation was necessary.

A cost-effective, one-step biological process for the cometabolic removal of chlorinated solvents using methanotrophic biomass does not seem to be feasible, although considerable removal efficiencies can be realized. If the groundwater is also contaminated with aromatic compounds, a certain level of removal of chlorinated solvents can be expected in the biofilm reactor.

ACKNOWLEDGMENTS

We would like to thank Sybe Hartmans, Wageningen Agricultural University, Department of Food Science, Division of Industrial Microbiology, for performing the in vitro experiments. The project was cofunded by The Dutch Agency for Energy and the Environment (NOVEM).

REFERENCES

De Bruin, W. P., M.J.J. Kotterman, M. A. Posthumus, G. Schraa, and A.J.B. Zehnder. 1992. "Complete Biological Reductive Transformation of Tetrachloroethene to Ethane." *Applied and Environmental Microbiology* 58(6):1996-2000.

Dolfing, J., A. J. van den Wijngaard, and D. B. Janssen. 1993. "Microbiological Aspects of Chlorinated Hydrocarbons from Air." *Biodegradation* 4:261-282.

Folsom, B. R., and P. J. Chapman. 1991. "Performance Characterization of a Model Bioreactor for Biodegradation of Trichloroethylene by *Pseudomonas cepacia* 64." *Applied and Environmental Microbiology* 57(6):1602-1608.

Hartmans, S., and J.A.M. de Bont. 1992. "Aerobic Vinyl Chloride Metabolism in *Mycobacterium aurum* L1." *Applied and Environmental Microbiology* 58:1220-1226.

Marsman, E. H., J.M.M. Appelman, L.G.C.M. Urlings, and B. A. Bult. 1994. "BIOPUR®, an Innovative Bioreactor for the Treatment of Groundwater and Soil Vapor Contaminated with Xenobiotics." In R. E. Hinchee, D. B. Anderson, F. Blaine Metting, Jr., and G. D. Sayles (Eds.), *Applied Biotechnology for Site Remediation*, pp. 391-399. Lewis Publishers, Boca Raton, FL.

Oldenhuis, R., R.J.L.M. Vink, D. B. Jansen, and B. Witholt. 1989. "Degradation of Chlorinated Aliphatic Hydrocarbons by *Methylosinus trichosporium* OB3b Expressing Soluble Methane Monooxygenase." *Applied and Environmental Microbiology* 55:2819-2826.

Shields, M. S., S. O. Montgomery, S. M. Cuskey, P. J. Chapman, and P. H. Pritchard. 1991. "Mutants of *Pseudomonas cepacia* G4 Defective in Catabolism of Aromatic Compounds and Trichloroethlene." *Applied and Environmental Microbiology* 57(7):1935-1941.

Speitel, G. E., R. L. Segar, and S. L. de Wijs. 1994. "Trichloroethylene Cometabolism by Phenol-degrading Bacteria in Sequencing Biofilm Reactors." In R. E. Hinchee, A. Leeson, L. Semprini, and S. K. Ong (Eds.), *Bioremediation of Chlorinated and Polycyclic Aromatic Hydrocarbon Compounds*, pp. 333-338, Lewis Publishers, Boca Raton, FL.

Degradation of Polycyclic Aromatic Hydrocarbons by a Marine Fluidized-Bed Enrichment

Esa S. Melin, Jaakko A. Puhakka,
Minna Männistö, and John F. Ferguson

ABSTRACT

Aerobic polycyclic aromatic hydrocarbon (PAH)-degrading bacteria from marine sediments were enriched and maintained in a fluidized-bed reactor (FBR). The FBR was continuously fed a mixture of naphthalene, biphenyl, 2-methylnaphthalene, 2,6-dimethylnaphthalene, acenaphthene, fluorene, and phenanthrene; and the effluent concentrations remained below 0.03 mg/L (detection limit) over a period of 6 months. In batch vial assays, the relative rates of PAH degradation were controlled by their solubilities and, in some cases, by their substitution patterns. The enrichment also degraded several other PAHs, including anthracene and pyrene. The results demonstrate that the predominant PAH constituents of coal-tar creosote can be degraded and that PAH-contaminated saline waters can be remediated by the marine enrichment.

INTRODUCTION

Remediation of contaminants in the marine environments is of concern because of the effects on marine biota. Estuarine and marine sediments accumulate PAHs (Shiaris 1989). Sources of PAHs in to marine sediment are, for example, oil spills, petrochemicals, coal tars, and combustion products. As hydrophobic compounds, PAHs adsorb onto suspended particles and finally settle. In Eagle Harbor, where sediment was collected for this study, a wood-preserving plant was located on shore. Nearby sediments were contaminated with creosote, and PAHs seeped through beach sands into the intertidal zone.

Microbial degradation of PAHs has been studied extensively and their metabolism is quite well known (Cerniglia 1984). PAH degradation in oxic marine sediments seems comparable to PAH degradation in terrestrial systems (Bauer & Capone 1985, 1988; Shiaris 1989). Numerous PAH-degrading bacterial species have been isolated from PAH-contaminated areas (e.g., Kästner et al. 1994). In this study, a mixed marine PAH-degrading culture was developed from marine sediment with a mixture of PAHs to simulate degradation of soluble creosote compounds and to study degradation of PAHs, both individually and in a mixture.

EXPERIMENTAL PROCEDURES AND MATERIALS

An aerobic, glass laboratory-scale fluidized-bed reactor (FBR) was used for enrichment and maintenance of marine PAH-degrading microorganisms. The reactor height was 51 cm, inner diameter 4 cm, liquid volume 600 mL; 100 g of Celite R-633 microcarriers (Manville) were used as growth support for biofilm. The initial bed expansion was set at 50% by liquid recycling (FB volume 310 mL). The reactor was aerated with pure oxygen. The artificial seawater medium (ASM) (Ofjord et al. 1994) was modified by replacing ammonia with 24 mg/L $NaNO_3$.

The FBR was inoculated with creosote-contaminated marine sediment (Eagle Harbor, Puget Sound, Washington). The upper layer of sediment material was slurried and allowed to settle. Supernatant with 3.7 g of total solids was used for inoculation (volatile solids [VS] 0.45 g). The reactor was kept on recycle for 4 days and then continuous feed was started at 5.5 h hydraulic retention time.

The reactor was fed PAHs including naphthalene (2.2 to 14 mg/L), 2-methyl-naphthalene (3.3 to 5.5 mg/L), and the following at 0.1 to 0.5 mg/L: biphenyl, 2,6-dimethylnaphthalene, acenaphthene, fluorene, and phenanthrene. Feed PAH concentrations constantly changed due to adsorption and depletion during feeding. In addition, the "salting-out" and simultaneous presence of several PAHs decreased their solubilities (Eganhouse & Calder 1976). The PAHs were dissolved into the ASM in a feed column (0.8 × 60 cm) prior pumping to the reactor. Feed PAHs were dissolved in pentane and mixed with 20 g of 100-μm glass beads. Pentane was evaporated and the column was charged with PAH-covered beads. The feed column was changed once a week.

Biodegradation tests were performed in 8-mL glass vials having Teflon™-lined screw caps. Pentane solution(s) of PAHs were pipetted into the vials, and the pentane was allowed to evaporate. Then, 3 mL of fresh ASM and 1 mL of suspended microorganisms were added. The suspension was obtained by vortexing biofilm-covered carrier material from the FBR. The vials were sealed and placed horizontally on a shaker table (200 rpm, 20°C). The killed controls contained 2% formaldehyde. Duplicate vials were analyzed by extracting the content of vials into hexane at each time point. The concentrations were calculated as if all PAHs were dissolved in liquid phase.

The hexane extracts were analyzed for PAHs using a Perkin Elmer gas chromatograph (GC) equipped with DB-5 column (J&W Scientific) and flame ionization detector (FID). The temperature program was isothermal at 125°C for 2 min, ramped first at 15°C/min to 250°C, then at 4°C/min to 280°C and held at 250°C for 4 min. The injector temperature was 225°C, detector 300°C. Protein was analyzed using the modified Lowry method (BioRad).

RESULTS

Enrichment of marine PAH degraders in the FBR was accomplished by continuously feeding seven PAHs. Following the first days, no PAHs (detection limit 0.03 mg/L) were seen in the effluent over 6 months of operation. Oxygen uptake and changes in ASM color also indicated rapid startup of biodegradation. Nonbiological losses of PAHs (volatilization and adsorption) were not determined. After 50 days on continuous feed, the reactor contained 15.3 mg volatile solids (VS)/g carriers.

Table 1 shows results of vial tests with individual PAHs. Some removal of all test PAHs was apparent, but 1,8-dimethylnaphthalene, fluoranthene, chrysene, and benzo[a]pyrene were only partially removed. With 1,2-dimethylnaphthalene, acenaphthene, and fluorene, transient accumulation of unidentified metabolites were observed in GC/FID. The Michaelis-Menten model (assuming no growth) was used to obtain maximum specific utilization rates (k) and half-saturation constants (K_s) (Table 1). Degradation rates of several PAHs slowed significantly at concentrations below 100 to 200 µg/L, which was not predicted by model.

The effect of substitution on PAH biodegradation is illustrated with naphthalenes (Table 1). The rates were comparable with naphthalene and 1- and 2-methylnaphthalenes. The half-life of 2-ethylnaphthalene was twice that of naphthalene. Degradation rates significantly varied with dimethylnaphthalenes. Location of methyl substituent in one ring or on both rings did not affect degradation. The 1,2- and 2,6-dimethylnaphthalenes were degraded at the same rate, but 1,4-dimethylnaphthalene was much slower, and 1,8-dimethylnaphthalene was not degraded after initial removal.

In PAH mixtures, the degradation rates (Table 2) and patterns changed. In most cases, the degradation was much slower than with the compound alone (Table 1); sometimes half-lives were an order of magnitude higher. The 1,4-dimethylnaphthalene and fluorene were removed faster in mixtures than alone.

DISCUSSION

The PAH-degrading microorganisms were readily enriched from contaminated marine sediment in the continuous-flow FBR with seven PAHs. The enrichment removed a variety of other PAHs, including anthracene and pyrene. Phenanthrene has been shown to enhance anthracene degradation in marine sediments, but not naphthalene (Bauer & Capone 1988). High-molecular-weight PAHs, including chrysene and benzo(a)pyrene, disappeared only marginally before degradation ceased.

Degradation of several PAHs was fitted to the Michaelis-Menten model using the nominal concentration of PAH in the vial. The results suggest that PAH degradation was not entirely dissolution controlled as has been observed for

TABLE 1. Degradation of individual PAHs by marine sediment enrichment in vial tests.

Substrate	Solubility (mg/L)	Protein Concentration (mg/L)	Concentration (mg/L)	Initial Half-life (h)	Reduction (%)	Time (h)	k (h⁻¹)	K_s (mg/L)
Naphthalene	31.3[a]	28.9	11.1	1.5	100	4.9	0.14	0.42
1-Methylnaphthalene	25.8[a]	20.5	7.5	1.4	100	4.6	0.15	0.79
2-Methylnaphthalene	24.6[a]	18.2	6.7	1.1	99.7	5.3	0.21	1.3
2-Ethylnaphthalene	8.0[a]	20.5	7.5	2.8	98.9	10.4	0.12	4.7
1,2-Dimethylnaphthalene		20.5	9.0	3.4	99.3	9.1	ND[d]	ND
1,4-Dimethylnaphthalene		22.6	5.2	20	90.1	52	ND	ND
1,8-Dimethylnaphthalene		10.3	8.5	NA[c]	20.0	287	ND	ND
2,6-Dimethylnaphthalene	1.3[a]	25.0	9.5	3.4	98.2	16.9	0.18	15
Biphenyl	7.5[a]	29.8	9.1	2.7	99.8	8.3	0.093	2.2
Acenaphthene	3.5[a]	27.8	9.9	25.2	100	63.7	ND	ND
Fluorene	2.0[b]	28.8	9.5	26.7	67.9	66.1	ND	ND
Phenanthrene	1.2[a]	8.8	3.9	1.29	98.3	5.9	0.21	0.66
		26.3	3.8	0.52	98.7	5.9		ND
Fluoranthene		23.0	3.8	NA	40.9	359	ND	ND
Anthracene	0.26[b]	11.3	4.1	11.7	100	148	ND	ND
	0.07[b]	34.1	4.1	8.3	100	146	ND	ND
Pyrene	0.14[b]	10.7	4.0	24.5	100	240	ND	ND
		32.1	4.0	12.0	100	110	ND	ND
Chrysene	0.002[b]	17.3	2.2	NA	11.6	852	ND	ND
Benzo[a]pyrene	0.003[b]	17.3	1.9	NA	13.1	855	ND	ND

(a) Eganhouse & Calder (1976).
(b) Mueller et al. (1989).
(c) NA = Not applicable. The degradation ceased before 50% removal of PAH was achieved.
(d) ND = The fit was not good or the resulting constants were unreasonably high.

TABLE 2. Degradation of a PAH mixture by marine sediment enrichment in vial test. The vials contained 32 mg protein/L.

Substrate	Initial Concentration (mg/L)	Half-life (h)	Reduction (%)	Time (h)
Naphthalene	9.6	20.3	100	31
2-Methylnaphthalene	13.3	22.6	100	45
Biphenyl	9.2	28.8	100	73
2,6-Dimethylnaphthalene	9.2	37.9	100	93
Acenaphthene	8.9	52.5	71.7	171
Fluorene	8.9	38.2	100	93
Phenanthrene	4.0	36.0	100	93

PAHs introduced as solids (Volkering et al. 1992; Tiehm 1994). In our tests, degradation of less-soluble PAHs slowed significantly at below 100 to 200 µg/L. A similar phenomenon has been observed with anthracene in low-organic-content soil (Gray et al. 1994). In addition to kinetic limitations, this is probably due to a decreased dissolution rate when the surface area of solid PAHs decreases (Volkering et al. 1992).

Alkyl substitution of the aromatic rings has been shown to decrease the PAH degradation rate (Elmendorf et al. 1994). In our study, 1- and 2-methylnaphtha-lenes degraded at the same rate as naphthalene. The degradation of other substituted naphthalenes was slower, and the susceptibility of dimethylnaphthalenes to degradation varied. The degradation of 1,4- and 1,8-dimethylnaphthalenes was most limited as earlier observed for 1,8-dimethylnaphthalene by a soil *Pseudomonas* sp. (Miyachi et al. 1993).

The degradation rates and patterns were significantly changed when the enrichment was exposed to a mixture of PAHs. This contrasts studies by Bauer & Capone (1988) and Heitkamp & Cerniglia (1989), who did not observe significant changes in individual PAH degradation rates in mixtures.

This study has shown that soluble PAHs in saline water can be degraded in an FBR. The degradation is affected by solubilities, substitution patterns, and the number of individual PAHs in mixtures.

ACKNOWLEDGMENTS

This research was funded by the Academy of Finland (E.S.M.) and Office of Naval Research (Grant N00014-92-J-1578).

REFERENCES

Bauer, J. E., and D. G. Capone. 1985. "Degradation and Mineralization of the Polycyclic Aromatic Hydrocarbons Anthracene and Naphthalene in Intertidal Marine Sediments." *Appl. Environ. Microbiol.* 50(1):81-90.

Bauer, J. E., and D. G. Capone. 1988. "Effects of Co-Occurring Aromatic Hydrocarbons on Degradation of Individual Polycyclic Aromatic Hydrocarbons in Marine Sediment Slurries." *Appl. Environ. Microbiol.* 54(7):1649-1655.

Cerniglia, C. E. 1984. "Microbial Metabolism of Polycyclic Aromatic Hydrocarbons." *Adv. Appl. Microbiol.* 30:31-71.

Eganhouse, R. P. and J. A. Calder. 1976. "The Solubility of Medium Molecular Weight Aromatic Hydrocarbons and the Effects of Hydrocarbon Co-solutes and Salinity." *Geochim. Cosmochim. Acta* 40:555-561.

Elmendorf, D. L., C. E. Haith, G. S. Douglas, and R. C. Prince. 1994. "Relative Rates of Biodegradation of Substituted Polycyclic Aromatic Hydrocarbons." In R. E. Hinchee, A. Leeson, L. Semprini, and S. K. Ong (Eds.), *Bioremediation of Polycyclic Aromatic Hydrocarbon Compounds*, pp. 188-202. Lewis Publishers, Boca Raton, FL.

Gray, M. R., D. K. Banerjee, P. M. Fedorak, A. Hashimoto, J. H. Masliyah, and M. A. Pickard. 1994. "Biological Remediation of Anthracene-Contaminated Soil in Rotating Bioreactors." *Appl. Microbiol. Biotechnol.* 40:933-940.

Heitkamp, M. A., and C. E. Cerniglia. 1989. "Polycyclic Aromatic Hydrocarbon Degradation by a *Mycobacterium* sp. Microcosms Containing Sediment and Water from a Pristine Ecosystem." *Appl. Environ. Microbiol.* 55(8):1968-1973.

Kästner, M., M. Breuer-Jammali, and B. Mahro. 1994. "Enumeration and Characterization of the Soil Microflora from Hydrocarbon-Contaminated Soil Sites Able to Mineralize Polycyclic Aromatic Hydrocarbons (PAH)." *Appl. Microbiol. Biotechnol.* 41:267-273.

Miyachi, N., T. Tanaka, T. Suzuki, Y. Hotta, and T. Omori. 1993. "Microbial Oxidation of Dimethylnaphthalene Isomers." *Appl. Environ. Microbiol.* 59(5):1504-1506.

Mueller, J. G., P. J. Chapman, and P. H. Pritchard. 1989. "Creosote-Contaminated Sites." *Environ. Sci. Technol.* 23(10):1197-1201.

Ofjord, G. D., J. A. Puhakka, and J. F. Ferguson. 1994. "Reductive Dechlorination of Aroclor 1254 by Marine Sediment Cultures." *Environ. Sci. Technol.* 28(13):2286-2294.

Shiaris, M. P. 1989. "Seasonal Biotransformation of Naphthalene, Phenanthrene, and Benzo(a)pyrene in Surficial Estuarine Sediments." *Appl. Environ. Microbiol.* 55(6):1391-1399.

Tiehm, A. 1994. "Degradation of Polycyclic Aromatic Hydrocarbons in the Presence of Synthetic Surfactants." *Appl. Environ. Microbiol.* 60(1):258-263.

Volkering, F., A. M. Breure, A. Sterkenburg, and J. G. van Andel. 1992. "Microbial Degradation of Polycyclic Aromatic Hydrocarbons: Effect of Substrate Availability on Bacterial Growth Kinetics." *Appl. Microbiol. Biotechnol.* 36:548-552.

Database on Microbial Degradation of Soil Pollutants

Dorothea Gleim, Herbert Milch, and Manfred Kracht

ABSTRACT

The factual database outlined in this paper has been developed to provide rapid reliable information on the biodegradability of soil pollutants and on possible bioremedial action. It contains information both on degradation in pure and mixed cultures and on degradation in soils.

INTRODUCTION

The project was initiated by the interdisciplinary DECHEMA[1] study group Environmental Biotechnology — Soil, which is composed of well-known scientists from industries, universities, and governmental institutes. This group agreed that there is an urgent need for a better transfer of scientific results on biodegradation of soil pollutants to people who are involved in bioremedial action. As a result, the group supported the setup of the factual database described below.

DESCRIPTION OF THE DATABASE

The database has been developed using the commercially available relational database management system DataPerfect 2.3. To facilitate the database access for prospective users, the data will be transferred into the FoxPro 2.6 database management system, which has enabled the development of suitable retrieval menus.

The data kept in the database have been retrieved from international scientific publications on biodegradation in pure and mixed cultures and from publications on biodegradation in soils. The format for input and storage of information consists of more than 50 fields, grouped into 8 files. These files are listed below together with the kind of information that can be entered.

---

[1] Deutsche Gesellschaft für Chemisches Apparatewesen, Chemische Technik und Biotechnologie e.V., Frankfurt am Main, Germany.

1. Compounds or mixtures of compounds. All frequently used names for a substance (systematic names, common names, well-known and long-established trade names, abbreviations) or a mixture are entered into the database and linked together by a common number. Single substances are also defined by the number of the Chemical Abstracts Services.
2. Classes of compounds. All substances in the list of compounds [1] are assigned to at least one class, such as polycyclic aromatic hydrocarbons.
3. Microorganisms. Microbial names, strain designation(s) and/or collection number(s), names of mixed cultures, and remarks on strains or mixed cultures (e.g., on the source).
4. Culture conditions. Special features of cultivation; oxygen relations (aerobic/anaerobic).
5. Metabolism. Metabolites (linked with the list of substances [1], the soil file [6], and the microorganisms file [3]); degradative pathways.
6. Soils. Soil properties (e.g., soil texture, pH, soil organic matter, cation exchange capacity); experimental scale (laboratory experiment, pilot study, full-scale experiment); experimental conditions (temperature, water content, oxygen relations, oxidation-reduction potential); bioremediation treatment (e.g., fertilization, tilling, liming, application of microorganisms with known degradation capabilities in pure culture).
7. Degradation data. Quantitative data on biodegradation of pollutants in soils (initial concentration, experimental time, decrease, losses caused by evaporation or leaching, biodegradation).
8. Bibliography. Conventional bibliographic descriptors of a literature source (author(s), title, editor, source, year, volume, number, pages).

At present the study comprises haloaliphatic compounds, polycyclic aromatic hydrocarbons, polychlorinated biphenyls, and polychlorinated dioxins and furans. In the first stage of the project until April 1994 the language of the database was in German. Since then, the project has been continued in English. The database should be available early in 1996 on diskettes (MS-DOS) from the DSM-Deutsche Sammlung von Mikroorganismen und Zellkulturen GmbH, Mascheroder Weg 1b, D-38124 Braunschweig, Germany.

ACKNOWLEDGMENTS

The project is financially supported by Germany's Federal Environmental Agency and the Federal Ministry of Research and Technology (UBA/BMFT project 1480743 A).

Effect of Sorption
and Substrate Pattern
on PAH Degradability

Bernd P. Ressler, Charlotte Kämpf, and Josef Winter

ABSTRACT

The effect of sorption and the substrate pattern on the degradability of polycyclic aromatic hydrocarbons (PAHs) during bioremediation of PAH-contaminated silt in a slurry reactor was investigated. Biological degradation of high-molecular-weight PAH compounds sorbed to silt and clay particles was enhanced in the presence of low-molecular-weight PAHs. In soil suspensions containing silt (30% w/w) contaminated with PAH compounds of different molecular weights, PAHs containing four aromatic rings were degraded more readily in the presence of naphthalene. Bioavailability of PAHs was correlated to the water solubility of different compounds; a significant limitation of bacterial growth and activity due to sorption of PAHs to the fine particles could not be observed.

INTRODUCTION

The silt and clay fraction of soils is known to be the bottleneck during in situ and ex situ bioremediation of contaminated soil because of its low k_f-value and high sorption capacity for hydrophobic soil contaminants, e.g., PAHs (Wefer-Roehl 1994). The increasing use of soil-washing plants has led to an increasing amount of contaminated fine-particle sludges as a residual fraction of the soil-washing process. Investigations on the bioremediation of PAH-contaminated soils have often stressed the influence of adsorption of PAHs to organic soil matter as a limiting factor.

In our studies, soil suspensions containing 30% (w/w) silt contaminated with PAHs containing two, three, and four rings were treated biologically in a slurry reactor. The influence of the sorption of PAHs to silt particles on the bioavailability and the effect of low-molecular-weight PAHs on the degradation of high-molecular-weight PAHs sorbed to silt particles was investigated.

EXPERIMENTAL PROCEDURES
AND MATERIALS

The silt used for experiments was the sieved fraction of an uncontaminated loess soil. The average particle diameter of this soil fraction was <63 µm, the organic carbon content was below 1%, and the clay content was 20%. For bio-remediation experiments, the silt was contaminated with seven different PAHs containing two (naphthalene), three (acenaphthene, fluorene, phenanthrene, and anthracene) and four (fluoranthene and pyrene) aromatic rings; dry silt was mixed with PAHs dissolved in acetone. Acetone was allowed to evaporate while the suspension was stirred continuously to achieve a homogeneous contamination of the silt. The silt (30% w/w) was then suspended in sterile mineral medium containing 0.2 mg/L KH_2PO_4, 0.8 mg/L K_2HPO_4, and 0.5 g/L KNO_3. The suspension was stirred in a slurry reactor (vol.: 6.0 L) at 350 rpm and 25°C.

In batches containing no silt, PAH crystals were added to the medium. Crystals with an average diameter between 100 µm and 50 µm were produced as follows: single PAH compounds were dissolved in acetone (0.1% w/v) and the solution was poured into a glass dish. Acetone was allowed to evaporate, leaving a layer of PAH crystals that could be easily removed and suspended into the mineral medium.

After the equilibrium concentration of PAHs in the liquid phase was reached (at t = 0), 10 mL of a mixed bacterial culture M1 (O.D.$_{578}$ = 1.0), isolated from a coal gas plant and cultured on PAH as the sole carbon source, was added to the soil suspension. Air was supplied by an electronically controlled valve that opened if the oxygen concentration dropped below 3 mg/L and closed at 7 mg O_2/L. The oxygen concentration in the soil suspension was measured with an O_2 electrode in intervals of 120 s. The PAH content of the silt and the liquid phase was measured by high-performance liquid chromatography (HPLC). Samples of the soil suspension (10 mL) were centrifuged (6,000*g, 10 min), 20 µL of the liquid phase was analyzed directly by HPLC, and 1 g of the pellet was dried at room temperature and extracted with 2*10 mL acetonitrile in closed glass vessels by ultrasonicating (50 W) for 2*30 min.

To account for an inhomogeneous distribution of suspended PAH crystals in batches without silt, the sample volume was 30 mL. Samples were centrifuged (10,000*g, 10 min) and 20 µL of the supernatant was analyzed directly by HPLC. The pellet was extracted with 2*10 mL acetonitrile in closed glass vessels by ultrasonicating (50 W) for 2*30 min. For each sample, three replications were made.

RESULTS

Bioavailability of PAHs

To investigate the effect of sorption of PAHs on the bioavailability, experiments were conducted with single PAH compounds, respectively. Experiments on the bioavailability of PAHs were conducted until the stationary phase of bacterial growth (measured by oxygen consumption) was reached. Figure 1 shows the maximum degradation rates by a mixed bacterial culture in batches containing single PAH compounds. The degradation rates of sorbed PAHs did not differ significantly from the degradation rates of nonsorbed PAHs. The bioavailability of PAHs was correlated mainly to the water solubility of different PAH compounds; a significant limitation of bacterial growth and activity due to sorption of PAHs to the fine particles could not be observed.

The concentrations of PAHs both in the aqueous phase and sorbed on the silt fraction during the experiments are given in Table 1. During logarithmic growth, the PAH concentrations in the aqueous phase dropped far below their water solubility, which indicated a direct bacterial uptake of dissolved PAHs. Naphthalene was degraded completely within 12 h; the degradation of acenaphthene ceased after 10 d.

Effect of Low-Molecular-Weight PAHs on the Degradability of High-Molecular-Weight PAHs

Microbial degradation of pyrene and fluoranthene could be enhanced in soil suspensions containing naphthalene as an additional carbon source. In cultures using a mixture of pyrene and fluoranthene as the sole carbon source, 2,150 mg/L O_2 were consumed during pyrene (fluoranthene) degradation from 400 (396) mg/kg to 155 (145) mg/kg within 8 days (Figure 2a). In cultures

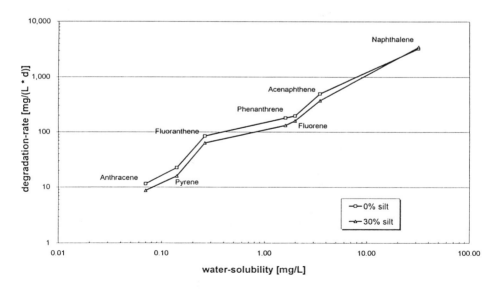

FIGURE 1. Correlation between water-solubility and degradation rate of sorbed and nonsorbed PAHs in suspensions containing 0% and 30% (w/w) of contaminated fine particles.

Biological Unit Processes for Hazardous Waste Treatment

TABLE 1. Partitioning of PAHs to the aqueous (A) and solid (S) phase (silt, PAH crystals) in mg/L at t=0 (t_0), during bacterial log-phase (t_{log}), and after stationary phase was reached (t_{stat}). The initial PAH concentration was 1,000 mg/L for naphthalene and 100 mg/L for the other PAHs; the initial concentration of dry silt was 284 mg/kg (2,840 mg/kg for naphthalene), resulting in a final concentration of 100 mg/L (1,000 mg/L for naphthalene) in batches containing 30% silt (w/w).

PAH	Phase	t_0		t_{log}		t_{stat}		PAH Recovery in Sterile Control at t_{stat}	
		0% silt	30% silt	0% silt	30% silt	0% silt	30% silt	0% silt	30% silt
Naphthalene	A	29.8	30.2	0.2	n.d.[a]	<0.1	n.d.	31.2	32.8
	S	914.7	893.0	—	—	0.8	n.d.	769.0	757
Acenaphthene	A	3.5	3.1	0.1	0.3	n.d.	n.d.	3.4	2.9
	S	94.1	91.0	—	—	n.d.	2.4	80.1	79.1
Fluorene	A	1.9	1.6	n.d.	n.d.	0.2	n.d.	1.7	0.8
	S	91.8	92.7	—	—	5.5	4.6	84.5	84.6
Phenanthrene	A	1.7	1.6	0.1	0.1	0.3	<0.1	1.7	1.2
	S	92.2	90.3	—	—	6.2	4.3	87.1	83.5
Fluoranthene	A	0.27	<0.1	n.d.	n.d.	<0.1	n.d.	0.28	n.d.
	S	98.0	96.4	—	—	12.3	10.9	96.1	92.4
Pyrene	A	0.14	n.d.	<0.1	n.d.	0.1	n.d.	0.11	n.d.
	S	99.2	94.1	—	—	19.5	14.2	93.5	89.5
Anthracene	A	0.05	n.d.	n.d.	n.d.	n.d.	n.d.	n.d.	n.d.
	S	98.0	90.2	—	—	23.1	17.8	92.4	90.0

n.d. = not detected.

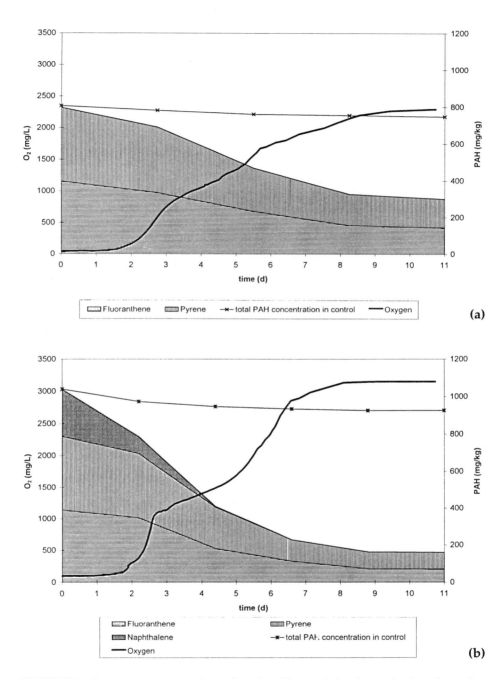

FIGURE 2. Oxygen consumption of a mixed bacterial culture during degradation of fluoranthene and pyrene in the absence (a) and presence (b) of
naphthalene.

containing 250 mg/L naphthalene as an additional carbon source, the same reduction of pyrene and fluoranthene was achieved within 6 days, requiring a total of 3,150 mg/L O_2 during two phases of intensive respiration and increased O_2 consumption (Figure 2b).

DISCUSSION

The results obtained lead to the conclusion that bacterial transformation and the degradation of high-molecular-weight PAHs are enforced in the presence of low-molecular-weight-PAHs. This might be due to enhanced bacterial enzyme synthesis during initial degradation of low-molecular-weight PAHs or/and the formation of metabolites that might fortify cometabolic processes. Bioremediation of PAH-contaminated silt seems to be limited mainly by the water solubility of PAH compounds and less by the sorption of PAHs to fine mineral silt particles if the particles are in direct contact with the liquid phase and if the desorption and diffusion of PAHs are not obstructed (Breure et al. 1992). Thus, in addition to the organic soil material such as, for example, humic matter, the microstructure and low permeability of silt are the limiting factors in reclamation processes, making the use of in situ remediation techniques for silty soils very difficult. Using bioreactor technology, reclamation of the silt fraction seems to be possible.

REFERENCES

Breure, A. M., A. Sterkenburg, F. Volkering, and J. G. van Andel. 1992. "Bioavailability as a rate controlling step in soil decontamination processes." In Dechema (Ed.), *International Symposium on Soil Decontamination using Biological Processes*, pp. 147-154. Reprints, Karlsruhe, Germany. 6-9 December 1992.

Wefer-Roehl, A. 1994. "Polycyclische aromatische Kohlenwasserstoffe in feinkörnigen Sedimenten: Adsorption und mikrobieller Abbau." Dissertation Inst. f. angewandte Geologie, Universität Karlsruhe, Werke Angewandte Geologie Karlsruhe Vol. 31, Karlsruhe, Germany.

AUTHOR LIST

Ahmadvand, Hassan
Arco Pipeline Company
P.O. Drawer 429
Colvis, NM 88102 USA

Albu-Cimpoia, Ruxandra
Biotechnology Research Institute
National Research Council—Canada
6100 Royalmount Avenue
Montréal, Québec H4P 2R2
CANADA

Andrilenas, Jeffrey S.
AGRA Earth & Environmental, Inc.
3232 West Virginia Ave.
Phoenix, AZ 85009-1502 USA

Arcangeli, Jean-Pierre
Technical University of Denmark
Dept. of Environ. Science & Engrg.
Building 115
DK-2800 Lyngby
DENMARK

Arneberg, Rolf
Biogénie, inc.
350 rue Franquet entrée 10
Sainte-Foy, Québec G1P 4P3
CANADA

Arvin, Erik
Technical University of Denmark
Dept. of Environ. Science & Engrg.
Building 115
DK-2800 Lyngby
DENMARK

Barbush, John A.
Waste Management, Inc.
2625 E. Broadway Street
Northwood, OH 43619-1062 USA

Basrai, Shabbir S.
Orange County Sanitation Districts
10844 Ellis Avenue
P.O. Box 8127
Fountain Valley, CA 92708-8127
USA

Belanger, David W.
CH2M Hill Engineering Ltd.
180 King St. S., Suite 600
Waterloo, Ontario N2J 1P8
CANADA

Belcher, David M.
ABB Environmental Services, Inc.
107 Audubon Road, Building 3,
   Suite 25
Wakefield, MA 01880 USA

Benazon, Netta
CH2M Hill Engineering Ltd.
180 King St. S., Suite 600
Waterloo, Ontario N2J 1P8
CANADA

Bender, Judith
Clark Atlanta University
Research Center for Science
   & Technology
Box 296
Atlanta, GA 30314 USA

Bishop, Dolloff F.
U.S. Environ. Protection Agency
Natl. Risk Mgmt. Research Lab
26 W. Martin Luther King Drive
Cincinnati, OH 45268 USA

Blackburn, James W.
Exxon Research & Engineering
Clinton TWP Rte 22E
Annandale, NJ 08801 USA

Bohner, Andrew K.
Envirogen, Inc.
4100 Quakerbridge Road
Lawrenceville, NJ 08648 USA

Bourbonais, Katherine A.
AGI Technologies
300 120th Ave. NE, Building 4
Bellevue, WA 98005 USA

Bower, Jr., David A.
Waste Management, Inc.
5701 Este Avenue
Cincinnati, OH 45232 USA

Braun-Lüllemann, Annette
Universität Göttingen
Forstbotanisches Institut
Büsgenweg 2
D-37077 Göttingen
GERMANY

Brown, Kandi L.
IT Corporation
1425 S. Victoria Court, Suite A
San Bernadino, CA 92408-2923 USA

Buehler, Verne T.
Envirex, Inc.
1901 S. Prairie Avenue
P.O. Box 1604
Waukesha, WI 53187 USA

Burick, Thomas
Envirogen, Inc.
4100 Quakerbridge Road
Lawrenceville, NJ 08648 USA

Cassidy, Daniel P.
University of Notre Dame
Center for Bioengineering &
    Pollution Control
Notre Dame, IN 46556 USA

Chen, Jian-Shin
University of Minnesota
500 Pillsbury Dr., S.E.
Minneapolis, MN 55455 USA

Cyr, Benoit
Biogénie, inc.
350 rue Franquet entrée 10
Sainte-Foy, Québec G1P 4P3
CANADA

Daun, Gregor
Fraunhofer Institut für
    Grenzflächen und
    Bioverfahrenstechnik
Nobelstrasse 12
D-70569 Stuttgart
GERMANY

Davila, Brunilda
U.S. Environ. Protection Agency
Natl. Risk Mgmt. Research Lab
5995 Center Hill Avenue
Cincinnati, OH 45268 USA

De Santis, Peter
BP Oil Company
295 SW 41st Street
Building 13, Suite N
Renton, WA 98055 USA

Deobald, Lee A.
Innovative BioSystems, Inc.
121 Sweet Avenue
Moscow, ID 83843-2386 USA

Deschamps, Stéphane
National Research Council—Canada
Biotechnology Research Institute
6100 Royalmount Avenue
Montréal, Québec H4P 2R2
CANADA

Devinny, Joseph S.
University of Southern California
Environmental Engrg. Program
KAP-224D
Los Angeles, CA 90089-2531 USA

Dooley, Maureen A.
ABB Environmental Services, Inc.
107 Audubon Rd., Bldg. 3, Suite 25
Wakefield, MA 01880 USA

Esler, Charles T.
AGRA Earth & Environmental, Inc.
7477 SW Tech Center Drive
Portland, OR 97223-8025 USA

Evans, Patrick J.
AGI Technologies
300 120th Avenue, NE, Building #4
Bellevue, WA 98005 USA

Farmer, Randall W.
University of Minnesota
500 Pillsbury Dr., S.E.
Minneapolis, MN 55455 USA

Feldt, Sabrina
BIOPRACT GmbH
Scheiblerstrasse 27
12437 Berlin
GERMANY

Ferguson, John F.
University of Washington
Dept. of Civil Engineering FX-10
Seattle, WA 98195 USA

Folsom, Brian R.
Envirogen, Inc.
4100 Quakerbridge Road
Lawrenceville, NJ 08648 USA

Frigon, Jean-Claude
National Research Council—Canada
Biotechnology Research Institute
6100 Royalmount Avenue
Montréal, Québec H4P 2R2
CANADA

Fruechtnicht, Antje
Umweltschutz Nord GmbH & Co.
Industriepark 6
D-27767 Ganderkesee
GERMANY

Gandee, John P.
Blasland, Bouck & Lee
189 W. Schrock Road
Westerville, OH 43081 USA

Gantzer, Charles J.
Membran Corporation
1037 10th Avenue SE
Minneapolis, MN 55414 USA

Germann, Gerry
Sinclair Pipeline Company
Route 4, Box 4A
Carrollton, MO 64633 USA

Glaser, John A.
U.S. Environ. Protection Agency
Natl. Risk Mgmt. Research Lab
26 W. Martin Luther King Drive
Cincinnati, OH 45268 USA

Gleim, Dorothea
DSM—Deutsche Sammlung von
    Mikroorganismen und
    Zellculturen GmbH
Mascheroderweg 1b
D-38124 Braunschweig
GERMANY

Govind, Rakesh
University of Cincinnati
Dept. of Chemical Engineering
620 Rhodes Hall
Cincinnati, OH 45221-0171 USA

Green, Roger B.
Waste Management, Inc.
5701 Este Avenue
Cincinnati, OH 45232 USA

Greer, Charles
National Research Council—Canada
Biotechnology Research Institute
6100 Royalmount Avenue
Montréal, Québec H4P 2R2
CANADA

Grey, Gary M.
HydroQual, Inc.
1 Lethbridge Plaza
Mahwah, NJ 07430 USA

Gromicko, Gregory J.
Groundwater Technology, Inc.
600 Clubhouse Drive
Moon Township, PA 15146 USA

Groshko, Veronica
Michigan Biotechnology Institute
P.O. Box 27609
Lansing, MI 48909-0609 USA

Guarini, William J.
Envirogen, Inc.
Princeton Research Center
4100 Quakerbridge Road
Lawrenceville, NJ 08648 USA

Guiot, Serge R.
National Research Council—Canada
Biotechnology Research Institute
6100 Royalmount Avenue
Montréal, Québec H4P 2R2
CANADA

Gundersen, Anders T.
Technical University of Denmark
Dept. of Environ. Science & Engrg.
Building 115
DK-2800 Lyngby
DENMARK

Hague, Keith
IT Corporation
312 Directors Drive
Knoxville, TN 37923 USA

Hahn, Hans H.
Universität Karlsrühe (TH)
Institut für Siedlungswasserwirt-
    schaft (ISWW)
D-76138 Karlsrühe
GERMANY

Hater, Gary R.
Waste Management, Inc.
5701 Este Avenue
Cincinnati, OH 45232 USA

Hawari, Jalal
National Research Council—Canada
Biotechnology Research Institute
6100 Royalmount Avenue
Montréal, Québec H4P 2R2
CANADA

Hayes, Thomas D.
Gas Research Institute
8600 W. Bryn Mawr Avenue
Chicago, IL 60631 USA

Heerenklage, Jörn
Technische Universität
    Hamburg-Harburg
Institute of Waste Management
Harburger Schlossstrasse. 37
D-21079 Hamburg
GERMANY

Heuermann, Elisabeth
Umweltschutz Nord GmbH & Co.
Industriepark 6
D-27767 Ganderkesee
GERMANY

Hickey, Robert F.
EFX Systems, Inc.
3900 Collins Rd., Suite 1011
Lansing, MI 48910 USA

Horn, William C.
Exxon Research & Engineering Co.
Clinton TWP Route 22E
Annandale, NJ 08801 USA

Hupe, Karsten
Technische Universität
    Hamburg-Harburg
Arbeitsbereich Abfallwirtschaft und
    Stadttechnik
Harburger Schlossstrasse 37
D-21079 Hamburg
GERMANY

Hüttermann, Alois
Universität Göttingen
Forstbotanisches Institut
Büsgenweg 2
37077 Göttingen
GERMANY

Irvine, Robert L.
University of Notre Dame
Center for Bioventing and Pollution
    Control
Notre Dame, IN 46556 USA

Jenkins, Joshua
Blasland, Bouck & Lee
189 W. Schrock Road
Westerville, OH 43081 USA

Jerger, Douglas E.
OHM Remediation Services Corp.
16406 U.S. Route 224 E
Findlay, OH 45840 USA

Jetté, Jean-François
National Research Council—Canada
Biotechnology Research Institute
6100 Royalmount Avenue
Montréal, Québec H4P 2R2
CANADA

Johannes, Christian
Universität Göttingen
Forstbotanisches Institut
Büsgenweg 2
37077 Göttingen
GERMANY

Johnson, Jaret C.
ABB Environmental Services, Inc.
107 Audubon Rd., Bldg. 3, Suite 25
Wakefield, MA 01880 USA

Kämpf, Charlotte
Universität Karlsrühe
Institut fur Hydrologie und
    Wasserwirtschaft (IHW)
Postfach 6980
76128 Karlsrühe
GERMANY

Knackmuss, Hans-Joachim
University of Stuttgart
Almandrin 31
D-70569 Stuttgart
GERMANY

Kopchynski, David M.
University of Minnesota
1609 Pleasant Street, #112
Lauderdale, MN 55108 USA

Kracht, Manfred
DSM—Deutsche Sammlung von
    Mikroorganismen und
    Zellculturen GmbH
Mascheroderweg 1b
D-38124 Braunschweig
GERMANY

Kudrjawizki, Felix
BIOPRACT GmbH
Scheiblerstrasse 27
12437 Berlin GERMANY
Deceased November 1994

Laakso, Gary L.
AGI Technologies
300 120th Avenue NE, Building 4
Bellevue, WA 98005 USA

Laford, H. Donald
URS Consultants, Inc.
1100 Olive Way, Suite 200
Seattle, WA 98101 USA

Lang, Craig
IT Corporation
312 Directors Drive
Knoxville, TN 37923 USA

Lee, Richard
Skidaway Inst. of Oceanography
10 Ocean Science Circle
P.O. Box 13687
Savannah, GA 31411 USA

Lee, Jean H.
AGI Technologies
300 120th Avenue NE, Building 4
Bellevue, WA 98005 USA

Lee, M. Kathryn
Exxon Research & Engineering Co.
Clinton TWP Route 22E
Annandale, NJ 08801 USA

Lei, Jiyu
Biogénie, inc.
350 rue Franquet entrée 10
Sainte-Foy, Québec G1P 4P3
CANADA

Lenke, Hiltrud
Fraunhofer Institut für
    Grenzflächen und
    Bioverfahrenstechnik
Nobelstraße 12
D-70569 Stuttgart
GERMANY

Lesky, Mark J.
Novacor Chemicals Inc.
690 Mechanic Street
Leominster, MA 01453 USA

Leuschner, A.
Remediation Technologies, Inc.
9 Pond Lane
Concord, MA 01742 USA

Loganathan, Bommanna G.
Skidaway Inst. of Oceanography
10 Ocean Science Circle
Savannah, GA 31411 USA

Lord, Denis
Biogénie, inc.
350 rue Franquet, entree 10
Sainte-Foy, Québec G1P 4P3
CANADA

Lüth, Joachim-Christian
Technical University of
    Hamburg-Harburg
Institute of Waste Management
Harburger Schlossstrasse 37
21079 Hamburg
GERMANY

Maiello, Joy A.
HydroQual, Inc.
1 Lethbridge Plaza
Mahwah, NJ 07430 USA

Maier, Walter J.
University of Minnesota
500 Pillsbury Dr., S.E.
Minneapolis, MN 55455 USA

Majcherczyk, Andrzej
Universität Göttingen
Forstbotanisches Institut
Büsgenweg 2
37077 Göttingen
GERMANY

Männistö, Minna
Helsinki University of Technology
Helsinki
FINLAND

Marsman, Eric H.
TAUW Milieu b.v.
P.O. Box 133
7400 AC Deventer
THE NETHERLANDS

McCauley, Paul T.
U.S. Environ. Protection Agency
Natl. Risk Mgmt. Research Lab
26 W. Martin Luther King Drive
Cincinnati, OH 45268 USA

McGuire, Patrick N.
Blasland, Bouck & Lee, Inc.
6723 Towpath Road
Syracuse, NY 13214 USA

Medina, Victor F.
RMT Inc.
20900 Swenson D., #100
Waukesha, WI 53186 USA

Melin, Esa S.
Tampere University of Technology
Inst. of Water & Environ. Engrg.
P.O. Box 527
33101 Tampere
FINLAND

Mercier, Pierre
National Research Council—Canada
Biotechnology Research Institute
6100 Royalmount Avenue
Montréal, Québec H4P 2R2
CANADA

Milch, Herbert
DSM—Deutsche Sammlung von
    Mikroorganismen und
    Zellculturen GmbH
Mascheroderweg 1b
D-38124 Braunschweig
GERMANY

Miller, R. Scott
AGRA Earth & Environmental, Inc.
7477 SW Tech Center Drive
Portland, OR 97223-8025 USA

Okada, Fusako
Marine Biotechnology Institute
Kamaishi Institute
3-75-1 Heita Kamaishi
Iwate 026
JAPAN

Peden, David H.
Cordova Chemical Co. of Michigan
500 Agard Road
North Muskegon, MI 49445 USA

Peterson, Lance E.
AGI Technologies
300 120th Avenue NE, Building 4
Bellevue, WA 98005 USA

Phillips, Peter
Clark Atlanta University
Research Center for Science & Tech.
Atlanta, GA 30314 USA

Pugh, Lucy B.
Earth Technology
5555 Glenwood Hills Parkway SE
P.O. Box 874
Grand Rapids, MI 49588-0874 USA

Puhakka, Jaakko A.
Tampere University of Technology
P.O. Box 527
SF 33101 Tampere
FINLAND

Rajan, Raj V.
Michigan Biotechnology Institute
P.O. Box 27609
Lansing, MI 48909-0609 USA

Ramaratnam, Mukund
Westates Carbon
2130 Leo Avenue
Los Angeles, CA 90040-1634 USA

Rediske, Richard R.
Grand Valley State University
Water Resources Institute
1 Campus Drive
Allendale, MI 49401 USA

Ressler, Bernd P.
Universität Karlsrühe
Inst. fur Ingenieurbiologie und
    Biotechnologie des Abwassers
    am Fasanengarten
Postfach 6980
76128 Karlsrühe
Baden-Wurteinberg
GERMANY

Rho, Denis
National Research Council—Canada
Biotechnology Research Institute
6100 Royalmount Avenue
Montréal, Québec H4P 2R2
CANADA

Ringpfeil, Manfred
BIOPRACT GmbH
Rudower Ch. 5, Geb 13.9 im IGZ
12489 Berlin
GERMANY

Rodríguez-Eaton, Susana
Clark Atlanta University
Research Center for Science & Tech.
James P. Brawley Dr. at Fair St. SW
Box 271
Atlanta, GA 30314 USA

Rogers, David F.
CPC International, Inc.
700 Sylvan Avenue
P.O. Box 8000
Englewood Cliffs, NJ 07632 USA

Saberiyan, Amy G.
AGRA Earth & Environmental, Inc.
7477 SW Tech Center Drive
Portland, OR 97223-8025 USA

Saha, Gautam
Clark Atlanta University
Research Center for Science & Tech.
James P. Brawley Dr. at Fair St. SW
Atlanta, GA 30314 USA

Samson, Réjean
École Polytechnique de Montréal
Chemical Engineering Dept.
P.O. Box 6079 Station Centre-Ville
Montréal, Québec H3C 3A7
CANADA

Sanschagrin, Sylvie
National Research Council—Canada
Biotechnology Research Institute
6100 Royalmount Avenue
Montréal, Québec H4P 2R2
CANADA

Sanseverino, John
IT Corporation
312 Director's Drive
Knoxville, TN 37923 USA

Scheible, O. Karl
HydroQual, Inc.
1 Lethbridge Plaza
Mahwah, NJ 07430 USA

Scheurlen, Dirk B.
CH2M Hill Engineering Ltd.
180 King St. S., Suite 600
Waterloo, Ontario N2J 1P8
CANADA

Schmid, Karl
Universität Karlsrühe (TH)
Institut für Siedlungswasserwirt-
    schaft (ISWW)
D-76128 Karlsrühe
GERMANY

Shah, Manish M.
Battelle Pacific Northwest
P.O. Box 999 M.S. P7-35
Richland, WA 99352 USA

Sheridan, Bill
AlliedSignal, Inc.
50 E. Algonquin Road
P.O. Box 5016
Des Plaines, IL 60017-5016 USA

Shi, Jing
Michigan Biotechnology Institute
3900 Collins Road
East Lansing, MI 48909-7609 USA

Shimomura, Tatsuo
EBARA Research Co. Ltd.
2-1 Honfujisawa 4-Chome
Fujisawa-shi 251
JAPAN

Smith, Tony
IT Corporation
312 Directors Drive
Knoxville, TN 37923 USA

Smock, Mark
Chester Environmental
600 Clubhouse Drive
Moon Township, PA 15108 USA

Solsrud, Tricia
Waste Management, Inc.
6207 Hempton Lake Road
Whitelaw, WI 54247 USA

Sonnenberg, Lucinda
Inst. of Paper Science & Technology
500 10th Street NW
Atlanta, GA 30318 USA

Stegmann, Rainer
TU Hamburg-Harburg
Institute of Waste Managemnt
Harburger Schlossstr. 37
21079 Hamburg
GERMANY

Stevens, David K.
Utah State University
Dept. of Civil & Environ. Engrg.
Logan, UT 84322-4110 USA

Steward, Kara J.
URS Consultants, Inc.
1100 Olive Way, Suite 200
Seattle, WA 98101-1832 USA

Stolpmann, Holger H. R.
Umweltschutz Nord GmbH & Co.
Industriepark 6
D-27767 Ganderkesee
GERMANY

Stormo, Keith E.
Innovative BioSystems, Inc.
121 Sweet Avenue
Moscow, ID 83843-2386 USA

Sunday, April
Michigan Biotechnology Institute
P.O. Box 27609
Lansing, MI 48909-0609 USA

Sutton, Paul M.
PM Sutton & Assoc., Inc.
8 Marvin Place
Bethel, CT 06801 USA

Thomas, Mark
IT Corporation
1425 South Victoria Court, Suite A
San Bernardino, CA 92408 USA

Tinari, Paul D.
World Envirotech Services &
 Technologies, Inc.
317 SW Alder #1195
Portland, OR 97204-2530 USA

Torres, Edward M.
Orange County
Sanitation Districts
10844 Ellis Avenue
P.O. Box 8127
Fountain Valley, CA 92728-8127
USA

Tzeng, J.-W. (Jim)
U.S. Environ. Protection Agency
Natl. Risk Mgmt. Research Lab
26 W. Martin Luther King Drive
Cincinnati, OH 45268 USA

Uchiyama, Hiroo
National Inst. for Environ. Studies
16-2 Onogawa
Tsukuba-shi 305
JAPAN

Urlings, Leon G.C.M.
TAUW Milieu b.v.
Handelskade 11
P.O. Box 133
7400 AC Deventer
THE NETHERLANDS

van Veen, Wiecher W.
TAUW Milieu b.v.
P.O. Box 133
7400 AC Deventer
THE NETHERLANDS

Vira, Alex
ABB Environmental Services, Inc.
Corporate Place 128
107 Audubon Road
Wakefield, MA 01880 USA

Voice, Thomas C.
Michigan State University
Dept. of Civil & Environ. Engrg.
A-124 Research Complex - Engrg.
East Lansing, MI 48824-1326 USA

Wagner, Dan
Michigan Biotechnology Institute
3900 Collins Drive
P.O. Box 27609
Lansing, MI 48909-0609 USA

Warrelmann, Juergen
Umweltschutz Nord GmbH & Co.
Industriepark 6
D-27767 Ganderkesee
GERMANY

Webster, Todd S.
University of Southern California
Environmental Engineering
KAP 2240
Los Angeles, CA 90089-2531 USA

Winter, Josef
Institut fur Ingenieurbiologie und
    Biotechnologie
IBA Am Fasanengarten
Universität Karlsrühe
Postfach 6980
76128 Karlsrühe
GERMANY

Wong, Arthur D.
Chester Environmental
600 Clubhouse Drive
Moon Township, PA 15108 USA

Woodhull, Patrick M.
OHM Remediation Services Corp.
16406 U.S. Route 224 E
Findlay, OH 45840 USA

Yagi, Osami
National Inst. for Environ. Studies
16-2 Onogawa
Tsukuba-shi 305
JAPAN

Yudelson, Jerry M.
World Envirotech Services
    & Technologies Inc.
317 SW Alder #1195
Portland, OR 97204-2530 USA

Zhao, Xianda
Michigan State University
Hazardous Substance Research
    Center
A-136 Research Complex - Engrg.
East Lansing, MI 48824-1326 USA

Zhou, Xin-Qiang
National Research Council—Canada
Biotechnology Research Institute
6100 Royalmount Avenue
Montréal, Québec H4P 2R2
CANADA

INDEX